Markus Båth: Introduction to Seismology

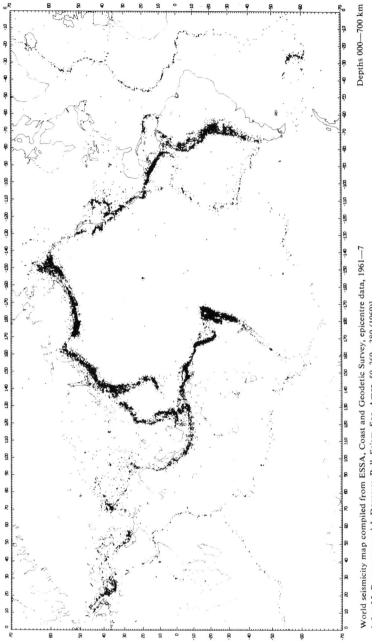

World seismicity map compiled from ESSA, Coast and Geodetic Survey, epicentre data, 1961—7 [after M. Barazangi and J. Dorman, Bull. Seism. Soc. Amer. *59*, 369—380 (1969)].

Depths 000—700 km

Markus Båth

Introduction
to Seismology

Second, Revised Edition

1979   Birkhäuser Verlag
Basel, Boston, Stuttgart

First published under
Markus Båth, Introduktion till Seismologin
by Natur och Kultur Stockholm
© 1970, Markus Båth and Bokförlaget Natur och Kultur, Stockholm

CIP-Kurztitelaufnahme der Deutschen Bibliothek

Båth, Markus:
Introduction to seismology / Markus Båth. – 2.,
rev. ed. – Basel, Boston, Stuttgart: Birkhäuser, 1979.
(Wissenschaft und Kultur; Bd. 27)
Einheitssacht.: Introduktion till seismologin ⟨dt.⟩
ISBN 3–7643–0956–3

English translation © 1973, 1979 Birkhäuser Verlag Basel
ISBN 3 – 7643 – 0956 – 3

Preface
to the First Edition

The data must be greatly amplified and strengthened.

BENO GUTENBERG (1959)

The purpose of this book is to give a popular review of modern seismology, its research methods, problems of current interest and results and also to some extent to elucidate the historical background. Especially in recent years, seismology has attracted much interest from the general public as well as from news agencies. The reasons for this are partly connected with recordings of large explosions (nuclear tests), partly related to earthquake catastrophes. This interest and the questions which people have asked us for the past years have to a certain extent served as a stimulus in the preparation of this book. I have aimed at answering the usual questions in a way which is both exact and easy to understand.

Seismology is an applied mathematical-physical science. In its presentation, I have aimed at giving the reader an understanding of those phenomena which can be considered as primary, rather than to dwell on long, detailed descriptions of various more or less secondary effects of earthquakes or just to list facts. In other words, I have aimed at giving the reader an apprehension of basic phenomena, which lie behind the more immediately visible ones. No doubt, several difficulties present themselves in giving an account of seismology using practically no mathematics. As a consequence, it was found necessary to communicate some results without proof, particularly where a mathematical treatment is unavoidable for a more thorough understanding of the procedures.

In general, we can state that the mathematical apparatus needed for the solution of seismological problems extends far above any high-school course. In fact, considerably more is needed of applied mathematics than is usually contained in a B. Sc. degree in the subject. On the other hand, it is true that high-school physics in many respects can very well lead to an understanding of seismological problems. This is then possible more in an intuitive way than by strict mathematical derivation. For instance, we may mention the importance of electricity theory for a correct apprehension of modern seismograph constructions or the importance of optics for an understanding of elastic wave propagation through the earth's interior. In this book I have therefore referred to high-school physics, rather than to advanced mathematical discussions. In this connection, it ought to be emphasized that seismology offers a number of instructive applications of school physics.

In the selection of items and in their treatment, I have had the use of the book constantly in mind: as a guide-book for high-school teachers in physics, geography and related subjects, as a reading-book for high-school pupils, as an introductory book on a university level, as an information source for construction engineers and others who are in need of knowledge of earthquake effects, as an introduction to the principles of seismic prospecting, and as an information book for an interested general public.

Chapters 1–7 comprise what we could term classical seismology. Thereafter, we have a background enabling us to understand trends in modern seismology. We discuss methods to improve seismological information, both with regard to observations in nature (Chapter 8) and by means of laboratory investigations (Chapter 9). Chapters 10–12 deal with three different branches of great current importance in seismology: earthquake prediction, nuclear test detection, and planetary seismology. Finally, Chapter 13 discusses education and practice in seismology.

The present book is essentially a translation of my book in Swedish entitled *Introduktion till Seismologin*, published by Natur and Kultur Co., Stockholm, in 1970. In working up the present English edition, several minor modifications of the Swedish version were undertaken, especially in the light of seismological developments during the last two years. Some smaller items of more limited Swedish or Scandinavian interest, such as the seismicity of Fennoscandia, have been excluded from this edition. On the other hand, many examples of seismological observations presented here are still based upon our own records in Sweden, of which we have the best knowledge for obvious reasons.

The author is grateful to Mr. R. J. BROWN, Ph. D., Uppsala, who has checked and improved the language, to Mrs. E. DREIMANIS, Uppsala, who has drafted most of the illustrations, and to Birkhäuser Verlag, Basel, for their interest and care in the production of this book.

November, 1971   MARKUS BÅTH

In reviewing the first English edition of this book it has been considered important not to change its scope, level or purpose, but rather to bring all information up-to-date. This concerns partly tabular material on large and destructive earthquakes (Chapter 5), partly data on nuclear explosions (Chapter 11). Fields in which recent significant developments have taken place have called for added sections. This concerns earthquake prediction (Chapter 10) and planetary, especially moon, seismology (Chapter 12). The seismology curriculum (Chapter 13) has been replaced by a somewhat changed, newer version, and the Literature Review has been up-dated by a selection of recent literature. Beyond this, only minor additions and modifications, corrections of a few misprints, etc have been done.

March, 1979   MARKUS BÅTH

Contents

10   Contents

Chapter 1

Scope and History of Seismology

1.1 *What is Seismology?*

The word seismology is formed from the Greek *seismos*=earthquake and *logos*=science. Sometimes one encounters the belief that seismology corresponds only to the exact translation of this word, i.e. the science of earthquakes. However, this is far from the truth. Like most other sciences, seismology has grown beyond its original boundaries. Even though the study of earthquakes still constitutes an essential part of seismology, also several other important branches have been added. The elastic waves emanating from an earthquake permit the most reliable studies and conclusions about the internal constitution of the earth, thanks to the records at the seismograph stations around the world. The physical properties of the earth's interior thus constitute another important branch of seismological studies. I shall here define seismology in two ways, which, however, agree by content:

1. Seismology
   a) the science of earthquakes, plus
   b) the physics of the earth's interior (essentially with regard to seismic wave propagation and the conclusions it permits concerning the internal constitution of the earth);
2. Seismology
   the science of elastic (seismic) waves, i.e.:
   a) their origin (earthquakes, explosions, etc.),
   b) their propagation through the earth's interior, and
   c) their recording, including the interpretation of the records.

Basic research in seismology is concerned with problems within the categories mentioned. In addition, there is applied seismology, where we can also distinguish between several branches, such as seismic prospecting, i.e. the search by seismic methods for economically significant occurrences of salt, oil, minerals, ores, furthermore depth-to-bedrock measurements for construction purposes, etc. The problem of distinguishing between earthquakes and explosions can be considered as another branch of applied seismology.

Seismology is a part of a much more comprehensive science, i.e. geophysics. By geophysics we mean physics applied to the earth, both the solid earth, the sea, the atmosphere and the ionosphere. By 'solid-earth physics' or geophysics in a more restricted sense we mean physics applied to the earth's interior. Just as the usual physics is divided into a number of smaller disciplines, corresponding to various physical phenomena, it is customary to split 'solid-earth physics' in a corresponding way. However, for natural reasons it is convenient to split geophysics in a slightly different manner, and within 'solid-earth physics' we list the following parts:

1. Seismology.
2. Volcanology (also part of geology): volcanoes, hot springs, etc.
3. Geomagnetism: the magnetic field of the earth.
4. Geoelectricity: the electrical properties of the earth.
5. Tectonophysics (common with geology): physics applied to geo-logical processes.
6. Gravimetry (also part of geodesy): measurement of gravity and its interpretation.
7. Geothermy: the temperature conditions in the earth's interior.
8. Geocosmology: the origin of the earth.
9. Geochronology: the dating of events in the earth's history.

This division is the internationally most usual one, corresponding to 'solid-earth physics', *Physik des festen Erdkörpers*, etc. Geodesy, which deals with the shape and size of the earth, is a science closely related to geophysics. Nowadays, it is usual to combine geophysics and geodesy with other sciences concerned with the natural properties of the earth, like geology and geography, into a still bigger unit, called geoscience. A schematic review is shown in Figure 1.

The term 'solid-earth physics' is not entirely satisfactory. 'Solid' has here to be taken as referring to those parts of the earth which remain when the liquid and gaseous parts (sea and atmosphere) are excluded. On the other hand, it cannot be taken as characterizing the earth's interior in a strict physical sense. The earth is solid in a physical sense, i.e. crystalline, only to a depth of around 80 km. In many universities one simply uses the term 'geophysics' to denote what has been listed under 'solid-earth physics'. Geochemistry is another related subject, which has not been listed separately here, as it forms a part of several of the subjects in Figure 1.

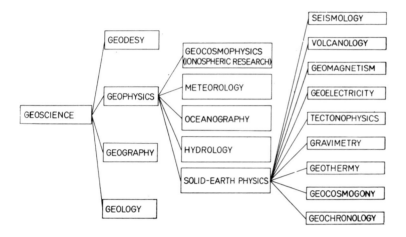

Fig. 1. Subdivision of geoscience, geophysics and solid-earth physics into different disciplines. For other subjects listed the subdivisions are left out.

Seismology, like geophysics in general, works on three parallel fronts: by observations in nature (including recordings of natural phenomena), by laboratory investigations and by theoretical investigations. The problems are often very difficult, mainly because the object of our studies — the earth's interior — is in general inaccessible to direct investigation. Instead we have to rely on indirect observations, made on the earth's surface or very near to it. It is clear that the interpretations of such observations can involve great difficulties. In fact, seismological problems consist, to a very great extent, of such interpretational questions. Still, there is no doubt that seismological observations (generally in the form of seismograph records) permit less ambiguous and more reliable interpretations than other geophysical observations, such as gravimetrical and magnetic. In order to get a picture of the earth's interior as reliable and complete as possible it is most suitable to combine as many different observations as possible. The interpretation of the internal constitution has to agree with all reliable observations, if it shall be accepted. Or as a famous geophysicist once expressed this: 'We have to remember that there is only *one* earth.'

Laboratory investigations may yield results of great significance to the interpretation of the observations. Thus, the behaviour of various materials (metals and minerals) at high pressures and high temperatures has been investigated in the laboratory. Moreover, it is possible to simulate earth processes in the earthquake regions as well as wave propagation through the earth's interior. Experiments of the last-mentioned types are usually referred to as model seismology. The advantage of laboratory investigations is that they permit variations of the parameters at will and thus having the specimen under much better control than in nature, even if it is not always possible to simulate nature in all details. Modern electronic computers offer another efficient method of simulating complicated natural phenomena.

In fact, theory and observations (recordings) have always played an important role in seismological research. Laboratory investigations, especially model seismology, have been added in more recent decades as a useful complement.

Obviously, seismology is an applied science. Above all, a thorough knowledge of mathematics and physics is essential in order that one might produce more significant contributions in this science. In addition, other subjects, such as geology, statistics, etc., also play a very great role.

Seismology first became an independent science around the turn of the century. The theoretical foundations, especially the theory of elasticity and wave propagation, had been developed much earlier, the elasticity theory particularly by Cauchy and Poisson as early as during the first half of the nineteenth century. Observations of earthquakes and their effects have been made in populated areas as far back as history goes. A certain instrument for earthquake observations, the seismoscope, already existed in China about one century after Christ. But the theoretical foundations and the observations were completely separated from each other until the end of the last century. Thanks to the construction of seismographs, it was then possible to combine the two disciplines. It is the operation of good seismographs which has led to the considerable advances in our knowledge of the earth's interior during the present century and which has made seismology a science.

1.2   *Development of the Theory of Elasticity*

We shall now scrutinize the development of seismology and we start with the theory of elasticity and strength of materials. This deals with the behaviour of bodies (especially solids) when they are subjected to forces, both how they are deformed and how they ultimately break under large enough stresses. The first mathematician who studied such problems was GALILEO in 1638. He investigated the behaviour of a beam attached at one end to a wall and loaded. He found that with increasing load the beam bends around an axis perpendicular to its length and situated in the plane of the wall. The problem to determine this axis is called Galileo's problem. Even though GALILEO did not give any mathematical relations between load and deformation, his works were pioneering in elasticity theory.

Two of the most important events in the further development of elasticity theory are the establishment of HOOKE's law in 1660 and the formulation of the general equations of elasticity by the French scientist NAVIER in 1821. HOOKE's law states that the deformation of a body is directly proportional to the applied stress. It forms the basis for the mathematical theory of elasticity, and also in the study of the earth's interior it is assumed to be valid. At least it still serves as a good first approximation to the elastic conditions in the earth. The main problems which occupied the elasticity scientists in this relatively early stage were a development of Galileo's problem and studies of vibrations of bars and plates, stability of columns, etc. NAVIER was the first to investigate the general equations both for equilibrium and for vibrations of elastic solid bodies. The elasticity theory and especially the problem of propagation of elastic waves through a medium attracted at about the same time two other French mathematicians of high standard: CAUCHY and POISSON. Through their studies the development of elasticity theory became closely linked with the problem of light propagation. In 1822, CAUCHY had presented most of the foundations of elasticity theory and later he extended his researches to crystalline bodies. In his research of wave propagation in an elastic medium POISSON found around 1830 two types of waves. At a greater distance from the source, these are effectively longitudinal and transverse, respectively, and their velocities have the ratio $\sqrt{3}:1$. This was later confirmed by STOKES in

England in 1849. This is the first time that we come across the waves *P* and *S*, now so well known in seismology. POISSON also studied the free radial vibrations of a solid sphere, a problem which was also studied later in England by LAMB and others.

In 1845, STOKES observed that the elastic resistance which an isotropic body exerts to an applied stress can be split into a resistance against compression or tension and a resistance against shear (by 'isotropic' we mean that the elastic properties are independent of direction). Compression or tension is due to normal stress, whereas shear is due to tangential stress. STOKES also defined two parameters or elasticity moduli to express these resistances. These two parameters are nowadays called the modulus of compressibility and the modulus of rigidity.

We have already mentioned POISSON's discovery of longitudinal and transverse waves through the interior of a solid elastic body. In 1887, Lord RAYLEIGH discovered still another type of elastic wave, which instead propagates along the surface of a body. The velocity or speed of propagation of this wave type was found to be lower than for the two first mentioned. These surface waves are also well known in seismology and are called Rayleigh waves, after their theoretical discoverer. Another important type of seismic surface wave, i.e. Love waves after the Englishman LOVE, was not found until in 1911.

The German scientist A. SCHMIDT published in 1888 a paper in which he discussed the propagation of waves through the earth's interior. He emphasized that in general the wave velocity must increase with depth in the earth and that because of this, the wave paths will be curved and not rectilinear. The curved wave paths will have to be concave towards the earth's surface. About the same time, KNOTT in England investigated the energy of reflected and refracted waves.

Among other problems, which were studied at an early stage of elasticity theory, we mention also conditions on impacts between two solid bodies. The early elasticity research workers devoted most of their efforts to basic research. However, their results have later been of the greatest significance in a number of applied branches, such as technology, seismology, etc. Thus, we have seen that the main types of seismic waves, nowadays regularly found on our seismograph records, had been discovered by mathematicians long before any seismic records had been obtained. For easier review, the main points in the development of the elasticity theory up to around the turn of the century are summarized in Table 1.

Table 1. Some historically important points in the earlier development of elasticity theory and seismology.

1638	GALILEO	Deformation of beams (Galileo's problem)
1660	HOOKE	Proportionality between stress and strain (Hooke's law)
1799	CAVENDISH	Determination of the earth's mean density
1821	NAVIER	General equations of elasticity
1822	CAUCHY	Foundations of elasticity theory
1830	POISSON	Longitudinal and transverse waves; vibrations of a sphere
1845	STOKES	Compressibility and shear modulus
1860	MALLET	World seismicity map
1874	DE ROSSI	The first more generally used intensity scale
1878	HOERNES	Classification of earthquakes
1880	GRAY, MILNE, EWING	Seismograph construction
1887	RAYLEIGH	Surface waves of Rayleigh type
1888	SCHMIDT	Wave propagation in the earth's interior
1897	WIECHERT	Iron-core hypothesis
1899	KNOTT	Reflection and refraction of elastic waves
1900	WIECHERT	Construction of the Wiechert seismograph
1900	MONTESSUS DE BALLORE, MILNE	World seismicity maps
1906	OLDHAM	Iron-core hypothesis verified seismologically
1906	GALITZIN	Construction of the Galitzin seismograph; application of electromagnetic induction
1906	REID	'Elastic rebound theory' – still the generally recognized theory for tectonic earthquakes
1909	MOHOROVIČIĆ	Earth's crust bounded below by a sharp discontinuity, the so-called Mohorovičić discontinuity
1911	LOVE	Surface waves of Love type
1913	GUTENBERG	Depth to the outer core determined as 2900 km
1922	TURNER	Indications of the existence of deep-focus earthquakes
1928	WADATI	Existence of deep-focus earthquakes established
1935	BENIOFF	Construction of the strain seismograph
1935	RICHTER	Earthquake magnitude scale
1936	LEHMANN	Discovery of the inner core

1.3 *Observations of Earthquakes*

Quite independent of the development of elasticity theory up to the end of the nineteenth century, there is a comprehensive body of literature on earthquake effects. Information on earthquakes exists at least as far back as 1800 B.C. Information of older date is, however, generally very scanty and does not satisfy modern scientific requirements on observations. First around the middle of the eighteenth century, descriptions of more scientific value began to be produced. Thus, the Lisbon earthquake on November 1, 1755, was studied in a more scientific way. Earthquakes in Calabria, Italy, in 1783, were studied by special scientific commissions. An earthquake in Cutch, India, in 1819 seems to be the first one for which clear effects of fault action were observed. An earthquake at Naples in 1857 was thoroughly investigated in the field, and this represents the first attempt to apply physical principles to such observations. Among other early earthquakes which were carefully investigated in the field, we may mention especially the Mino-Owari earthquake in Japan in 1891 with great fissures on the surface, likewise the San Francisco earthquake in 1906 (which gave rise to a still valid theory for earthquakes, the 'elastic rebound theory', due to REID) and finally the Kwanto earthquake in Japan in 1923, when large parts of Tokyo and Yokohama were devastated. Chapter 5 below gives a summary of the more important earthquakes.

Most of the earlier reports on earthquakes deal only with various effects, such as on constructions, topography, etc. In most cases such descriptions call for a critical attitude on the part of the reader. But, to an equally high degree, it is necessary for a present-day observer to be critical of his own observations. To observe earthquake effects in a scientifically satisfactory way is not so simple as it may seem. It is very common that secondary effects are observed and described in detail, whereas more primary effects, such as faulting, are given much less attention. Still, the latter are of most interest and value in any scientific study of earthquakes. From the engineering point of view, it is of importance to investigate any effects in relation to the structural details of buildings or other constructions, and not simply describe the degree of destruction.

In 1878, R. HOERNES in Germany proposed a classification of earthquakes, which is still valid:

1. Collapse earthquakes, caused by collapse of cavities in the earth's interior.
2. Volcanic earthquakes.
3. Tectonic earthquakes.

Tectonic earthquakes are due to folding in the earth's crust, formation of mountain ranges, etc., in general, motions within the solid earth, and these are the only earthquakes which are of greater significance on the earth as a whole. Shocks of the categories 1 and 2 were earlier overestimated in importance, but they are generally only of local significance and nearly always very small.

It was obvious from earthquake observations at an early stage that effects were considerably stronger on soft, moist ground than on dry material, and that the amplitudes were smallest on solid rock. These observations have later been confirmed by direct recordings. One problem, which early attracted considerable attention, was the direction in which pillars had fallen. It was often believed that such directions could give important information about the direction to the earthquake source. However, it was found that in equally many cases the direction of fall was perpendicular to the direction of the source. It depends obviously on what wave causes the pillar to fall: if it is P, or the longitudinal wave, the pillar will fall in the direction of the source, or in the opposite direction, but if it is S, or the transverse wave, which made the pillar fall, the direction of fall may be perpendicular to the source direction. In addition, this may be modified because of special conditions.

In earthquake reports, observations of sound and light play a certain role. While sound observations probably derive from the elastic waves (sometimes noticed even before the main earthquake) the light effects are more difficult to explain. A conceivable reason may be electric charges (friction between adjacent sides in the fault) and following discharges, or (more trivially) short-circuiting in the electric net.

At sea earthquakes can be noticed on ships because of the wave which can traverse the water from a shock below the sea bottom. This is possible only for the longitudinal wave, and there are several reports telling that this can cause the same sensation as if the ship had hit a hard rock. Another sea effect consists of the so-called tsunamis (or tidal waves, as they are also, but erroneously, called). These are waves on the sea surface, several hundred kilometres in length but not particularly high. They propagate with a velocity of about 220 m/sec when the water depth is 5 km, and more slowly in shallower water. On the open sea they do not present any danger, but when they hit a coast, particularly in narrow passages, they rise in height and may cause much damage to ships and harbours.

On the basis of direct earthquake observations, the first maps of the distribution of earthquakes over the earth were made around the middle of the nineteenth century. A map by MALLET (England) from about 1860 was the most reliable, even up to around 1900, when MONTESSUS DE BALLORE (France) and MILNE (England) independently constructed seismic world maps. They both noticed that the seismic activity was concentrated to areas with tectonic and volcanic activity in late geologic time. It is quite clear that such seismicity maps to a certain extent are also population maps, and therefore they cannot satisfy modern requirements. Already around the middle of the nineteenth century, numerous investigations of possible periodicities of earthquakes were started, but hardly any of these gave any definite results.

In order to express earthquake effects (the so-called macroseismic observations) in a quantitative way, intensity scales were introduced as early as the 1870's. Thus, DE ROSSI in Italy proposed between 1874 and 1878 the first more commonly used intensity scale. In 1881 FOREL in Switzerland proposed a similar scale, and soon thereafter they joined their efforts into the Rossi-Forel scale. This scale which had 10 degrees has later been modified and modernized. About this we shall learn more in Chapter 4. By expressing the observations in an intensity measure, the foundation was laid for the combination of all observations for a given earthquake into an intensity map. Such a map shows by special curves (isoseismals) the geographic distribution of the intensity. Such maps permit some conclusions about the depth to the earthquake source (focus). If the intensity decreases rapidly outward from its maximal value, then this indicates a shallow focus, whereas if the intensity decreases only slowly this suggests a greater focal depth. On the basis of this principle it was concluded in the 1880's that some earthquakes in the Eastern Mediterranean Sea with exceptionally large macroseismic areas also had focal depths in excess of normal—conclusions which have later been confirmed by seismic records.

1.4   *Early Knowledge of the Earth's Interior*

It is correct to say that our knowledge of earthquakes was quite imperfect up to the beginning of the present century, but this is true to an even higher degree of our knowledge about the earth's interior. The interior of the earth has been a very popular field for various speculations and free imagination as far back as history goes. The first more scientific opinions were probably based on the observation that volcanoes from time to time spew out molten lava, which was taken as evidence that the interior of the earth was red-hot and in a molten condition. This opinion was considered proved, when it was found over a century ago that the temperature grows with depth in the earth. However, other opinions were also proposed, for instance by POISSON, who did not believe in a gaseous central part in the earth with temperatures of hundreds of thousands of degrees, and who warned against extrapolation to great depth of observations made near the surface. Others maintained that the lava only came from local cavities near the earth's surface and did not represent any global condition. By means of observations of the tidal effect in the solid earth, Lord KELVIN claimed in 1863 that the earth as a whole is more rigid than glass. This opinion has been confirmed later, the only difference being that steel offers a better comparison.

As early as in the eighteenth century a rough idea had been reached about the value of the mean density of the earth, among others by the English scientist CAVENDISH in 1799. As this density exceeded the density of rocks on the surface, the conclusion was that the density must increase with depth in the earth. It was assumed that it increases to a maximum in the centre of the earth. In 1897, the German geophysicist WIECHERT found by theoretical calculations that the earth's interior consists of a mantle of silicates, about 1500 km thick, surrounding a core of iron. This was the first proposal of the so-called iron-core hypothesis, which still has a majority of adherents. The existence of the earth's core was confirmed by the English seismologist OLDHAM in 1906. The depth to the boundary of the core was revised to its present value (2900 km) in 1913 by the famous German, later American geophysicist, GUTENBERG.

1.5   *Installation of Seismographs: Seismology Becomes a Science*

Even though earthquake observations were expressed quantitatively in an intensity scale, the purely descriptive reports were too far from the mathematical foundations of the elasticity theory to permit a unification of the two disciplines—even in spite of the fact that basically the phenomena which were observed and which were studied theoretically were closely related. The unifying link consisted of the seismographs, by which exactness was introduced into earthquake study. These turned seismology into a science, not only one of a descriptive nature but a mathematical-physical science. If we should mention any breakthrough in the study of earthquakes and the earth's interior, then the installation of seismographs is no doubt the most important.

Nowadays, it is equally difficult to conceive seismology without seismographs as astronomy without telescopes. While telescopes already existed around 1600, the first useful seismographs date back to the time between 1880 and 1890, i.e. remarkably late. In 1880 seismographs were constructed in Japan by the Englishmen GRAY, MILNE and EWING. They were mainly intended for recording of Japanese earthquakes. It took until 1889 before the first record of a distant earthquake was obtained. This earthquake occurred in Japan and the record was written in Potsdam near Berlin.

After this, the development was rapid and two very famous seismologists published detailed descriptions of seismographs which they had constructed. One of these was due to WIECHERT in Germany who in 1903 gave a detailed account of his mechanical seismograph, constructed in 1900; the other was due to the Russian Prince GALITZIN who in 1911 reported his seismograph. In this apparatus he applied in 1906 electromagnetic induction with galvanometric recording on photographic paper. Both these papers have been of fundamental importance in the later development of seismographs and their theory. Several seismographs from the beginning of the century are still in operation at various stations, especially in Europe where significant development took place early, and such seismographs may still give good records. Because of their low magnification of ground motion, by modern standards, the records of these older instruments are still among the most useful for the biggest earthquakes. The modern seismographs generally have a magnification which is too high to give really good records of the largest shocks.

Soon after the initial studies of the first seismograms, the identification of longitudinal and transverse waves and of Rayleigh waves was clear. Thus, in the 1890's OLDHAM in England and WIECHERT in Germany asserted independently that longitudinal and transverse waves and surface waves exist in the seismograms. Through this discovery the link with elasticity theory was established, and this initiated a very fruitful period in seismology. The decades around the turn of the century are characterized by the new combination between earthquake observations and the elasticity theory. Efforts to explore the earth's internal constitution by means of seismic records were pursued with much enthusiasm. On the whole, it can be maintained, that in this stage the research workers were more inclined to suspect sharp discontinuity surfaces in the interior, as soon as some observation seemed to call for such a surface. Later, several such discontinuity surfaces disappeared, i.e. they were not necessary to explain the observations, and they were for the most part replaced by zones of more gradual transition. However, in the course of time, the pendulum swings, and recently mantle discontinuities begin to appear again. One discontinuity surface which has remained in spite of all reinterpretations and has been strongly confirmed is the sharp limit of the earth's core. Its depth was placed at 2900 km by GUTENBERG in 1913, a depth value which later investigations have not been able to modify considerably. During the 1930's the inner core was discovered, having its surface or rather zone of transition at a depth of around 5000 km. Another well-known and well-established discontinuity is the base of the earth's crust. This was found in 1909 by A. MOHOROVIČIĆ by means of seismic records of earthquakes in Croatia. After him, this surface is called the Mohorovičić discontinuity or, for brevity, the Moho. The epoch 1910–1940 is characterized by an intensified study of earthquakes and the earth's interior by means of seismograms. At the end of this period, seismology had arrived at fairly accurate values of the internal properties of the earth, such as the elastic wave velocities (calculated from observed travel times of P and S waves), and moreover the distribution within the earth of density, pressure, gravity and elastic parameters. Another discovery in this period (by the English seismologist TURNER in 1922 and the Japanese seismologist WADATI in 1928) concerned deep earthquakes. It was found that some earthquakes occurred at depths of several hundred kilometres. The deepest (about 720 km below the earth's surface) have been found in the Indonesian archipelago.

In a study of the history of seismology one cannot avoid observing that national considerations have played a certain role in the judgment of the achievements of various seismologists. This is more true for the breakthrough years from 1890 to 1910 than for any other epoch. A famous American seismologist once stated this very clearly at a conference in the 1950's. With regard to a certain seismological result, he said that the remarkable thing in this connection is that the name of the discoverer is pronounced so differently in different countries: in England they pronounce it as MILNE, in Germany as WIECHERT, and in Russia as GALITZIN.

With regard to the period after World War II, it is correct to say that the picture of the earth's interior from about 1940 has stood up very well. It has not been subject to any major changes, but has rather been improved in detail, thanks to more numerous and more accurate observations. It is not until after the installation of special so-called array stations during the 1960's, that reason has been found to somewhat modify our earlier knowledge about the variation of the elastic wave velocities with depth in the earth. Seismological research work after World War II will occupy us to a great extent in the following chapters. Already at this stage, we can emphasize one very important thing, namely the extension of the seismic recordings to a considerably larger range of period of the seismic waves than earlier, both towards shorter and longer periods. This has been made possible thanks to newer seismograph constructions and to the installation of particularly sensitive stations.

It is hardly any exaggeration to characterize seismology today as a popular science. The reasons are quite obvious. Above all, the possibilities of detecting explosions, which are offered by seismic methods, have contributed considerably to increased interest in this science, not the least from the military and political points of view. Seismologists have been faced with a number of questions to which they have not had any ready-made answers. This explains why in some countries, especially in the USA, basic research in seismology has since late 1950's received considerably more support than ever before in its history. This research has covered a wide range of basic interests in seismology, but the ultimate goal was to improve detection and identification capabilities. The American VELA Project has spent enormous funds on seismological research, which has been of great benefit to many institutes, even outside the USA itself. For instance, many institutes in different European countries have conducted seismological research on American funds during the 1960's. This research has been completely free and open, available to anybody through publication in international journals. Around 1971, this phase of research has been largely terminated and followed by a stage of applications, done inside the USA.

UNESCO has shown a very great interest in seismological problems, especially after the earthquake catastrophes during the spring of 1960 (Agadir in Morocco, Lar in Iran, Chile). Accordingly, UNESCO sent in the early 1960's four different seismological missions to different countries (Southeast Asia, South America, Mediterranean countries including the Middle East and to Africa). Each mission consisted of a few experts whose job it was to investigate various conditions in connection with earthquake risks for people and buildings and to suggest various measures to be taken to eliminate the risks as much as possible. An accurate mapping of seismicity is required in order to facilitate judgment on the risks at different places, and this includes the installation of more and better seismograph stations. Special instructions for building construction, so-called building codes, are still missing in numerous earthquake countries, and it was part of the missions' duties to suggest such codes in applicable parts, where they are still non-existent. In several cases, UNESCO has also sent experts to study earthquake effects immediately following the occurrence of a catastrophe. Numerous tasks are involved in the duties of such an expert group, such as observation of geological and topographic effects, damage to buildings and other types of construction, as well as installation of temporary stations in the area to record aftershocks, etc.

Moreover, the International Geophysical Year (IGY) in 1957–8 meant a considerable improvement in the observation net, especially in the southern hemisphere, including Antarctica, and also an intensification of the global seismological cooperation. During the 1960's the international Upper Mantle Project (UMP) has similarly contributed to an increase in seismological resources to attack big problems and has also stimulated closer cooperation between different branches of geophysics and related sciences.

After the termination of the UMP in 1970, the International Geodynamics Project provides for another broad, multi-disciplinary attack on major geophysical, geochemical and geological problems.

A notable feature of the development of seismology in the last five-year period is the enormously increased demand for accurate information on earthquake risk in connection with various engineering undertakings, in particular for the planning and construction of nuclear power plants, as well as for a number of other applications, e.g. in mining operations. Earthquake insurance is another important applied field.

Finally, we can make the following commentary on the development of seismology, which is probably typical of most subjects within science. Earlier it was possible by relatively simple means to reach results of surprising accuracy. When later, one wants to improve these results, really large problems are encountered, which require major input even for relatively modest output. This is illustrated simply by the phases A and B, respectively, in Figure 2. One example is furnished by the development

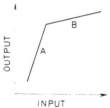

Fig. 2. Schematic relation between output and input in earlier (A) and later (B) stages of the development of seismology.

of seismographs. All facilities up to about a decade ago were relatively inexpensive in installation and operation, yet they have given all the fundamental results of present-day seismology. Nowadays, enormous funds have to be spent on large stations, but the results are not in the same proportion to the input as earlier. The condition is similar within seismological theory. The simple elasticity theory has carried us very far in the exploration of the earth's interior. For any further progress, even of minor extent, considerably more involved mathematical discussions are necessary.

Chapter 2
Seismographs

2.1 Fundamental Principles

Some definitions. First we shall define some fundamental concepts: a *seismoscope* is an apparatus or device which only indicates that an earthquake has occurred, but which does not record anything; a *seismograph* is an instrument which gives a continuous record of the motion of the ground, i.e. a *seismogram*; a *seismometer* is a seismograph whose physical constants are so well known that the true ground motion can be calculated from the seismogram. However, seismometer has also another and more usual meaning. Electromagnetic seismographs (see below) consist partly of an electromagnetic pick-up (a sensor), usually a pendulum instrument, partly of a galvanometer with a recording device. In such an installation, the pick-up is usually termed a *seismometer*, whereas the whole apparatus is called a *seismograph*. In addition, we have to observe that nowadays numerical characteristics of practically all seismographs ('seismograph constants') are so accurately known as to permit an accurate calculation of the ground motion. Also for this reason, it is advantageous to use the latter definition of the word seismometer.

The pendulum principle. When an earthquake of sufficient strength occurs, elastic waves are radiated from the earthquake source through the whole of the earth. A seismograph at some point on the earth's surface records the waves which arrive at or pass this point. Obviously everything in connection with the earth, such as buildings, the seismograph pier, the frame of the seismograph, etc., takes part in this motion. However, it is required to have some fixed point, not taking part in this motion, just to be able to record the motion. This fixed point corresponds to the seismograph pendulum, which has been made more or less independent of its surrounding by special suspension. For instance, in a vertical-component seismograph of the simplest kind, consisting of a weight suspended by a vertical spiral spring, it is the frame, where the spring is attached, which moves when hit by a seismic wave, whereas the weight remains still. A completely statical pendulum is an ideal case; in reality, even the pendulum gets its own small motion.

Figure 3 illustrates in principle a vertical pendulum and a horizontal pendulum of the simplest kind. The principle of the stationary pendulum mass is that for a displacement of frame, including recorder, upwards

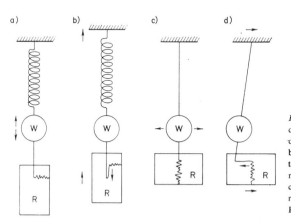

Fig. 3. Principles of a vertical pendulum (a) and a horizontal pendulum (c) in their simplest forms. b) and d) show the relations between ground displacement and recorded displacement for a vertical and a horizontal displacement, respectively. W = pendulum mass, R = record.

(Fig. 3b), the recording pen traces a displacement downwards. The corresponding holds for a horizontal displacement in the case of the horizontal pendulum (Fig. 3d). These considerations are of fundamental importance in the transcription of a given record into the corresponding ground motion.

In principle, the pendulums in Figure 3 are seismographs, but of a very unsatisfactory construction. There are three main objections which can be made against them:

1. Damping. The pendulums in Figure 3 are practically undamped.

2. Period. The formula for the free period of a pendulum tells us that pendulums of the construction shown in Figure 3 can only have short periods, if the size of the instruments should be kept within reasonable limits.

3. Magnification. The pendulums in Figure 3 work with practically no magnification of the ground motion.

In the following, we shall see how we can eliminate these shortcomings.

Evidently, any instrument containing a pendulum, i.e. a mass movable in relation to its frame, can be considered as a kind of seismograph. This is true, for instance, of gravimeters, tilt meters, magnetometers, etc. It is important to observe that such instruments can operate as seismographs upon the arrival of seismic waves. In many cases, such records have been interpreted, often erroneously, as due to some real variations of gravity or of the earth's magnetic field, caused by the earthquake or the waves radiated from it.

Practically all physical principles have been applied in different seismograph constructions. Many of these have now only a curiosity interest. We shall limit our discussion to constructions which have shown to be of scientific value, i.e. such which have well-defined properties.

Translation, rotation, deformation. The general motion of the particles in a solid body can be divided into three kinds: translation, rotation and deformation. For a complete recording we should evidently need seismographs which react to all three kinds of motion, or still better, seismographs for each one of the three motions: translation seismographs, rotation seismographs and deformation seismographs. In fact, nearly all seismographs are of the first kind, since the translatory motion has always been the one to attract the greatest attention. Rotation seismographs have been constructed, but rotational effects are considered to be of significance only at or near the earthquake source and to be of interest only in connection with the influence of the ground motion on various constructions. Deformation seismographs or strain seismographs, as they are usually called, have also been constructed (see Section 2.3), but they still only exist in a very limited number.

In order to get a complete picture of the translation, it is necessary to record the ground motion in three components, generally in the north-south (N), east-west (E) and vertical (Z) directions. Other directions have been recorded from time to time. In this connection we want especially to mention so-called *azimuthal seismographs* which Russian seismologists have installed. In these, a series of seismographs are arranged with their direction of vibration along the generators of a cone with vertical axis. By comparison of amplitude and phase of the different records, it is possible to determine the nature of the waves and their direction of arrival. A similar development in the USA is termed *triaxial seismometer*.

Every seismic wave consists of a whole spectrum, and to be able to reproduce this fairly well, it is necessary in addition to record in different period ranges, from a few tenths of a second up to several minutes. Finally, parallel instruments with different sensitivity are needed to a certain extent, because of the great variation among the amplitudes of the incoming signals. Thus, we easily see that a well-equipped station comes up to a dozen instruments, recording simultaneously.

Damping. These days, all seismographs are equipped with damping devices. As we shall see, the record of an earthquake consists of a series of onsets of different waves, which have different velocities and therefore do not arrive at the same time but at certain time intervals. In order to be able to observe all these sufficiently well, it is necessary that the motion provoked by one phase has been adequately damped before the next phase arrives. In time readings on records, it is only the first swing of each phase which is measured. A damped pendulum (preferably with critical or aperiodic damping) also gives a record which best corresponds to the ground motion. Figure 4 illustrates different degrees of damping.

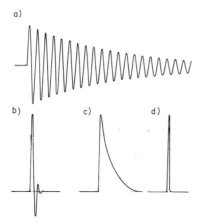

Fig. 4. Different degrees of damping: a) the pendulum motion when the damping mechanism is removed, b) underdamped vibration (with damping mechanism), c) overdamped vibration, d) critical (aperiodic) damping.

The period problem. Seismometer periods as long as 10 sec or more certainly entail some constructional problems. For one thing the instrument has to be made as compact as possible, for another the equilibrium position has to be as stable as possible, even if the restoring force is small. For example, for a vertical pendulum of the simplest kind, i.e. a weight hanging by a vertical helical spring (Fig. 3a), one has to use a spring of about

25 m length in order to achieve a free period of vibration of 10 sec. The length of the spring increases with the square of the free period. By special suspensions, such difficulties can be avoided. For a vertical pendulum a suspension system as shown in Figure 5a can be used. The pendulum mass

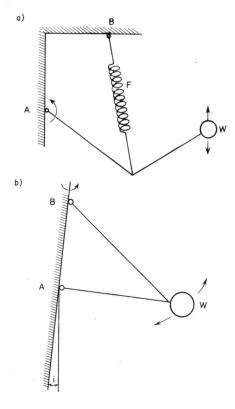

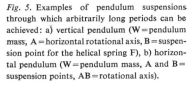

Fig. 5. Examples of pendulum suspensions through which arbitrarily long periods can be achieved: a) vertical pendulum (W = pendulum mass, A = horizontal rotational axis, B = suspension point for the helical spring F), b) horizontal pendulum (W = pendulum mass, A and B = suspension points, AB = rotational axis).

W is attached to a bar, movable vertically around an axis A and suspended by the helical spring F. This is attached to the frame at B. A variation of this suspension principle was suggested by LaCoste in 1935. Nowadays, helical springs of 'zero length' are used, which means that the real length under load is equal to the extension.

For horizontal seismographs a suspension, which can be characterized as a 'hanging gate', is frequently used (Fig. 5b). If the axis AB is exactly vertical, then the pendulum boom and the pendulum are in equilibrium in any position, in other words the free period becomes infinitely great. By tilting AB, a fixed equilibrium position and a finite period are achieved. The more AB is tilted, the shorter the period will be and the greater the restoring force, i.e. the force that restores the pendulum to its equilibrium position after deflection.

Transmission of pendulum motion to record. Magnification. Practically all translation seismographs presently in use are pendulum instruments. This means that the basic principle is the same for all of them. Nonetheless, different types may deviate from each other by the manner in which the pendulum motion relative to its frame is transmitted to the record. We can distinguish two main types for this transmission:

1. Direct-recording seismographs.
2. Electromagnetic seismographs.

In seismographs of older date (for instance, the Wiechert) the motion is transferred from the pendulum to the recording pens in a purely mechanical way and simultaneously magnified (Fig. 6). In addition, the record-

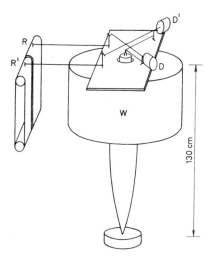

Fig. 6. Schematic picture of the Wiechert seismograph. W = inverted pendulum (1000 kg at Uppsala), at its lower end supported by a Cardan suspension, DD' = cylinders for air damping, RR' = pens recording on smoked paper, one for the E-component and one for the N-component.

ing is mechanical (stylus on smoked paper). These things entail that the seismograph becomes rather bulky and that friction is introduced both in the mechanical transmission and above all in the mechanical recording.

In order to overcome the friction as much as possible it is necessary to use large pendulum masses, for instance 1000 kg in many Wiechert seismographs. In certain other Wiechert seismographs, pendulum masses of as much as around 20 tons are used. Such instruments were installed in the beginning of the century at several places, especially in Central Europe.

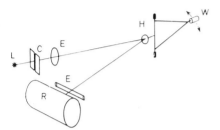

Fig. 7. The principle of a mechanical-optical seismograph. W = pendulum (horizontal), L = light source, C = slit, E = lenses, H = mirror, R = recording drum.

Another type of direct-recording seismograph uses optical recording with small mirrors attached to the pendulum, which reflect a light beam from a light source onto the photographic paper on a recording drum (Fig. 7). This principle is used in the English seismograph Milne—Shaw, among others.

Other seismograph types (e.g. Galitzin) are electromagnetic, in which one or several coils are attached to the pendulum such that they are movable in the field of a permanent magnet. The magnet is attached to the frame and the coils are connected to a galvanometer. On the arrival of seismic waves, the coil is set in motion in relation to the magnet, and an electromotive force is induced, whereby the galvanometer shows deflections. These are recorded optically on photographic paper. In this construction, friction is avoided completely. One of the best seismographs at present is the electromagnetic type constructed by BENIOFF (Pasadena) as early as in 1931. It has great magnification for periods less than 1 sec. The common principle for all electromagnetic seismographs is the use of the induction of an electromotive force by the variation of the magnetic flux through a coil. In most types (e.g. Galitzin) this is achieved by relative motion of coil and magnetic field, in the Benioff type by variation of the magnetic reluctance.

As distinct from the mechanical and mechanical-optical seismographs, the electromagnetic seismographs thus transform the originally mechanical vibrations into corresponding electrical vibrations, both in analogue form.

Time measurements. When a record of an earthquake has been obtained, the first task is to determine the arrival times of various phases, and moreover, periods and amplitudes for the different waves. As the earthquake waves exhibit periods from less than one second up to several minutes, it is not possible to achieve the same good reaction to all these waves with one and the same instrument. The reaction is naturally best for those earthquake waves, whose periods agree with the free period of the seismograph. This period is somewhat modified with the addition of damping to the system. The common Wiechert seismograph has a period around 10 sec. If we want a very good record of the short-period, preliminary phases (periods around 1—2 sec), as is nowadays generally required, then short-period seismographs are used. However, these will in general not show the longer period waves. In modern highly sensitive instruments (like the Benioff) the recording speed (drum speed) is usually 60 mm/min for the short-period instruments, which for sharp onsets permits a time measurement to 0.1 sec accuracy. For long-period records, a drum speed of 15 or 30 mm/min is generally used. For reliable time measurements, we need reliable time marks on the record (usually each minute), a constant rotation of the recording drum, and an exactly known correction of the clock that activates the time marking.

In general, it is true that it is not particularly difficult to construct an apparatus that records seismic waves. But what may be far more difficult is to construct an apparatus with well-defined properties and to achieve accurate timing. With no exact time, the seismic records are as a rule useless. Today, quartz crystal clocks are generally used on the seismograph stations. With good construction and with proper adjustment, such a clock can keep its correction unchanged within 0.1 sec for several weeks. Frequently, such a clock is made to accomplish the following three operations:

1. Deliver minute and hour marks to the recording apparatus.

2. Deliver electric voltage with well-controlled frequency to the motors driving the recording drums.

3. Automatically turn on and off a radio receiver for direct recording of a time signal on the records.

Amplitude measurements. As a rule, amplitude determination is considerably more complicated and less reliable. The amplitudes on a record can be conveniently measured to 0.1 mm accuracy, but the problem is to recalculate these amplitudes into the corresponding amplitudes of the ground motion. Only the latter are of interest and only these can be used for comparison between different places with different instruments or for comparison between vibrations of different periods recorded by one and the same instrument. In order to penetrate deeper into these questions, we need a long mathematical treatment. Just as the motion of an ordinary simple pendulum is governed by a special equation, a differential equation, so also is the case with a seismograph. For a direct recording seismograph, there is *one* equation, and this contains terms corresponding to the damping of the instrument and corresponding to external forces (i.e. the ground motion affecting the instrument). For an electromagnetic seismograph, there are two vibrating systems coupled to each other—seismometer and galvanometer—and then we have to solve a system of *two* differential equations. Fully exact solutions have offered great difficulties, and in many cases one has to be content with an approximate solution.

As already emphasized, the pendulum is not absolutely stationary when seismic waves arrive. At some point the pendulum has to be attached to its frame, and therefore it cannot entirely be prevented that the pendulum also makes some slight motion of its own. What we record is always the relative motion of pendulum and frame. This record would give an exactly true picture of the ground motion if the pendulum were absolutely still (while the frame, naturally, faithfully follows the ground motion). As this ideal case cannot be perfectly achieved, the record obtained will be more or less deformed, compared to the incoming wave motion. Also the relative periods of the ground motion and of the pendulum are of importance for the behaviour of the seismograph. If the pendulum period is much longer than the ground period, the deflections on the record are proportional to the *displacement (amplitude)* of the ground and the seismograph is an *amplitude meter*. Conversely, if the pendulum period is much shorter than the ground period, then the deflections are proportional to the ground *acceleration* and the seismograph is an *acceleration meter* (or *accelerograph*). Finally, when the two periods are about equal, the deflections are proportional to the *velocity* of the ground motion. However, this subdivision has no decisive importance in practice, because the three quantities, amplitude, velocity and acceleration, are mathematically connected to each other. Therefore, a seismograph which is said to record one of these quantities, also records the other two. In pure seismology, the ground motion is generally expressed by its amplitude (plus its period), while acceleration is of more concern to engineering seismology.

In the theoretical solution of the differential equations for a seismograph, it is customary to assume the ground motion to be simple harmonic (for instance, a simple sine curve of some given period). This is obviously a great simplification, as the ground motion deviates considerably from this simple type. However, in two ways it is possible to get mathematical expressions which resemble reality much better. One method is to replace the simple sine term by a summation over a large number of sine terms, with different periods. According to FOURIER's theorem it is possible in this way to represent any shape of curve or any ground motion. Each simple sine vibration has a certain period and is assumed to prevail without beginning or end. Another method, which may even better approximate the often impulsive onsets in the beginning of a record (*P* and *S*), is to assume a mathematical expression of a corresponding type. In course of time, several such investigations have been published.

2.2   *Pendulum Seismographs*

In this and in the following section we shall learn how the basic ideas described above have been applied in seismograph constructions. In fact, there are often a great variety of constructions which agree in principle, and therefore we shall limit the discussion to the most representative and most commonly used types.

Mechanical seismographs. We have already mentioned the Wiechert seismograph, which is one of the best-known types (Fig. 6). As an example, I might mention the Wiechert seismograph that was installed at Uppsala, Sweden, back in 1904. It has a pendulum of 1000 kg and it records the two horizontal components (E and N) but not the vertical component. This seismograph is still in operation, but not only out of pious considerations. By its relatively low magnification (maximum around 240 times at a period of about 8—9 sec), the instrument is still the very best one for the largest earthquakes. For these the motions are so strong that the records of the modern, highly sensitive instruments become almost unreadable.

The corresponding vertical-component seismograph of the Wiechert type was also constructed in the beginning of the century and is still operative at some stations. It was in general more difficult to keep in good operation. Another widely used type of mechanical seismograph was constructed by MAINKA in Germany.

Mechanical-optical seismographs. The Milne–Shaw seismograph from 1915 is another one of the early seismograph types, especially common in England and its former colonies. It can yield a magnification of the ground motion of up to 350 times for a period of 10–12 sec. It is also direct-recording, but in an optical way. A mirror is attached to the pendulum or other parts connected to it, and this reflects a light beam towards the photographic paper on a recording drum (Fig. 7). The magnification which is achieved mechanically (by means of levers) in the Wiechert seismograph, is here accomplished by the optical path length.

Another type of optical direct-recording seismograph is the torsion seismograph. Of these, the best-known one was constructed in 1925 in USA by ANDERSON and WOOD, and it is usually called the Wood–Anderson torsion seismograph. The pendulum consists of a metal cylinder (gen-

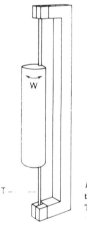

Fig. 8. The suspension (T) of the stationary mass (pendulum W) in the Wood-Anderson torsion seismograph. In this case the pendulum mass is of the order of a few grams. The height of the instrument is usually around 36 cm.

erally made of copper), which is attached to a vertical suspension wire along one of its generators. Instead of a cylinder, a metal plate can also be used. Upon arrival of seismic waves, the cylinder (or the plate) rotates slightly around the wire, and a mirror, attached to the suspension, is used for photographic recording of the displacements (Fig. 8). Two variations of this instrument were constructed. One of them had a free vibration period of 0.8 sec and a maximum magnification of 2800 (usually with an optical path length of 1 m), the other had a period of 6 sec and a maximum magnification of 800. The short-period version proved to be a stable and suitable instrument, particularly for the recording of local shocks, in which more high-frequency waves dominate. It was earlier a very much used instrument, among other places, in a station network in southern California. Its records also played a decisive role in the original formulation of the magnitude scale (see Chapter 4). The long-period variation of the torsion seismograph is most suited to recording of distant earthquakes. One drawback of these instruments is that they record only the horizontal components of the ground motion, but not its vertical component. Another drawback is their limited magnification, especially as regards the short-period type, which is insufficient for detection of the small ground motions, which is nowadays considered necessary. During the 1950's some improved versions of the torsion seismograph with photocell-type amplifier were constructed, notably in Pasadena in California.

Moving-coil seismographs. GALITZIN's application in 1906 of electromagnetic induction to the transmission of the pendulum motion (relative to its frame) to the records had a radical significance for later developments. The majority of the seismographs constructed since then and presently in operation at seismograph stations around the world is based on the pendulum principle plus electromagnetic induction. In course of time the original Galitzin seismographs have been improved and modified. At some stations, e.g. Kiruna, Sweden, Galitzin seismographs of the original type are still in operation. This is a set of three seismographs, one for each component, and in spite of their age they still operate properly, provided the instrumental parameters are under control. Their maximum magnification falls within the same period range as for the Wiechert or Milne-Shaw, and the instruments are intended only for recording of distant earthquakes.

The electromagnetic seismographs meant a great step forward compared to the mechanical ones. Practically no friction exists in the electromagnetic types, and this eliminates the large pendulum masses. Also, much higher magnifications are possible. Moreover, the electromotive force, recorded through a galvanometer, is proportional to the *velocity* and not the displacement of the pendulum, relative to its frame. This means that such seismographs are less sensitive to drift of the equilibrium position than mechanical seismographs may be.

In 1929, GUTENBERG modified a Galitzin seismograph by lowering the seismometer period to 3 sec and increasing the magnetic field by using a narrower air gap. In this way, a much better recording was achieved of short-period waves (such as of local earthquakes and of P from distant earthquakes). The maximum magnification was kept at 4400 and was in fact limited only because of the background noise (microseisms).

In the 1940's, SPRENGNETHER in St. Louis, USA, constructed two seismograph types, of which one deviates from the original Galitzin only in mechanical details, while the other has almost the same characteristics as GUTENBERG's modification. Other developments of moving-coil seismographs have been made by BENIOFF (USA), WILLMORE (England), HILLER (Germany), GRENET and COULOMB (France), KIRNOS and KHARIN (USSR), JOHNSON and MATHESON (USA), and others. The Grenet–Coulomb type, generally called 'Grenet', is among other places used at several of the stations in Sweden (Fig. 9).

Fig. 9. Short-period vertical-component seismometer of type Grenet-Coulomb. M = magnet, P = coil, F = helical spring, B = suspension of the spring in the frame, W = pendulum mass (of the order of 1—2 kg), A = rotational axis (horizontal). The suspension corresponds to the one shown in Figure 5a. Photo H. BORG.

A number of seismographs, often of the moving-coil type, have been constructed for special applications. Among these we want to mention seismographs with small magnification (usually only up to 100 times) to be used in earthquake regions. Such instruments, so-called *strong-motion seismographs*, are of great significance in engineering seismology. Another important development consists of *portable seismographs*, which find much application for recording of aftershocks and microearthquakes in temporary installations, as well as for recording of explosions for prospecting work. In the latter application, the seismometers are usually termed *geophones*.

We have seen that the earlier seismograph types (Wiechert, Milne–Shaw, Galitzin, and others) had their maximum magnification of the ground motion at a period around 10 sec. This was sufficiently good to produce fine records of stronger, distant earthquakes, showing both preliminary phases (*P* and *S*) and surface waves. In this way, these instruments were well suited to form the bridge between theory and observations that was so essential for the development of seismology (Chapter 1). However, a requirement gradually emerged to extend the period range considerably, partly towards shorter periods (around 1 sec and below) to be able to record local shocks and also to get more exact information on *P*-waves from distant events, partly towards longer periods (of the order of several minutes) to be better able to record long-period surface waves from distant earthquakes. The first attempts to extend the range towards shorter periods are marked by the Wood–Anderson torsion seismograph and GUTENBERG's modified Galitzin. Later and more efficient developments in this direction were made

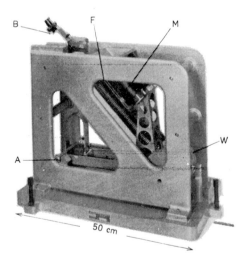

Fig. 10. Press-Ewing vertical-component seismometer. W = pendulum mass (of the order of 7 kg), M = magnet, F = helical spring, B = suspension for the helical spring. A = horizontal rotation axis. Photo Sprengnether Co., St. Louis, USA.

by the seismologists whose names were given above. These developments commenced in the 1930's. The extension towards longer periods is a considerably later development, in the 1950's. One of the most significant constructions in this connection is the Press–Ewing seismograph or the Columbia seismograph, as it is also called (Fig. 10).

Originally, PRESS and EWING used a seismometer period of 15 sec and a galvanometer period of about 75 sec. Later, seismometer periods of 30, 60 sec or even longer have been used. Long-period vertical-component seismometers are sensitive to drift of the equilibrium position, especially due to variations in the atmospheric pressure. Such effects can be avoided by enclosing the seismometer in an airtight case or by applying a compensator for variations in air pressure (Archimedes' principle!). In addition, long-period seismometers (both vertical- and horizontal-component) are often sensitive to air currents (convection) in the seismometer vault (particularly when these are heated), and therefore they must be protected by special covers from direct impact of such currents. Another method to achieve high sensitivity for long periods, without using seismometers and galvanometers of very long periods, with ensuing problems, is to use *overdamped* seismometers and galvanometers of more medium periods.

Reluctance seismographs. The reluctance seismograph implies a variation of the principle of the moving-coil seismograph. The principle of a

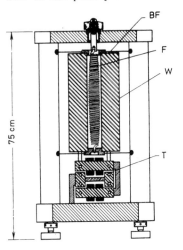

Fig. 11. Schematic cross-section of BENIOFF's variable-reluctance seismometer (vertical component). BF =flat springs which restrict the pendulum to a purely vertical motion and at the same time to a certain extent determine the free period, F = helical spring, W = pendulum mass (about 100 kg), T = transducer (shown in detail in Fig. 12).

reluctance seismometer, constructed by BENIOFF, is clear from Figure 11–12. M is a permanent magnet, O pole-pieces (flux distributing members) of soft iron, N armature of soft iron, P coils. The magnet M is attached to the seismometer pendulum, while the armature N and the coils P are fixed to the frame, or vice versa. Upon motion of the magnet M in relation to

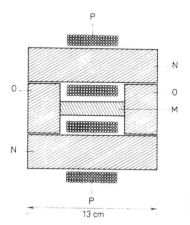

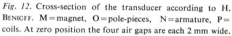

Fig. 12. Cross-section of the transducer according to H. BENIOFF. M = magnet, O = pole-pieces, N = armature, P = coils. At zero position the four air gaps are each 2 mm wide.

the armature, the air gaps between N and O are varied and thus the magnetic reluctance (= magnetic resistance) is varied with a corresponding variation of the magnetic flux through the magnetic circuit. In this way, an electromotive force is induced in the coils P, and this is then recorded via a galvanometer. The part shown in Figure 12 is called a *transducer* and is in effect nothing other than a modification of an ordinary telephone receiver.

In a usual construction of BENIOFF's reluctance seismograph, the transducer contains eight coils. Of these, four are connected in parallel to a short-period galvanometer (usually of free period 0.2 or 0.7 sec), while the other four coils are connected in series to a long-period galvanometer (usually of a free period of 80–100 sec). In this way, it is possible to pick out two quite different spectral ranges from one and the same seismometer. In the Benioff seismometers, pendulum masses around 100 kg are used. The pendulum (Fig. 11) is restricted in its relative motion to a vertical direction by flat springs above and below the pendulum mass. By altering the tension in these springs it is possible to vary the free period of the pendulum within certain limits. The helical spring in the vertical-component seismometer is made of a special nickel alloy with very low temperature coefficients for elasticity and expansion. The pendulum remains stable and operative even if it should be exposed to temperature variations up to 55 °C. The corresponding horizontal-component seismometers have a similar construction, only with the difference that special flat springs restrict the pendulum motion to a horizontal direction.

The development of very useful variable-reluctance seismographs is due especially to the availability of magnetic material of much higher quality than earlier, which makes it possible to use much stronger magnetic fields.

Electrostatic seismographs. Only a few electrostatic seismographs have been constructed. Their common principle is to use the variable capacity in a condenser with one plate attached to the pendulum and the other plate attached to the frame. Such instruments have been constructed by BENIOFF (USA) and GANE (South Africa). The condenser (with variable capacity) is part of an electronic circuit, which in turn is connected to a short-period galvanometer for recording in the usual way. These electrostatic seismographs are useful only for the recording of short-period waves, and so far they have found only limited application for special purposes. The seismographs installed on the moon (first in July 1969) are of a capacitor type, with a resonant period of 15 sec for the long-period types, and 1 sec for the short-period one.

A number of constructions of electronic seismographs have been proposed. For instance, the application of various electric filters offers a possibility to separate out different period ranges from a given seismometer. In general, such constructions involve much more complicated electric circuits than the simple electromagnetic seismographs, often with less reliable operation as a consequence.

Recording methods. Of the various recording methods, the photographic one has dominated ever since GALITZIN's days. This is still the case. In general, this method is very reliable in operation, while the drawbacks consist in the developing, etc., of the records before they can be read. Figure 13 shows a typical interior of a photographic recording room. Instead of recording on photographic paper, sometimes film recording (usually 35 mm film) is applied. In order to eliminate the photographic

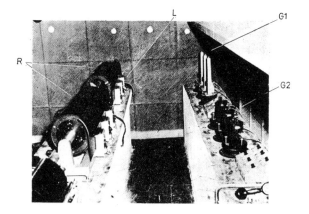

Fig. 13. The interior of a pho-tographic recording room. R = two so-called triple drums, each with provision for three records of size 30×90 cm, L = light sources, G1 = long-period galvanometers (free period 100 sec), G2 = short-period galvanometers (free period 0.7 sec). According to the U.S. Coast and Geodetic Survey (now the U.S. Geological Survey).

work, as well as the need for darkrooms for recording and developing, instruments with visible recording have been introduced in recent years. They have the further advantage of permitting an immediate inspection of the records. There are several different types available on the market, such as ink-writing recorders, hot-stylus recorders, etc. In order to operate these, an amplifier has to be installed between the seismometer and the recorder. Therefore, the operation may be more expensive and less dependable than the photographic one.

Quite another principle for the recording is applied in the so-called *digital seismographs*, developed at some institutes in the USA, among others. In modern research large electronic computers are constantly used. As a rule, the ordinary records cannot be used directly with computers, but have to be digitized first (i.e. the curve has to be represented at equal intervals by numerical values, corresponding to the continuous curve). This transcription of the record is frequently a laborious job, at least when more sophisticated digitizers are not available. Therefore, in the digital seismographs this trouble is eliminated and one may get directly a series of numbers (or some other equivalent data form) instead of the continuous record. Another newer development, which also facilitates the application of computers, consists in the recording of seismometer outputs on magnetic tape.

Calibration. Frequently, electromagnetic seismographs are adjusted to critical (aperiodic) damping, both for the seismometer and the galvanometer. In this way the picture of the ground motion will be as true as possible, and moreover, the mathematical treatment will be considerably simplified. In modern electromagnetic seismographs (as for instance, Benioff, Grenet) the damping is achieved by electric resistances in the circuit connecting the seismometer with the galvanometer. By the shunt shown in Figure 14 it is possible to reduce the magnification to a value which is

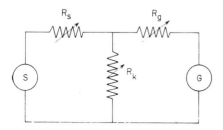

Fig. 14. A standard circuit between an electromagnetic seismometer (S) and a galvanometer (G). R_s = resistance for regulating the seismometer damping, R_g = resistance for regulating the galvanometer damping, R_k = bridge resistance for regulating the sensitivity (magnification).

suitable with regard to the always prevailing background noise. Nowadays, there is no problem reaching high magnifications. Instead the problem consists just of the microseismic disturbances in the ground, which always put a limit on the magnification that can be used. For instance, with a modern short-period instrument, with a galvanometer directly connected to a seismometer, the magnification would be too big and of no use. Instead the magnification has to be reduced by using a bridge as shown in Figure 14. The magnification has to be adjusted so as to give the optimum signal-to-noise ratio.

In seismographs of older construction, the damping was achieved in various ways, as for instance by air damping (Wiechert), oil damping (in an early version of the Benioff variable-reluctance seismograph), by eddy currents (Foucault currents) in a copper plate movable in a magnetic field (Galitzin, Sprengnether, Wood–Anderson). The Wiechert and Galitzin seismographs usually work in a slightly underdamped condition (Fig. 4).

Besides time measurements on seismograms, also amplitude measurements are of essential importance, not only for the calculation of magnitudes of earthquakes (see Chapter 4), but also for the study of wave propagation through the earth's interior. In the long run, it is not unlikely that amplitude measurements can furnish more complete information about the earth's interior than time measurements alone can do, even if the amplitude measurements are considerably more difficult to interpret. As only the ground amplitudes and not the recorded amplitudes ('trace amplitudes') are of interest (except in special cases concerning only relative measurements), then we are faced with the problem of calibrating our seismographs. A physicist, accustomed to accurate laboratory measurements, will certainly be surprised at the relative lack of accuracy in the determination of the true ground motion. The reasons are to be found partly in insufficiencies in the seismograph theories, as hardly any theory has been proposed taking due account of all factors, partly in the complications of the ground motion itself. On the other hand, it should be emphasized that the accuracy of the measurements is in general sufficiently great for most seismological purposes.

In principle, the magnification is determined by imparting to the seismometer pendulum a known displacement, e.g. by means of an electromagnetic hammer (Galitzin) or by means of weights placed on the pendulum and lifted (Benioff, Grenet, Wiechert), and measuring the corresponding displacement on the record. One difficulty, especially for high-sensitive short-period seismographs, is that even a very small pendulum displacement, not measurable with commonly available techniques, will produce a very large displacement on the record. As an example of this, we may mention that on the Benioff reluctance seismometers at Uppsala a pendulum displacement of only 0.01 mm produces a displacement of about 400 mm on the record. In such cases, the pendulum displacement is not *measured* but instead it is *calculated*, which is possible from known values of the pendulum mass, the seismometer free period and the mass of the test weight used. This impulse method has to be accompanied by measurements of the free periods and the damping of both the seismometer and the galvanometer. Having this, it is possible by means of theoretically deduced formulas (i.e. solutions of the differential equations for the whole system) to calculate the magnification curve or response curve, as it is also called. A few typical response curves are shown in Figures 15 and 16.

In order to avoid the detailed determinations of each constant one by one and to avoid the often imperfect theoretical assumptions, which have to enter the procedure just described, other and more direct methods have been developed to determine a response curve. One such method

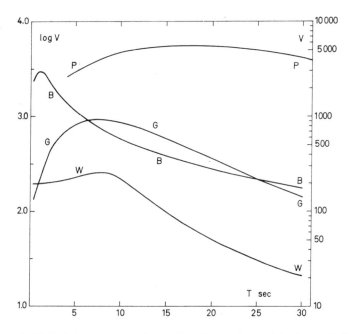

Fig. 15. Typical response curves for amplitude for some long-period seismographs. BB=Benioff N (Up), WW=Wiechert N (Up), GG=Galitzin N (Ki), PP=Press-Ewing N (Um). V=dynamic magnification, T=period. Cf. Table 2.

uses a *shaking table.* In this method, the seismometer is placed on such a table which is set into vibration (usually simple harmonic vibration) with known amplitude and period. The oscillation is recorded and this gives immediate information on the magnification at the period used. The experiment is repeated with another period of vibration of the shaking table and after a sufficient number of such tests it is possible to construct a complete response curve. In some constructions it is also possible to generate wave shapes more similar to those encountered on seismograms. This shaking-table method has the advantage that it completely avoids all theoretical difficulties, but, on the other hand, there may be a number of other problems with this method, especially of a technical nature. For instance, the shaking table should not have any free vibrations of its own within the investigated period range. For tests of horizontal-component seismographs it is very significant that the table does not tilt in any way under the load. As a consequence, shaking tables have been mostly applied for testing short-period seismographs, especially those used in seismic prospecting. However, more recent constructions have permitted testing with periods of several minutes.

In other calibration methods, a device is used by which the seismometer pendulum is set in motion of known amplitude and period, without

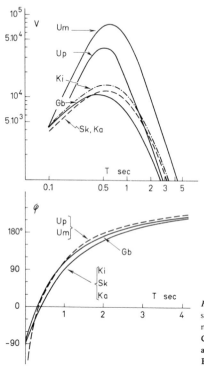

Fig. 16. Response curves for amplitude (*V*) and phase shift (*φ*) for some short-period vertical-component seismographs at Swedish stations: Benioff at Up and Um, Grenet at the other stations. The station abbreviations are explained in Table 2. After E. S. HUSEBYE and B. JANSSON (1966), modified.

any need to move the seismometer to any shaking table. There are several different implementations of this method, for instance, by attaching to the pendulum boom a small magnet, movable in a fixed coil. By connecting an electric oscillator to the coil, the magnet and thus the pendulum are set into vibrations which are recorded. In another calibration method, developed by the British seismologist WILLMORE, the seismometer coil itself is used to put the seismometer into forced oscillations, where the seismometer coil is part of a Maxwell resistance bridge. Many modern seismographs are equipped with calibration coils which make it possible to test them easily every day. The recorded test pulses provide a continuous check on the operation. In other constructions, use is made of a condenser, of which one plate is attached to the pendulum boom and the other is fixed. By connecting the condenser to an electric voltage oscillator, the air gap in the condenser is varied in a fully controllable way.

An exactly known response curve is a prerequisite to our being able to make full use of a given record. Only a magnification which were completely independent of the vibration period could furnish a fully true picture of the ground motion. This ideal case cannot be achieved by pendulum instruments, even though this condition is sometimes aimed at, for instance, by connecting a short-period Benioff variable-reluctance seismometer to a very long-period galvanometer. In other cases, the ideal is instead a response curve which is as selective as possible, i.e. it has a sharply limited maximum around some period range which one wants to emphasize and only small magnification outside this range. This is generally the principle followed in the short-period seismographs, which are suitable for the short periods in near events and for *P*-waves from distant earthquakes. On the other hand, they do not in general record the more long-period transverse waves or surface waves, except for larger shocks.

From Figure 15 we see that the Galitzin has a larger magnification than the long-period Benioff within the period range of 6–25 sec, but smaller outside this range. As a consequence, short-period *P*-waves (periods less than 6 sec) and long-period surface waves (periods over 25 sec) are better recorded by the Benioff, while *S*-waves (generally with periods of 10–20 sec) are better recorded by the Galitzin. The maximum magnification of both the Galitzin and Wiechert falls within the range of the usual microseisms (period range about 4–10 sec), which is a certain drawback.

However, the motion shown on the seismic records deviates from the ground motion not only because of the magnification. There is also a phase shift between the recording and the ground motion. Figure 16 shows some response curves both for amplitude and for phase shift as functions of the period of the ground motion. The value of the phase shift depends partly upon the properties of the seismograph (free period, damping), partly upon the period of the ground motion. As the ground motion is in general composed of a large number of different periods, we understand that the resulting record must be quite distorted in relation to the incoming vibration. But an exact knowledge of the response characteristics of the seismograph, both concerning amplitude and phase shift, permits a calculation of the true ground motion.

2.3   *Other Seismograph Types*

We have on purpose dealt quite extensively with pendulum seismographs, as these still play a dominant role in seismology. Among seismograph types, which are not based upon the pendulum principle, the most important is the *strain seismograph*, constructed by BENIOFF in the 1930's.

While the pendulum seismographs react to translation in the ground, the strain seismograph reacts to deformation in the ground. In principle, this instrument consists of a long bar (R in Fig. 17) which is attached to

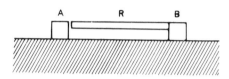

Fig. 17. The principle of a strain seismometer. A, B = two piers firmly attached to bedrock at a mutual distance of 20—50 m, R = rigid tube of steel or fused quartz attached to B and extending very nearly to A. A transducer for conversion of mechanical vibration into electric current is mounted in the air gap at A.

the ground at one end B, while the other end extends to another point A, also attached to the ground. As such a system is hit by a seismic wave, a relative motion occurs between the two points A and B. As a consequence, the small air gap at A between R and A will vary, and this variation can be measured. The usual construction is to provide this end, A, with a transducer, of the same type as in the variable-reluctance seismometer, and then recording can be made in the usual way via a galvanometer. In some installations, provisions have also been made for direct visual readings of the air gap by means of a microscope; such readings can reveal longer-period deformations in the earth. By using a quartz tube (R) it has been possible to measure permanent deformations of 10^{-9}, i.e. an extension (or contraction) of 1 mm per 1000 km. For non-permanent deformations, the sensitivity is about 10 times as high.

In actual constructions, the bar R is made of quartz because of its low temperature coefficient. Earlier a steel bar was used, but with less satisfactory results. The length of the bar R is usually of the order of 20–50 metres. In order to get a well-measurable relative displacement of the two points A and B, the bar should exceed a certain minimum length; on the other hand, the construction problems and the installation put an upper limit on the suitable bar length. The strain seismographs, constructed by BENIOFF, were made exclusively for measurements of the horizontal component, but not the vertical component, of the relative displacements. On the other hand, he has successively used a combination of two of his strain seismographs, oriented at right angle to each other.

BENIOFF installed his strain seismographs in some mountain tunnels in southern California. They are very sensitive to temperature variations (in spite of the quartz bar) and they have even been able to record the daily expansion and contraction of the mountain where they are installed. Only deep underground installations have some prospect of leading to useful recordings. In connection with the International Geophysical Year 1957–8, BENIOFF installed similar instruments also in South America (Peru). Strain seismographs have been in use also in Japan and in Germany. They are still rather rare, which among other things is due to difficulties in finding suitable localities for their installation.

The period response of a strain seismograph depends only upon the galvanometer, as the seismometer has no free period of its own. By varying the galvanometer properties a possibility is offered to vary the seismograph response within wide ranges. For instance, in combination with a galvanometer of 0.8-sec period, a response curve is obtained which corresponds to the Wood–Anderson short-period torsion seismograph. As the superiority of a strain seismograph compared with a pendulum seismograph increases with the period, it is customary to connect the strain seismometer to very long-period galvanometers, for instance, of 70 or 180 sec or even up to 8 min (in Pasadena). It was with such a system with a 180-sec galvanometer that BENIOFF was able to record a ground motion of 57 min period from the great Kamchatka earthquake of November 4, 1952. This constituted one of the strongest impulses to continued theoretical and observational studies of the free vibrations of the earth, generated by large earthquakes. Such seismographs have also been used to record the daily tides in the solid earth as well as to investigate the slow motions in the solid earth, which give rise to earthquakes in seismically active regions.

It is frequently an advantage to be able to use much longer base lines (up to several kilometres). In such cases, a strain seismograph as described here can naturally not be used. Instead, corresponding strain or deformation meters have been developed, which use light or microwaves between two points. In this case, the atmospheric influence on the wave propagation enters as a special factor to correct for. Recently, it has been possible to eliminate such influences from the observations by a sophisticated combination of double laser rays. The laser reflector method is based on the interference principle. Two laser rays are made to interfere with each other, and shifts in their interference pattern can be measured to a high accuracy. Such devices permit higher accuracy in strain measurements than BENIOFF's strain seismometer. Accuracies of one part in 10^{12} have been attained, i.e. a change of length of 1 mm per 10^6 km ($=25$ times the earth's circumference). Contrary to BENIOFF's strain seismometer, the length of a laser strain meter is practically unlimited, and therefore its sensitivity can be made very high. In addition, a laser strain meter has a linear response, as there is no dependence on the frequency of the ground motion. The method no doubt provides accurate means for measuring small motions of large scale. Examples are the relative motions across a fault, which have significance in connection with earthquake prediction (Chapter 10).

A number of other physical phenomena has been applied in the construction of seismographs and, even more commonly, for construction of static pressure meters. This concerns such phenomena as, for instance, that minerals exposed to pressure show certain electric and magnetic properties (piezoelectricity, magnetostriction). Thus, the Swedish professor NILS HAST has constructed a measuring sonde, based on magnetostriction, which can measure the absolute pressure in solid rock with high accuracy. Originally developed for stress measurements in cement and concrete, and later applied to measurements of stability in mines, the method has proved more and more to be of very great geophysical significance.

2.4   *Some Lines of Development*

The present network of operating seismograph stations in the world consists of nearly 1500 stations (about 1460 listed in 1974). However, they are very unequally spaced. While USA, incl. Hawaii and Alaska, have about

480 stations and Japan about 155, there are only very few in large areas, especially in the great oceans. In recent years, much effort has been devoted to improvement of the station networks, partly by standardizing instruments, partly by extending observations to new areas. The latter item includes installation of unmanned stations in remote parts of the continents and of stations on the sea bottom. Several experiments with the operation of ocean-bottom seismographs have been made, particularly by the Russians and the Americans. But still very much remains to be done, before the earth is covered by an equally spaced station network.

Among instrumental improvements, I may mention the so-called array stations (or multiple stations). They consist of a system of seismo-meters arranged in some regular geometric pattern over an area extending over a few kilometres to hundreds of kilometres. Such stations, built mainly for the detection of nuclear tests, now exist in the USA, Canada, Scotland, Sweden, Norway, India, Australia, and in a few other places. In combination with recordings on magnetic tape and the use of big computers, such stations offer a quicker and more complete analysis of the recorded waves. However, special research has shown that even more tradi-tional station networks can be treated as great array stations, as long as the signals on the different stations show enough similarity that array techniques can be applied. Tests of this kind on the Swedish network were very promising, probably because of the similar ground conditions at the different stations. Recordings of seismometers placed in deep bore holes (so-called bore-hole seismographs) is another method which can be used with advantage to avoid the disturbing background noise on the surface. A suitably placed bore-hole seismometer could reach a higher signal sensi-tivity than a considerably more expensive array station on the surface or near to it.

As a consequence of more numerous and more sensitive stations, the data flow to the world centres has increased enormously. This has led to some reorganization of such centres, and since 1960 big computers are regularly used for the calculation of earthquakes and other seismic events.

In Chapter 8 we shall describe in more detail how we are trying to improve seismic records and their sensitivity. In Chapter 12 we shall learn something about moon seismographs, which have been developed especially for the observation of seismic phenomena on the moon.

It is a correct statement that the development during the last decades has not particularly been focussed on new constructions of seismometers. Those already in existence and described above have proved to be practically unsurpassed. Rather, the development has been concerned with the location and distribution of these seismometers with those modifications of the construction that this can entail. The general trend of the instrumental development has been toward more sensitive, stable, and rugged equipment that can be set up quickly and can operate reliably under the most severe field conditions. This development has been paralleled by improved recording technique and improved data handling.

2.5   A Seismograph Network

Most countries have now well-developed seismograph networks, usually operated by some central institution. As an illustration of such a network, I shall here briefly describe the one which is operated by the Seismological Institute in Uppsala, Sweden, and with which I am most familiar for obvious reasons. The present seismograph network in Sweden is equipped with modern, high-sensitive instruments (Table 2). The recording is made photographically via galvanometers except in the following cases: Wiechert EN (Uppsala) records mechanically with a stylus on smoked paper; Press–Ewing ENZ (Uppsala) recorded in 1962–8 on magnetic tape in parallel with the photographic recordings; Press–Ewing Z (Uppsala) with visible recording has an ink writer with an amplifier. The Wiechert seismograph of 1904 has been in practically continuous operation since then and has up to now (1977) produced an archive of nearly 53 000 records. It is still the best instrument for the largest earthquakes, just because of its lower magnification.

All records from the network are sent weekly to the institute in Uppsala, nearly all undeveloped. At Uppsala, the records are developed and analysed for our bulletins (see Chapter 8) and stored to be used for special research. With presently (1977) 27 parallel records per day (each record measuring about 30×90 cm in size), our archive increases by about 9850 papers per year. In order to provide for an additional file, these records are put on microfilm.

The long-period seismographs are best suited to recording of distant

Table 2. The Swedish network of seismograph stations, operated by the Seismological Institute at Uppsala.

Station; geographic coordinates; height; ground	Seismograph		Seismometer period sec	Galvanometer period sec	Maximum dynamic magnification	Start of operation
Uppsala (Up):	*Short-period:*					
59°51.5′ N, 17°37.6′ E;	Benioff	E	0.9	0.7	80 000⎫	
14 m; granite	Benioff	N	0.8	0.7	80 000⎬	1955
	Benioff	Z	1.0	0.7	40 000⎭	
	Grenet-Coulomb	Z	1.4	0.7	13 510	1951 (1969)
	Medium-period:					
	Benioff	E	0.9	83	3 030⎫	
	Benioff	N	0.8	92	3 290⎬	1955
	Benioff	Z	1.0	88	3 670⎭	
	Wiechert	E	11	—	240⎫	1904
	Wiechert	N	10	—	230⎭	
	Long-period:					
	Press-Ewing	E	15	103	2 270⎫	
	Press-Ewing	N	15	98	2 250⎬	1957
	Press-Ewing	Z	15	100	1 770⎭	
	Press-Ewing (visible)	Z	15	100	1 200	1963
Kiruna (Ki):	*Short-period:*					
67°50.4′ N, 20°25.0′ E;	Grenet-Coulomb	Z	1.4	0.7	13 310	1951
390 m; porphyry	*Medium-period:*					
	Galitzin	E	12	12	770⎫	
	Galitzin	N	12	12	800⎬	1951
	Galitzin	Z	10	10	860⎭	
	Long-period:					
	Press-Ewing	Z	15	100	3 800	1963 (1971)
Skalstugan (Sk):	*Short-period:*					
63°34.8′ N, 12°16.8′ E;	Grenet-Coulomb	Z	1.3	0.8	12 040	1956
580 m; gneiss						
Umeå (Um):	*Short-period:*					
63°48.9′ N, 20°14.2′ E;	Benioff	E	1.0	0.7	75 000⎫	
16 m; mica gneiss and	Benioff	N	1.0	0.7	75 000⎬	1962[1]
pegmatite (WWSSN)	Benioff	Z	1.0	0.7	75 000⎭	
	Long-period:					
	Press-Ewing	E	15	100	5 500⎫	
	Press-Ewing	N	15	100	5 500⎬	1962
	Press-Ewing	Z	15	100	5 500⎭	
Uddeholm (Ud):	*Short-period:*					
60°05.4′ N; 13°36.4′ E;	Benioff	Z	1.0	0.7	75 000	1967[2]
240 m; granite						
Delary (De):	*Short-period:*					
56°28.2′ N, 13°52.2′ E;	Grenet-Coulomb	Z	1.4	0.7	12 990	1967
150 m; gneiss						

Table 2. The Swedish network of seismograph stations, operated by the Seismological Institute at Uppsala (continued).

Station; geographic coordinates; height; ground	Seismograph		Seismometer period sec	Galvanometer period sec	Maximum dynamic magnification	Start of operation
Göteborg (Gb): 57°41.9′ N, 11°58.7′ E; 66 m; gneiss	*Short-period:* Grenet-Coulomb	Z	1.4	0.5	10 530	1958[3]
Karlskrona (Ka): 56°09.9′ N, 15°35.5′ E; 11 m; granite	*Short-period:* Grenet-Coulomb	Z	1.5	0.7	11 590	1961[3]
Hedemora (Hd): 60°17.3′ N, 15°56.9′ E; 124 m; gneiss-granite	*Short-period:* Grenet-Coulomb	Z	1.4	0.7	(15 000)	1969[4]
Kungsör (Ku): 59°24.1′ N; 16°08.0′ E; 63 m; gneiss	*Short-period:* Grenet-Coulomb	Z	1.4	0.7	(15 000)	1969[4]

[1] Umeå operated a Grenet-Coulomb Z seismograph in 1960—2, until the world-wide standardized equipment of the U.S. Coast and Geodetic Survey was installed in 1962.
[2] Uddeholm operated a Grenet-Coulomb Z seismograph in 1966—7.
[3] The operation of Göteborg and Karlskrona was discontinued in 1968, and replaced by the more sensitive station at Delary.
[4] Hedemora and Kungsör operated for one year (1969—70) together with Uppsala as a triangular array.

earthquakes, and then especially for the more long-period waves, as the transverse waves and surface waves. For such recordings, the net does not need to be particularly dense, and our three stations with such apparatus (Uppsala, Kiruna, Umeå) fulfill the need very well. According to international recommendations, fully equipped stations (like those we have at Uppsala, Kiruna and Umeå) should be placed at intervals not exceeding 1000 km. As is evident from the map in Figure 18, our net fulfills this requirement very well. On the other hand, a denser net of short-period stations is needed, partly for a more complete recording of the short-period components of *P*-waves from distant events, partly for recording of near events, i.e. those which occur within Scandinavia and its nearest surroundings. This requirement is well fulfilled together with our supplementary stations (Skalstugan, Uddeholm, Delary).

The present network extends over more than 1300 km and exhibits also a relatively even distribution over the country. Its extent and regularity

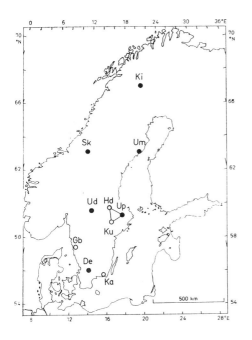

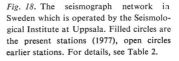

Fig. 18. The seismograph network in Sweden which is operated by the Seismological Institute at Uppsala. Filled circles are the present stations (1977), open circles earlier stations. For details, see Table 2.

permit good locations not only of near events but also of those at greater distance. It should also be emphasized that our network ties in very well with corresponding networks in the neighbouring countries. Seen from a global viewpoint, there is no doubt that Sweden, like all the Nordic countries, is today well equipped with seismograph stations.

The combination of highly sensitive instrumentation and good bedrock has placed our network among the most sensitive ones in the world. As an example of the sensitivity of the network, I quote here the number of events recorded during the first six months of 1970:

Month	Number of events
January	377
February	371
March	458
April	804
May	538
June	524

The total amounts to 3072 events, i.e. on the average 16.8 events per day, and this is approximately a normal figure. The majority of the events are distant earthquakes but even a number of Scandinavian events (both explosions and earthquakes) is recorded.

When large earthquakes take place, these are generally followed by a great number of aftershocks. Then, the number of recorded events is considerably higher than just mentioned. One example is provided by August, 1969, when after a Kurile Islands earthquake on August 11, we recorded during the following day (i.e. for 24 hours) no less than 219 aftershocks. During the month of August, 1969, we were also able to identify by means of our network 172 more aftershocks than reported from anywhere else.

For one year, August 1969 to August 1970, the Seismological Institute in Uppsala also operated a triangular array station. The three stations (Uppsala, Hedemora and Kungsör) formed an almost equilateral triangle with a side of about 100 km length. The recording was made centrally in Uppsala over the telephone lines.

In addition to the stations listed in Table 2, the Seismological Institute at Uppsala has at different times operated about a dozen temporary stations, partly in collaboration with the Research Institute of National Defence, Stockholm. These stations have been located at places scattered over the whole of Sweden and they have operated only for a few months each. This has been done mainly for investigations of signal sensitivity (Chapter 8).

Chapter 3
Seismic Waves

3.1 *Main Types of Seismic Waves and Fundamental Laws*

By seismic waves we mean every motion that can be observed on a seismogram, with the exception of direct disturbances of the instruments. The seismic or elastic waves, which arise through the sudden rupture in an earthquake source or by an explosion, propagate through the whole of the earth's interior or along its surface layers. The waves are recorded by seismograph stations the world over, provided that the released energy has been big enough. The seismic waves are of two main types:

1. Body waves, which propagate through the interior of the earth. These consist of two types:
a) longitudinal waves or *P*-waves;
b) transverse (or shear) waves or *S*-waves.

2. Surface waves or guided waves, which propagate along some surface. These consist of the following types:
a) Love (*L*) and Rayleigh (*R*) waves which follow the free surface of the earth;
b) Stoneley waves, which are related to Rayleigh waves, but follow a discontinuity surface in the earth's interior;
c) channel waves, which propagate along some layer of lower velocity in the earth's interior.

While the body waves can be considered as 'free waves', i.e. they have freedom to propagate in practically every direction through the earth's interior, the surface waves are 'bound waves', i.e. they are bound to some surface or some layer during their propagation.

Besides in their way of propagation, the different wave types also differ both concerning particle motion and propagation velocities. *P* is longitudinal, i.e. particles hit by this wave oscillate back and forth around their equilibrium position in the same direction as the wave propagates. *S* is transverse, i.e. the particle motion is confined to a plane perpendicular to the direction of propagation. For simplicity, we split the *S*-wave motion into a horizontal component (*SH*) and a vertical component (*SV*). Love waves (*L*) have a particle motion which agrees with *SH*, i.e. transverse horizontal. For the Rayleigh wave (*R*), the particle motion is elliptic, with the plane of the ellipse vertical and lying in the plane of propagation. The rotation in the ellipse is retrograde. With regard to speed of propagation, this is highest for *P* and decreasing in the following order *P–S–L–R*. As a consequence, at some distance from the source, *P* is recorded first (*P=* =primary), followed by *S* (*S*=secondary), *L* and finally *R*. Figure 19

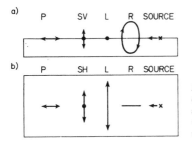

Fig. 19. Sketch to demonstrate the propagation of the direct waves in a slab: a) seen from one side, b) seen from above. The arrows indicate the particle motions and the waves have been arranged in order of their propagation velocities.

shows schematically the particle motion and the propagation of the different waves.

We came across the different wave types already in Chapter 1. In fact, it is the elasticity theory which can give complete information about these waves, in other words, about the behaviour of materials upon compression or dilatation (in the *P*-wave) and upon shear stresses (in the *S*-wave). The conditions can be expressed in an exact mathematical form. In most cases, however, simplifying assumptions are justified and acceptable in the study of the conditions of the earth's interior. Thus, it is assumed that:

1. Relative displacements between adjacent particles are infinitesimally small.

2. The material is perfectly elastic, i.e. the stress is a homogeneous linear function of the strain and vice versa; usually, a generalized form of Hooke's law is applied.

3. The material is isotropic, i.e. the elastic parameters are independent of direction (the same in all directions).

4. External forces, such as gravity, friction, etc., can be neglected.

Under certain circumstances it may be necessary to abandon one or the other of the above-mentioned simplifying assumptions. In such a case, we may be faced with the problem of finite displacements (especially in the near vicinity of the source) instead of infinitesimal ones; or we may have to deal with time effects of the material exposed to stress (in the simple elastic case, the deformation follows the application or the removal of stresses instantaneously, but this is often not the case with real materials, which exhibit a certain lag). Such time effects are often so difficult to treat in an exact mathematical way that laboratory models may be necessary to give more reliable results. If the material is not isotropic, but has elastic properties which are different in different directions, the material is said to be *anisotropic* or *aeolotropic*. Such conditions certainly exist in some parts of the earth's interior and they play a certain role in the wave propagation. External forces, such as gravity, have in general only negligible effects on the wave propagation, while friction (absorption) can be of great significance in certain studies.

In spite of the assumptions underlying the simple elasticity theory, this has proved to be remarkably useful in the treatment of wave propagation through the earth. In fact, it is only in more special cases that it is found necessary to drop one or another of the simplifying assumptions, and then mostly just to investigate this particular effect. The major part of seismology, including our knowledge of the earth's interior, is based on these simple assumptions. And investigations have confirmed that the results are nevertheless remarkably accurate. For the most part additional investigations deal just with improvements of the picture already arrived at by these simple means. Seismology is certainly not unique in this situation: that a first-order theory can lead very far, but for even modest further progress, an enormously larger input is required. See the diagram in Figure 2.

In the simple elasticity theory, the elastic properties of the material are characterized by only two independent parameters. A number of these have been defined in the literature, which have certain relations to each other. We shall choose $k = incompressibility$ *modulus* or *bulk modulus* and $\mu = modulus$ *of rigidity* or *shear modulus*. If we denote the density of the material by ϱ, then the wave velocities v_P and v_S for P- and S-waves, respectively, are given by the following simple equations:

$$v_P = \left(\frac{k + \frac{4}{3}\mu}{\varrho}\right)^{\frac{1}{2}} ; \quad v_S = \left(\frac{\mu}{\varrho}\right)^{\frac{1}{2}} \tag{1}$$

For the earth's crust and the upper part of the earth's mantle, we have with good approximation that $k = \frac{5}{3}\mu$ (corresponding to a POISSON ratio $= \frac{1}{4}$, i.e. a ratio of $\frac{1}{4}$ between the lateral contraction and the longitudinal extension of a cylinder). Substituting this into equation (1), we find that for the upper parts of the earth $v_P : v_S = \sqrt{3} : 1$ (cf. Chapter 1).

Now, we can get an explanation for the order in which the different waves arrive (Fig. 19). The P-wave has always a higher velocity than the S-wave according to equation (1) and therefore it always arrives before S in the seismogram. The P-wave, which corresponds to the sound wave through the earth, has the highest velocity among all seismic waves. No reliable velocity difference has been observed between SH and SV. Nor is any such difference to be expected in elastically isotropic media, while with anisotropy a velocity difference may exist. The Love (L) wave is an SH-wave, as we have seen, but it arrives later than S. The reason for this is that during its propagation it is bound to the surface layers, where the velocity is lower than in the interior parts of the earth, through which the S-wave travels. Finally, the Rayleigh (R) wave arrives later than the Love wave. This can also be proved theoretically. In the simplest case, with Rayleigh waves propagating along the surface of a homogeneous medium, it can be shown that their velocity is $=0.92 v_S$. Thus, they have a lower velocity than the Love wave would have, and this result is not changed if instead we assume a layered structure. Incidentally, Love waves require a layered structure for their existence, whereas Rayleigh waves can also exist on the surface of a homogeneous medium.

During their propagation through the earth's interior, the body waves follow the same laws which hold for any other wave propagation, for instance, in optics. The most important relation is SNELL's refraction law. In optics, it is generally written as $n \sin i =$ constant along a given wave path, where $n=$ the refractive index and $i=$ the angle between the incident wave direction and the normal to a discontinuity surface. This angle is generally termed *angle of incidence*. In seismology, it is more suitable to work with wave velocity v instead of refractive index n, and then the law reads $(\sin i)/v =$ constant along a given wave path. On account of the curvature of the earth and thus the ensuing curvature of all discontinuity surfaces and layers in the earth, we cannot use this simple form of SNELL's law but we have to introduce the radius or the distance r from the earth's centre as well. By simple geometrical considerations, we find that the formula for plane parallel layers has to be modified into the following expression, when we deal with concentric, spherical layering:

$$\frac{r \sin i}{v} = \text{constant along a given wave path} \tag{2}$$

The expression in equation (2) is called the *wave parameter*. It has a characteristic value for each wave path, but is different for different wave paths. SNELL's law is a direct consequence of FERMAT's principle, which says that a wave propagates along a path which corresponds to a stationary time. This means that the travel time along the path should be a minimum or a maximum or correspond to an inflexion point. A consequence of this is that each seismic wave corresponds to a stationary time, but different waves can correspond to different kinds of stationarity (minimum, maximum, inflexion).

The conditions in the earth's interior are more complicated than for light waves also by virtue of the existence of two types of motion (P and S) instead of only one. For instance, when a P-wave impinges upon a discontinuity surface between two solid media, there are four emergent waves created, i.e. two reflected (P and S) and two refracted (P and S). Evidently the wave propagation becomes in this way rather complicated. In addition, there may be diffraction phenomena under certain circumstances. It has to be emphasized that the wave propagation is governed by exact mathematical laws; in other words, it is a good example of mathematics at work. In this way, the wave propagation from an earthquake contrasts sharply to the often chaotic phenomena that earthquake effects may exhibit.

Figure 20 illustrates different cases at a discontinuity surface separating two different media. In order that reflection or refraction should take place at the surface, it must exhibit a contrast in the wave velocities. From equation (1) we thus see that a contrast is required in the elastic properties and/or the densities between the two adjacent media, in such a way that

the wave velocities are changed. If the two media differ only in other properties, then the surface of separation has no influence on the wave propagation. In a study of Figure 20, we must remember that the *P*-wave,
being longitudinal, is able to propagate through any medium (whether

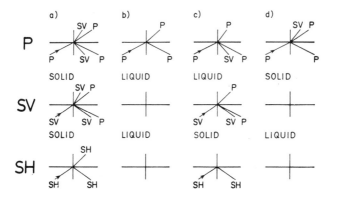

Fig. 20. Reflection and refraction of *P*, *SV* and *SH* at boundaries between different media: a) solid-solid,
b) liquid-liquid, c) solid-liquid, d) liquid-solid. In the figure, 'solid' and 'liquid' have been indicated only for
SV (middle figure). Cases where no wave paths have been drawn, cannot exist (incidence from below assumed
in all cases).

solid, liquid or gaseous), whereas the transverse wave (*S*) propagates only
through solid media. Moreover, we have to keep the particle motions in
mind (Fig. 19), from which it follows that, for example, an incident *SH*-
wave can only give rise to *SH*-waves (by reflection or refraction). Similarly,
an incident *P*-wave gives rise only to *P*- and *SV*-waves, but not to *SH*-
waves, and an incident *SV*-wave gives rise to *P*- and *SV*-waves, but not to
SH-waves. From equation (2), we also find that the angle of incidence i is
smaller for *S* than for *P*, because v_S is smaller than v_P. The atmosphere
can as a rule be treated as a vacuum, i.e. without any wave propagation.

3.2   Body Waves from Distant Earthquakes

Concerning the wave propagation through the earth's interior, there are
two facts which we have to take into consideration:

1. The earth is a sphere and all layers have the same curvature (as a
first approximation).

2. The properties of the earth vary with depth and in general the wave velocities increase with depth.

Item 1 has a dominating significance, while item 2 is to be considered at most as a correction to the results that item 1 leads to. This means that we can get a very good apprehension of the wave propagation by studying the conditions in a *homogeneous* sphere. This simplifies the discussion considerably. In some modern wave propagation studies by famous theoretical seismologists, it is also customary to assume a homogeneous sphere, as its wave propagation approximates well to the real earth. We have to understand this in such a way that the homogeneous earth provides for a qualitative approximation, which will greatly assist us in understanding the principles of the wave propagation in the real earth. On the other hand, there may be quite large divergences quantitatively, i.e. when numerical applications are made, between the homogeneous and the real earth.

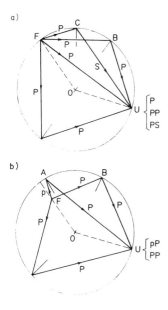

Fig. 21. Principles of the wave propagation through a homogeneous sphere: a) surface focus, b) deeper focus. F = focus, U = station.

Thus, we begin with a homogeneous sphere and assume an earthquake source (also called *focus*) located on the free surface at F (Fig. 21). Waves radiate from F in all directions through the sphere along rectilinear paths. Figure 21 shows some of the waves reaching the station U. Besides the direct waves *P* and *S*, there are those which have been reflected at the earth's surface from underneath. The incident *P*-wave gives rise to a reflected *P*-wave, denoted *PP*, and to a reflected *S*-wave, denoted *PS*. Similarly, by reflection against the earth's surface, the *S*-wave gives rise to a reflected *S*-wave, denoted *SS*, and a reflected *P*-wave, denoted *SP*. This illustrates how the notation for the seismic waves is built up. *One* symbol (*P* or *S* in the cases just mentioned) is used for each part of the wave path. In general, the *S*-wave leaving the focus F consists both of *SH* and *SV*. However, if *SV* were missing in the direction from F to U, then there would be no *SP*-wave at U. According to equation (2), the incidence angle at the reflection point is bigger for *P* than for *S*. While *PP* and *SS* have a common reflection point B midway between F and U, the reflection points for *PS* and *SP* must be shifted from B. The reflection point for *PS* is located at C, i.e. closer to F than to U, and the reflection point for *SP* is closer to U than to F. Applying equation (2) and assuming $v_P : v_S = \sqrt{3}$, we find easily for the homogeneous sphere in Figure 21a and for FU = 180°, that CB = 30°, whereas for FU = 130° we find that CB = 45°. We also find that for *PS*, as distinct from *PP*, there is a certain minimum distance FU, below which *PS* cannot be obtained. For the homogeneous sphere, this minimum distance is 110°. For the real earth, the minimum distance for *PS* is about 44°.

Then we can repeat the process, with two successive reflections against the earth's surface. From an original incident *P*-wave, we thus get *PPP*, *PPS*, *PSP* and *PSS*, and an original incident *S*-wave gives rise to *SSS*, *SSP*, *SPS* and *SPP*. We have now three parts (from F to the first reflection point, between the two reflection points, and from the second reflection point to U), and as a consequence three letters are used in the symbol for each wave. Obviously, we can continue this operation with three reflections and form the corresponding symbols. However, it is quite rare that more than two reflections from the earth's surface are clearly readable in seismograms. In addition, we can consider wave propagation, including reflections, along the greater arc between F and U in Figure 21.

After this, we shall study the wave propagation from a deeper focus (Fig. 21b), instead of a surface focus, but still for a homogeneous sphere. Then it can easily be shown that the law of reflection is fulfilled only at *three* points of the earth's surface between F and U. Limiting ourselves to the smaller arc FU, the angle of incidence is equal to the angle of reflection at only two points on the periphery, but not at any more points (Fig. 22).

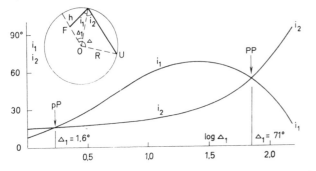

Fig. 22. Diagram demonstrating that the two reflection angles i_1 and i_2 are equal at two points on the smaller arc FU. The curves have been constructed under the assumption that $\varDelta = 150°$ and $R/(R-h) = 1.1$.

One of the reflection points (A in Fig. 21b) is located near the focus and obliquely above it, whereas the other (B) is nearer to the midpoint between F and U. The first reflection point corresponds to the optical case, while the other one is generally not observed in optics because of the limited extent of mirrors. Waves leaving F in an upward direction (above the horizontal plane through F) are denoted by lower-case letters: *p* for longitudinal waves and *s* for transverse. Besides the waves reflected at B and already dealt with in the preceding paragraph, there are now also waves reflected at A. An incident longitudinal wave at A gives rise to the reflected waves *pP* and *pS*, and a transverse wave gives *sS* and *sP*. Just as in the case of a surface focus, the reflection points for the transformed waves (*pS, sP*) do not coincide with those of the non-transformed waves (*pP, sS*). Also in the case of a deeper focus, it is possible to have several reflections at the earth's surface, with the first reflection at A. Such waves are then logically denoted by *pPP, pPS, pSP, pSS, sSS, sSP, sPS, sPP*. These waves are frequently observed on seismograms. The deeper F is, the later, clearly, is the arrival of *pP* in relation to the direct *P*-wave. For deeper earthquakes, we get a greater number of phases than for a surface focus, and in addition the recorded phases are usually sharper for the deeper earthquakes.

An important observation, made already at a relatively early stage, is that the direct P-wave becomes weaker or disappears at distances $\Delta > 103°$ and does not reappear until at $\Delta \gtrsim 144°$. This was explained by the earth's core. The direct P-wave is tangential to the core boundary at $103°$ distance and at greater distances it is hidden by the core. At $144°$ a P-wave reappears, being the one which has traversed the core with two refractions. The disappearance of the P-wave at $103°$ made it possible to calculate the depth to the core boundary. GUTENBERG, who made this discovery already in 1913, in this way calculated this depth to be 2900 km. Later calculations have fully confirmed GUTENBERG's result. The depth has been determined to 2898 km with an error of only 4 km; the boundary is thus a very sharp discontinuity. See Chapter 7 for recent revisions of the core depth. The range from $103°$–$144°$ is a so-called *shadow zone*, but weak phases are observed even within this range.

While P-waves are able to traverse every part of the earth's interior, it has still not been possible to observe any waves which have passed through the core (below 2900 km depth) as transverse waves, in spite of several efforts to find these on records. This fact has been taken as a strong suggestion that the core is liquid; in any case it behaves as a liquid in relation to the seismic waves.

The results regarding the earth's core are based on careful analyses of seismograms. But even in this case it is possible to make a simple and quite accurate picture of the wave propagation by means of a homogeneous sphere, inside which we place another homogeneous sphere concentrically, the latter corresponding to the core (Fig. 23). The properties have to be different in the two spheres. The wave propagation from F will now

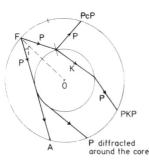

Fig. 23. Principles of the wave propagation in the case of two concentric, homogeneous spheres.

be disturbed by the core in several different ways. FA denotes the limiting ray which is tangential to the core (and in the real earth corresponds to a distance of 103°), behind which the core shadow begins. If we let the angle i at F decrease further, we get waves which hit the earth's core. This is a boundary between a solid medium (the mantle) and a liquid medium (the core). We get reflected waves, and in their notation a c is inserted between the symbols to avoid confusion with reflections at the outer surface. Thus, PcP, PcS, ScS and ScP are waves which have been reflected from outside against the core boundary. But in addition we have refracted P-waves which penetrate through the core. A P-wave within the core is not denoted by P but by K (from German $Kern$=core). When it has passed through the core and reached its opposite side, there is again a partitioning into a reflected P- (or K-) wave and refracted P and S in the mantle. Then we can immediately form the following symbols for waves which have passed through the core: PKP, PKS, SKS, SKP, and here we have to remember that K is always a P-wave. Instead of PKP the notation P' is sometimes used.

If in a homogeneous sphere, the incidence angle i at the focus F is gradually decreased, we will reach greater distances from F, until when $i=0°$ we get the wave which have left F vertically downwards (to the distance 180°). If we now place another homogeneous sphere concentrically inside the first one, we shall find interesting deviations from this simple scheme. For the calculations illustrated in Figure 24, I have assumed that

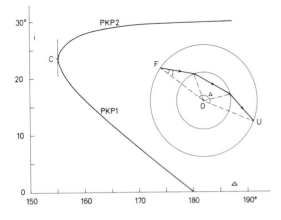

Fig. 24. The origin of a caustic C with ray propagation through two concentric, homogeneous spheres. In the calculations for this figure, we have assumed that $(v_P)_{mantle}$ $:(v_P)_{core}=1.2$ and the core radius has been assumed to be half the earth's radius.

$(v_P)_{\text{mantle}} : (v_P)_{\text{core}} = 1.2$ and that the core radius is equal to half the earth's radius and I have applied the wave-path equation (2). I want to emphasize again that the intention in this and other model calculations concerning the wave propagation is not to approximate the earth numerically, but simply to bring the wave propagation back to simple principles, well known from school physics. In the case mentioned (Fig. 24), $i = 30°$ corresponds to the ray that is tangential to the earth's core. Then letting i gradually decrease, we find that the distance to U first decreases from 187° down to 155°, where it 'turns' and increases again up to 180°, the latter distance corresponding to $i = 0°$. The implication of this is that every point U situated more than 155° from F will receive not one, but two *PKP*-waves, which have propagated along different paths. Exactly at 155° these two waves coincide. This means a great concentration of energy at a distance of 155° in our model (which can easily be seen by simple geometrical considerations). We say that a *caustic* is formed on the surface of the sphere at this distance. The two *PKP*-waves use to be distinguished by the notation *PKP1* and *PKP2* for the first and second arrivals, respectively.

The conditions in the real earth correspond to the phenomenon just described for our model. But the properties of the real earth deviate from those of the model chosen, and so does the location of the caustic. For the real earth, we find it at a distance of about 144° from the source. But in principle the phenomenon is nothing other than what we know from school optics as the minimum deviation for light passing through a prism. Figure 25 illustrates the caustic phenomenon in a striking way. The existence of a shadow zone and a caustic is a direct consequence of a low-velocity layer at the inside of the outer-core boundary (see further Chapter 7).

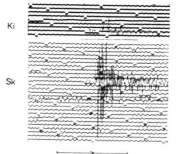

Ki

Sk

1 min

Fig. 25. PKP from an earthquake in the Kermadec Islands (north of New Zealand) on July 14, 1957, as recorded at Kiruna (Ki) at a distance of 140° (inside the shadow zone) and at Skalstugan (Sk) at a distance of 145° (outside the shadow zone).

This is the most marked low-velocity layer in the earth, but corresponding effects, though less pronounced, exist also from other less marked low-velocity layers at other depths in the earth.

According to the simple geometrical wave propagation, there should be a shadow in the range of $103°–144°$. However, weaker signals are observed also within this range, which can be explained by three different effects:

1. Diffraction of the P-wave around the core boundary. Such waves have been observed up to a distance of about $165°$. These waves are as a rule of very long period (around 20–30 sec), and therefore they are best recorded by long-period seismographs (Fig. 26). The corresponding dif-

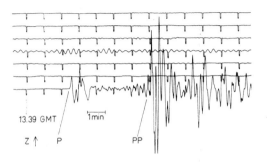

13.39 GMT   1min

Z ↑   P   PP

Fig. 26. Earthquake at New Guinea (magnitude $M=7.7$) on May 28, 1968, as recorded by Press-Ewing Z at Uppsala (distance 108°), showing P (diffracted around the core boundary) and PP.

fracted short-period P-waves are observed only to much shorter distances (around $119°$).

2. Diffraction at the caustic at $144°$. Just as in optics, a caustic is always accompanied by diffraction phenomena, which give rise to weak waves within the shadow zone. Theoretical calculations show that the diffraction at the caustic at $144°$ can explain weak signals down to a distance of about $139°$.

3. As the phenomenon, mentioned under 2., proved to be insufficient to explain the often relatively weak but clear signals within the shadow zone even down to distances of about $110°$, it was in the 1930's found necessary to postulate still another core, inside the outer core and concentric with it. The boundary of the inner core is located at about 5000 km depth. It was assumed to have higher wave velocities than the outer core. In this way it could divert sufficient energy to explain the waves observed

within the shadow zone. The existence of the inner core has later found full confirmation from numerous observations.

In fact, the boundary of the inner core is more complicated with several layers, extending altogether over a few hundred kilometres in depth. Therefore, it is more correct to talk about a transition zone between the outer and the inner core than a simple discontinuity surface. Recent investigations suggest that some layering exists also at the outer-core boundary. See further Chapter 7.

The inner core made it necessary to further supplement the wave notation. A wave which has traversed the earth's inner core as a longitudinal wave is denoted by *I* (this symbol refers to the part of the wave path that falls within the inner core). As concurring evidence indicates that the inner core, as distinct from the outer core, is solid, there should also be a possibility of transmission of transverse waves through the inner core; these phases are denoted by *J*. Thus we can form the following wave notations: *PKIKP, PKIKS, SKIKS, SKIKP, PKJKP, PKJKS, SKJKS, SKJKP*. Naturally, the symbols *I* and *J* have always to be surrounded by *K*. Much effort has been spent in searching for waves with *J* characteristics, but still with no reliable results. This may be due to small amplitudes of the corresponding waves. As mentioned above the boundary of the inner core is rather a transition layer and not as sharp as the outer core boundary. A consequence of a gradual transition to the inner core may be that the *K*-wave is only insignificantly transformed into a *J*-wave, while nearly all its energy passes into an *I*-wave. For similar reasons, there are not many completely reliable observations of waves reflected from the outside of the inner core (denoted *PKiKP*, as an example).

Without difficulty we can build further with the symbols already introduced. For example, *PKKP* is a *P*-wave which has been reflected once against the inside of the core boundary. *PKPPKP* or *P'P'* is a *PKP*-wave which has been reflected once against the earth's surface. Also several such reflections have been observed in seismograms, for instance *P'P'P'*. As *P'* is strongest at a distance of $144°$, we would also expect that the strongest reflections are obtained at about this distance. The consequence of this is that the strongest *P'P'*-waves will be observed at distances around $72°$, and the strongest *P'P'P'* also around $72°$. While *P'P'* is regularly observed in seismic records, *P'P'P'* is more seldom seen. On records in

Sweden it has been observed only a few times and then from earthquakes in Mexico. $P'P'$ arrives about 18–20 minutes after the PKP-phase. It resembles a new P- or PKP-phase and earlier it was often misinterpreted as such. However, $P'P'$ in general differs from its parent PKP-phase, partly by somewhat longer periods, partly by a less defined onset of the wave (moreover, it is frequently doubled on the records).

The next step in developing the model, we have built up by means of homogeneous spheres, is to introduce the velocity distribution that we know exists in the interior of the real earth (Chapter 7). Figure 27 gives a picture of the corresponding wave propagation. Since the wave velocities generally increase with depth in the earth, the wave paths will be curved and concave towards the free surface, according to equation (2). We have seen that the homogeneous sphere models lead us remarkably far into the understanding of the wave propagation in the earth, at least in principle. The velocity distribution in the real earth entails at most modifications of a quantitative nature in the picture we have arrived at.

However, in real seismograms there are a number of minor features which are not incorporated in the picture so far presented. Such features have nevertheless attracted much attention, especially in recent years by means of array-station records, and they have given useful information about factors of significance for the wave propagation. Particularly noteworthy are the *precursors* or early arrivals, i.e. smaller-amplitude waves arriving a few seconds and more ahead of the main wave. These have been observed for PP in the distance range of 90°–110° and for PKP. Such precursors are nowadays usually interpreted as due to *scattering* phenomena. Scattering — earlier only looked upon as a disturbance — has certainly been raised recently in importance as a wave-shaping factor.

3.3   Body Waves from Near Earthquakes

For distances less than 10°, other complications enter into the wave propagation, this time depending upon the regional structure of the earth's crust. Figure 28a shows a vertical section through a typical continental crust of the earth. OO is the earth's surface, CC is the Conrad discontinuity between the upper layer (granite) and the lower layer (basalt), and finally MM is the Mohorovičić discontinuity which marks the base of the earth's

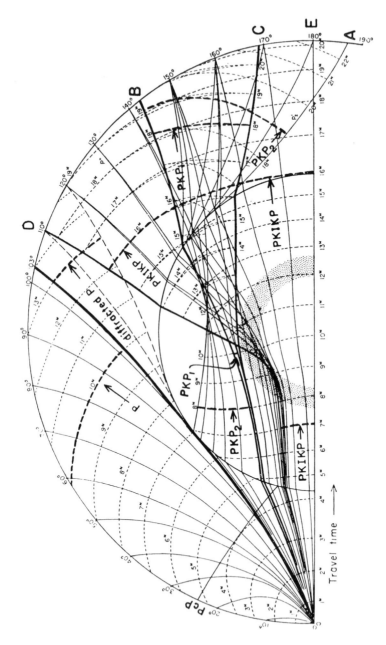

Fig. 27. Vertical section through half the earth, showing the propagation of the longitudinal waves from a source in the left corner of the figure. The outer-core boundary is marked by a circular arc at a depth of 2900 km and the transition to the inner core is marked by a shaded zone at about 5000 km depth. After B. GUTENBERG.

crust. Below MM we have the earth's mantle with ultrabasic rocks. The layer thicknesses given in Figure 28a correspond approximately to average conditions on the continents.

From an earthquake source F in Figure 28a we have the following direct waves propagating to the station U:

Pg, Sg = longitudinal and transverse waves, respectively, through the granitic layer (sometimes the notation $\bar{P}$, $\bar{S}$ is used for these waves);

P^*, S^* = longitudinal and transverse waves, respectively, which have followed CC (sometimes also denoted by Pb, Sb);

Pn, Sn = longitudinal and transverse waves, respectively, which have followed MM.

Evidently, it is necessary to further amplify the wave notation we learnt in the preceding section, when we have to deal with distances less than about $10°$. Waves propagating along the boundary between two media with different velocities travel with the higher of the two adjacent velocities. In addition to the *direct* waves mentioned above, there are also *reflected* waves, for instance, waves reflected from CC and MM.

For a typical crust under the deep ocean, we can make a corresponding picture by excluding the granitic layer altogether and by diminishing the thickness of the basaltic layer to about 5 km (Fig. 28b). However, for

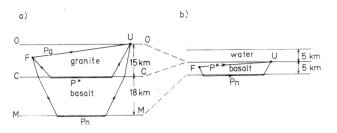

Fig. 28. Principles of the wave propagation through a continental crust (a) and an oceanic crust (b). F = source, U = receiver, OO = earth's surface, CC = Conrad discontinuity, MM = Mohorovičić discontinuity.

obvious reasons the near-source wave propagation is much more seldom observed in the case of the oceanic crust than for the continental one. The deficiency of such observations will be eliminated by the installation of more ocean-bottom seismographs. The wave propagation shown in Figure 28 is also directly applicable to explosions, by placing the focus F at or near the surface instead of at some depth. By means of controlled explosions as wave sources the structure of both the continental and the oceanic crust has been extensively investigated, especially since early 1950's (Chapter 7).

In Figure 28 we have assumed the velocities to be constant within each layer and as a consequence the wave paths to be rectilinear within each layer. As distinct from the discussion of the earth's deep interior in the preceding section, this assumption is not only a good model in the treatment of the crust; beyond that, it corresponds very well to reality. In fact, an assumption of layers with constant velocities is generally made as working hypothesis in the evaluation of actual records over short distances. The formulation of mathematical relations between travel times of the different waves, wave velocities, layer thicknesses and focal depths can be made easily by means of the refraction law and the geometry of Figure 28. Such relations are used in the interpretation of records of near earthquakes or near explosions (Chapter 7). Proceeding into more detail, it is necessary to admit velocity variations with depth, even within layers, such as downwards increasing velocity, imbedded low-velocity layers, etc. Moreover, the granitic layer has a thin surficial low-velocity layer (thickness 1.4 km according to a recent field investigation in central Sweden), apart from possible sedimentary layers at the surface.

In continental earthquakes it has often been observed that Sg has the largest amplitudes. This has also been used as a guide in the identification of the different waves in a record. On comparison of results from different areas, some variation has been found in the wave velocities, often apparently greater for Sg than for other crustal waves. This has then been taken as evidence of real structural variations from place to place. However, such conclusions are partly erroneous. It has been shown that in records of continental earthquakes at short distances there are two Sg-phases, denoted $Sg1$ and $Sg2$. The $Sg1$-wave is the proper Sg with largest amplitudes, while $Sg2$ probably belongs to the surficial granitic layer. In the analysis of records of earthquakes as well as of explosions over continental paths it is necessary to pay attention both to $Sg1$ and $Sg2$, and not to mix them in the false belief that these are only one wave. There is a corresponding doubling of Pg into $Pg1$ and $Pg2$; cf. Table 3. A typical record of a near earthquake is shown in Figure 29.

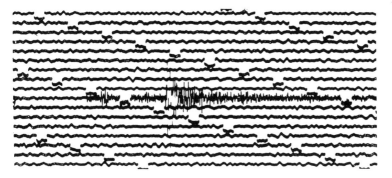

Fig. 29. Typical record of a relatively near earthquake. The shock occurred in Västergötland, Sweden, at 58.4°N, 14.1°E at 22 35 19 GMT on September 3, 1968, and the record shown here was obtained at Uddeholm at a distance of about 180 km. The first phase (*Pg1*) and the largest amplitudes (*Sg1*) are easy to recognize (cf. Table 3). There is 1 minute between successive time marks and time advances from left to right.

In records at short distances (less than 10°) it often happens that the *P*-waves are missing and that the record consists only of *S*-waves, usually *Sn* and *Sg1*. This is true especially for weaker events. Sometimes, only *Sg1* is seen. As all the waves are of short period, only short-period seismographs are suitable for their recording.

Summarizing the discussion in this and the preceding section we arrive at the following review:

1. Distances 0°–10°: the records are complicated because of the regional structure of the earth's crust.

2. Distances 10°–100°: the records are relatively simple and easy to interpret, as the wave propagation is dominated by the mantle.

3. Distances 100°–180°: the records again become complicated, this time because of the outer and the inner core of the earth.

3.4   Surface Waves (Fundamental Mode)

In the beginning of this chapter, we denoted Love waves by *L* and Rayleigh waves by *R*. An alternative notation for Love waves is *LQ* or *Q* (from the German *Quer–Wellen*, while *L* in the combination *LQ* refers to long, i.e. long waves). Similarly, Rayleigh waves are often denoted by *LR* instead of just *R*. Sometimes only the symbol *L* is used to denote the beginning of 'long waves', i.e. surface waves, without making any distinction between Love and Rayleigh waves.

Among the surface waves, the Love and Rayleigh types dominate on the seismograms. Stoneley waves are not observed as a rule, as they are restricted to internal discontinuity surfaces in the earth and the layers adjacent to such surfaces. Love and Rayleigh waves follow the earth's free surface and the layers just beneath it during their propagation. Their properties are thus defined by the earth's crust and the upper part of the mantle.

There are two main results which indicate that this part of the earth is layered and not homogeneous:

1. Existence of Love waves. It can be shown theoretically that Love waves cannot exist on the surface of a medium which is homogeneous. They require at least one outer layer or a velocity which increases continuously with depth.

2. Velocity dispersion. For both Love and Rayleigh waves it is true that the whole wave group does not arrive at the same time at a station, but as a rule the longest waves travel with the highest velocity and thus arrive first, followed by shorter and shorter waves. This phenomenon is called *velocity dispersion* or simply *dispersion*.

The dispersion depends on the layering in the upper parts of the earth, where velocities in general increase with depth. This is illustrated in Figure 30. During their propagation, the surface waves extend to some

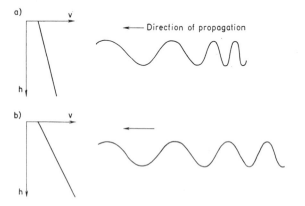

Fig. 30. The influence of a vertical velocity gradient on the dispersion of surface waves: a) smaller gradient— smaller dispersion, b) larger gradient—larger dispersion. v = wave velocity, h = depth below the earth's surface.

depth in the earth, say, to a depth of the same order as the wave-length. Rayleigh waves with periods of 60 sec would thus be noticeable to depths of around 200 km, while those of 20 sec would affect depths only to about 70 km. If we assume the velocity to increase continuously with depth as in Figure 30, it is obvious that the longer waves should propagate with a higher velocity than the shorter waves. This is called *normal dispersion* (the opposite case, when the short waves arrive first, is called *inverse dispersion*). The more rapidly the velocity increases with depth the greater will be the difference in propagation velocity be for different periods, i.e. the greater is the dispersion. In Figure 30 this is illustrated by two cases with different velocity gradient and a correspondingly different dispersion. Instead of assuming a continuous velocity variation with depth, we could have considered several superimposed layers each with a constant velocity, but increasing downwards. This latter model is more common in applications, as being more apt to the real earth. In practice, the procedure is the reverse: from a given dispersion curve, obtained from records, one should calculate the corresponding structure. Here, as in any similar reversal of a procedure, one has to consider carefully if the interpretation is unambiguous or not. In most cases a given dispersion curve can be interpreted in terms of more than one structure. If dispersion curves are available both for Love and for Rayleigh waves, this limits to a certain extent the number of alternative interpretations.

Especially earlier, the arrival time of the beginning of the long waves (*L*) was frequently reported in seismological bulletins. Obviously, such information has no great significance, not even if one distinguishes between *L* and *R*, since also the period has to be given. In addition, an analysis of the whole wave train is necessary for a full understanding (this is, however, beyond the routine duties of a seismological bulletin).

A consequence of the dispersion phenomenon is that we have to distinguish between two different velocities for the propagation of surface waves: phase velocity and group velocity. By *phase velocity* or *wave velocity* we mean the velocity with which a certain phase, e.g. a maximum, propagates. By *group velocity*, on the other hand, we mean the velocity with which a whole wave group propagates. The two velocities are not equal when there is dispersion, and this can be understood by realizing that a certain phase (e.g. a maximum) can propagate in relation to the wave group to which it belongs. A new wave appearing at the front of the wave group can be followed through the wave group until it disappears at the rear of the group; in such a case, the phase velocity is less than the group velocity. Or conversely, a particular wave may be seen to move through the wave group from its rear end until it disappears at the front of the group; in the latter case, the phase velocity is greater than the group velocity. This latter case holds for waves on a water surface, which incidentally offer one of the best visual examples of surface waves. For body waves, no clear indication of dispersion has been observed, and then the phase velocity and the group velocity are equal.

For surface waves the amplitudes are largest at or near the earth's surface and they decrease (roughly exponentially) with depth. The earth's surface is an antinode for the vibration and at greater depth there are one or more nodes, where the amplitude vanishes. Conversely, a given earthquake would generate the biggest surface waves if it is located near the surface, whereas if located at or near a node it would give only insignificant surface waves. This is also confirmed by observations in nature. This is an immediate application of a well-known principle, which holds for every kind of wave motion. In order to generate vibrations, the onset should be made at an antinode and not at a node of the vibration in question. In seismology this fact is generally termed RAYLEIGH's *principle.* Most earthquakes are located in the vicinity of the earth's surface (i.e. within the upper 30 km) and therefore they give rise to clear surface waves; in fact, the surface waves then have the largest amplitude on the whole record, far exceeding the amplitude of the body waves. A typical case is shown in Figure 31. However, with increasing depth of focus the surface waves become smaller and smaller, and may be relatively insignificant compared with the body waves. This fact provides a reliable means to distinguish at first sight between shallow (or so-called normal) earthquakes and deep earthquakes.

In seismological bulletins of older date, in a few exceptional cases still, we find in connection with surface waves some further symbols, such as *C* and *F*. By *C* (from Latin *cauda*) is meant the tail of surface waves which follows their maximum displacements. This tail can continue for several hours on the records of stronger earthquakes. It is not a phase in the usual sense and, in fact, it is not covered by the usual theory of seismic surface waves. Instead C is usually explained by the circumstance that part of the surface-wave energy has not followed the great circle arc from source to station, but has been subjected to lateral reflections and refractions. These in turn are explained by lateral variations of the structure

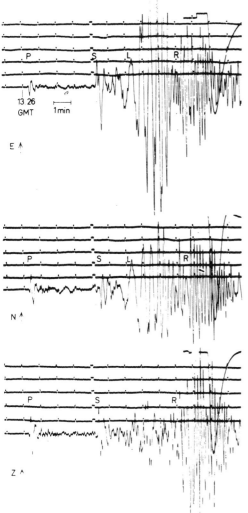

Fig. 31. Records obtained at Uppsala by three long-period seismographs (Press-Ewing) of an earthquake in Turkey (39.2°N, 28.4°E, focal depth 37 km) on March 25, 1969 (origin time 13 21 32.4 GMT). Magnitude $M = 6.1$. The arrows indicate the direction of ground motion, i. e. upwards on the records corresponds to eastward, northward and upward ground motion, respectively.

of the earth's crust and upper mantle, which are especially pronounced in transition zones between continents and oceans. As some parts of the surface-wave train will thus have propagated a longer path they will also arrive later at any station. Alternative explanations consider the *cauda* waves to be due to standing vibrations of crustal layers, generated by the surface waves. By *F* (from Latin *finis*) the time was indicated for the termination of any visible motion on the seismogram. Such an indication has no real significance, as it depends very much on the sensitivity of the seismograph used. We have to remember that the phenomenon at the earthquake source lasts only a very short time of the order of seconds. The fact that an earthquake record at some distance from the source can last for several hours depends exclusively on various wave propagation effects, particularly dispersion, and has nothing to do with the duration of the earthquake as such. For a given seismograph (of a given sensitivity) the time interval *F–L* provides a certain measure of the earthquake magnitude (Chapter 4), and has in some cases been used for such calculations.

Thanks to the installation of more long-period seismographs in recent years a greater number of records of long-period surface waves, so-called *mantle waves*, have been obtained. These can be both of Love-wave type (usually denoted *G* after GUTENBERG) and of Rayleigh-wave type (*R*). They frequently have periods of 8–10 minutes, which corresponds to wave-lengths of more than 2000 km. Evidently, nearly the whole of the earth's mantle must take part in these vibrations simultaneously. As the wave-lengths are an appreciable fraction of the earth's radius, it is necessary to take the earth's curvature into consideration in their study. A typical feature of these mantle waves is that they can be observed repeatedly, as they are passing around the earth. After the direct waves, which have travelled the shortest distance from source to station, those mantle waves arrive that have travelled along the greater arc from source to station. Some time later waves arrive which have again gone the direct way, but which in addition have travelled once around the earth, and so on. How the mantle waves are denoted is evident from Figure 32. An exceptional case with very clear mantle waves was offered by a strong earthquake in Chile in May 1960 (Fig. 33). At Uppsala mantle waves were recorded up to 60 hours after the earthquake. These must have been able to encircle the earth about 20 times before they were too weakened to be recorded. *G*20 or *R*20 have

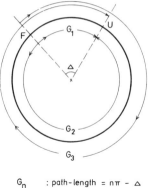

G_n : path-length $= n\pi - \Delta$

$G_{(n+1)}$:   "   " $= n\pi + \Delta$

$n = 0, 2, 4, \ldots$

Fig. 32. Explanation of the notation used for mantle waves of Love-wave type: $G1$ has travelled the shortest path (Δ) from the focus F to the station U, $G2$ has travelled along the greater arc from F to U, etc. A corresponding notation is used for mantle waves of Rayleigh type: $R1, R2, \ldots$

travelled a distance which is about equal to the distance from the earth to the moon. In many other cases observations have been made up to $G8$ or $R8$ and over. Occasionally, mantle waves are also reported in seismological bulletins of older date, but naturally only for the strongest earthquakes. In these bulletins they were usually denoted by W (from German *Wieder-kehr–Wellen*); $W2$ corresponds to $R2$ and $W3$ corresponds to $R3$.

Closely related to the mantle waves are the free vibrations of the earth, which are generated by stronger earthquakes and which have also been recorded a number of times in the last decades (e.g. Kamchatka in 1952, when BENIOFF suspected a wave with a 57 minute period, and notably Chile in 1960). In the interval between these two earthquakes, the theory had been worked out, especially by PEKERIS and his group in Israel, and the Chilean earthquake in 1960 gave a brilliant confirmation of the theories developed. Another early observation refers to torsional oscillations from an Alaskan earthquake in 1958 (BÅTH, 1958). Since 1960, free vibrations have been recorded and studied from several earthquakes, especially Alaska on March 28, 1964, and the Aleutian Islands on February 4, 1965. A useful summary has been published by DERR (1969). The free vibrations are of different kinds: spheroidal (including radial vibrations) and torsional (without radial component). See Figure 34. Like all free vibrations these are standing waves. The spheroidal waves arise through mutual interference of propagating Rayleigh waves, and the torsional vibrations

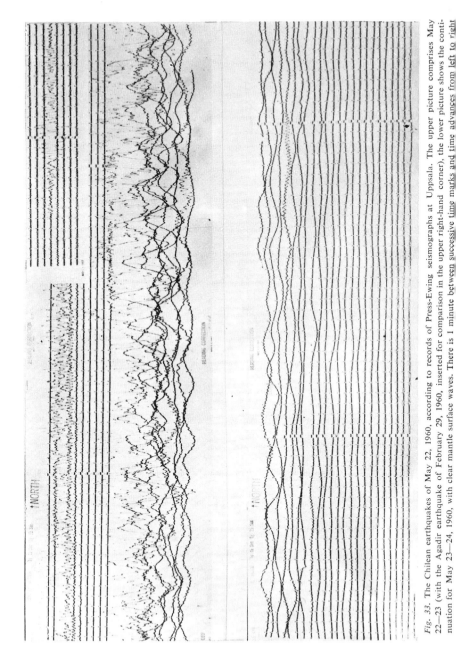

Fig. 33. The Chilean earthquakes of May 22, 1960, according to records of Press-Ewing seismographs at Uppsala. The upper picture comprises May 22—23 (with the Agadir earthquake of February 29, 1960, inserted for comparison in the upper right-hand corner), the lower picture shows the continuation for May 23—24, 1960, with clear mantle surface waves. There is 1 minute between successive time marks and time advances from left to right

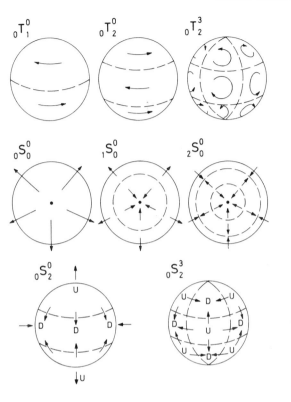

Fig. 34. Lowest modes of the free torsional and spheroidal vibrations of the earth. Mode notation is explained in Table 4c). Nodal surfaces are marked by dashed lines, and arrows indicate the direction of motion. U=upward motion, D=downward motion. After BARBER (1966), modified.

arise from Love waves in the same way. Not only fundamental vibrations but also those of higher modes ('overtones') have been investigated theoretically and observed in records. The longest theoretically calculated periods amount to 44 min for the torsional vibrations and 56 min for the spheroidal vibrations. Spheroidal vibrations of the earth's core have been calculated to have periods up to about 100 min. The observations of the earth's free vibrations have not caused any essential change of the picture of the earth's interior which had already been arrived at by other means. Observed and theoretically calculated periods agree within 1%. Comparisons have been made both with BULLEN's and with GUTENBERG's earth models (Chapter 7), and a somewhat better agreement was found with the latter model (including a low-velocity layer in the upper mantle). Observations of torsional oscillations have only been of limited help in discriminating between different models, whereas those of spheroidal oscillations provide for better discrimination. Free oscillations are recorded by seismographs, tiltmeters and gravimeters. Comparison between these records has proved to be elucidating. Torsional oscillations, lacking radial component, imply unchanged density of the earth, i.e. no gravity effects are produced. Therefore, gravimeters are only able to record spheroidal vibrations, whereas seismographs record both types. The outer core, being fluid, cannot make any torsional, but only spheroidal vibrations. Even though core oscillations produce only small amplitudes at the earth's surface, further detailed observations of these could lead to a more reliable determination of the density at the earth's centre as well as to a solution of the problem of the solidity of the inner core. There is no doubt that the observations of the earth's free vibrations rank among the most significant seismological discoveries in recent time. They are a brilliant example of what the combination of advanced theoretical developments and improved observational techniques can lead to.

3.5   Surface Waves (Higher Modes) and Channel Waves

From physics we know that a vibrating string, fixed at both its end points A and B (Fig. 35a), is able to perform a fundamental vibration, in which the wave-length is equal to double the length of the string. But, in addition, the string can at the same time perform a number of vibrations of higher modes (overtones). Mathematically we have the following simple relation between the length l of the string and the wave-length λ:

$$l = n \frac{\lambda}{2} \tag{3}$$

where $n=1$ for the fundamental vibration, $n=2$ for the first overtone, $n=3$ for the second overtone, etc.

Similar conditions prevail for the surface waves in the earth, and we can illustrate this simply with a Love wave or SH-wave propagating in a layer (Fig. 35b). In order that a wave should be able to propagate to a greater distance *constructive interference* is required. This means, for instance, that the down-going wave AB should be in phase with CD which propagates in the same direction. Just as in optics we can express this

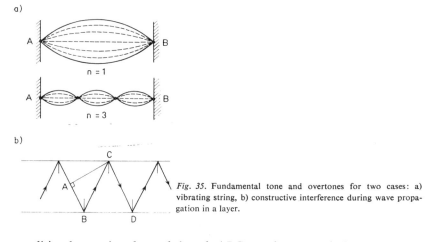

Fig. 35. Fundamental tone and overtones for two cases: a) vibrating string, b) constructive interference during wave propagation in a layer.

condition by putting the path length ABC equal to one whole wave-length or an integral multiple of one wave-length. But in addition we have to take account of possible phase shifts on the reflections at B and C. Assuming the phase shifts at the two reflections to be ε_1 and ε_2 radians, respectively, we then get the following condition

$$\overline{ABC} - (\varepsilon_1 + \varepsilon_2)\frac{\lambda}{2\pi} = n\lambda \tag{4}$$

where ε_1 and ε_2 have to be inserted with their signs, positive or negative. For $n=1$ we again have the fundamental vibration, $n=2$ corresponds to the first 'overtone', $n=3$ to the second 'overtone', etc. Exactly as for the string, the 'overtones' have shorter and shorter wave-lengths the bigger n is. A similar discussion can be made for Rayleigh waves, but not in such an easily apprehensible way as for Love waves. The overtones are called higher-mode surface waves. Equation (4) permits a simple deduction of the relation between the phase velocity and the period, i.e. the *dispersion equation* or the *period equation*, as it is also called.

The plate in Figure 35b can correspond to the earth's crust or some layer in the crust or the upper mantle. In nature, however, we have to deal with a series of such layers with ensuing complications in the wave propagation. The thickness of the plate plays a certain role in the constructive interference and it also enters into equation (4). The consequence is that if this thickness is altered during the wave propagation, especially if this happens rapidly (within a short distance), then equation (4) will not hold any more. This means that the constructive interference is destroyed and the waves are not able to propagate any further but are very much weakened or completely disappearing. This agrees with observations that higher-mode surface waves are not able to cross the transition between a continental and an oceanic structure, also that they are strongly reduced in passages across mountain ranges, where the crust is usually thicker.

The study of the higher-mode waves is relatively new and was not started until the 1950's, in spite of the fact that they had been recorded on seismograms ever since the turn of the century. The studies have consisted partly of theoretical calculations of disperion curves for assumed structures, partly of observations. Extensive observations of these waves, at among other places the Seismological Institute at Uppsala, showed that the first overtone of the Rayleigh waves ($2^{nd} R$) is the one most frequently found, but only over continental paths. Wave paths, which at any point have crossed an oceanic structure, do not show the higher modes. On the other hand, observations have been made at some other places of such waves propagating along purely oceanic paths. It is just this transition between the continental and the oceanic structures which constitutes a barrier to these waves. There is no doubt that ocean-bottom seismographs will be able to furnish extremely interesting information about the propagation of these waves over oceanic structures. As already mentioned,

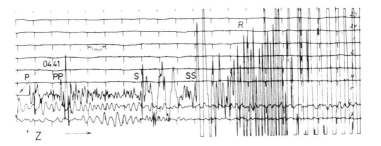

Fig. 36. An earthquake in Sinkiang, China, on November 13, 1965, recorded by a long-period vertical-component Press-Ewing seismograph at Kiruna. The distance is 41.5° and the magnitude M is 7.0. *PP* signifies the *P*-wave that has been reflected once from the earth's surface about midway between the epicentre and the station; *SS* is the corresponding reflected *S*-wave. The higher-mode Rayleigh waves are exceptionally pronounced.

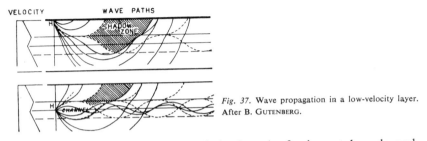

Fig. 37. Wave propagation in a low-velocity layer. After B. GUTENBERG.

the higher modes have shorter periods than the fundamental mode and, moreover, they have higher velocities. Therefore, they arrive ahead of L and R, but like the fundamental modes they exhibit dispersion with a wave train which lasts for several minutes. See Figure 36.

The so-called channel waves are another but related type of waves. In Figure 37 we have depicted a layer with decreased velocity (a low-velocity layer). The refraction law, equation (2), requires that every wave path which is not too steep towards the channel axis should be refracted back to the channel, and this holds both above and below the channel axis. Obviously, such a channel preserves the energy once fed into it in a very efficient way, and in fact such channels are the most capable guides known in nature. Such low-velocity layers are well-known from the atmosphere and the seas (for sound waves). An explosion in the ocean can emit part of its energy into such a channel in the sea. A longitudinal wave travels along the channel until it strikes a coast, where it is transformed into the usual land-propagated seismic waves. Earthquakes near the sea bottom are also able to generate longitudinal waves through the ocean, but in this case they generally propagate by multiple reflections between the bottom and the surface. Longitudinal waves through the ocean are termed T-waves (T as in tertiary, corresponding to P=primary and S=secondary). As their velocity during the water propagation is only about 1.5 km/sec, they arrive several minutes after the corresponding P-wave which has passed through the earth's interior. See Figure 38.

In the earth's interior there are similar channels or low-velocity layers. The most significant is situated at the inside of the outer core boundary at a depth of 2900 km but concerns only the P-wave. However, there are no sources (earthquakes) in this layer which otherwise would be able to feed much energy into it. But waves which have been multiply reflected

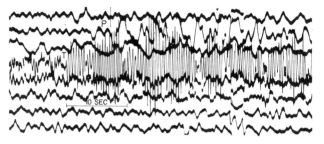

Fig. 38. Typical case of a so-called *T*-phase, recorded by Benioff Z in Bermuda from a North Atlantic earthquake (10.6°N, 43.7°W, depth 25 km, magnitude $M = 6.8$) at a distance of 3231 km, on March 17, 1962. The travel time of *T* is 35 min 21 sec, yielding a velocity of 1.52 km/sec through water. The upper trace shows *P*, the lower *T*.

in this layer are recorded, e.g. *PKKP, PKKKP,* etc. These may be considered as a kind of leaking channel wave.

Among low-velocity layers nearer to the surface, for which it is maintained that channel waves have been observed, the following are the most important:

1. A layer in the upper mantle of the earth. This layer has been generally accepted since around 1960, but was originally proposed as early as 1926 by GUTENBERG. *Pa* and *Sa* (*a* stands for asthenosphere, a name for the upper mantle) are longitudinal and transverse waves, respectively, which have been explained as channel waves in this layer. As a rule they have periods of 10 sec and 20 sec, respectively, and they are often recorded by long-period instruments. Alternatively, these waves have been interpreted as higher-mode surface waves.

2. Layers in the earth's crust, both in the granitic and the basaltic media. Such low-velocity layers, suggested during the 1950's, especially by GUTENBERG, did not find general acceptance. However, waves are observed which have been characterized as the corresponding channel waves (*Li, Lg1, Lg2,* see Table 3). Crustal low-velocity layers have experienced a certain revival in the last decade, as several investigators deem them necessary to explain their observations.

Most of the channel waves mentioned here have been found to bear relation both to the higher modes of surface waves and to the waves found on records of near earthquakes (see above, this chapter). Thus, it has been possible to explain most or all of the channel waves in terms of higher

Table 3. Corresponding phases in records of near and distant events (earthquakes, explosions).

Near events $\Delta < 10°$	Distant events $\Delta > 10°$	Group velocity[1] km/sec
Pn	Pn	7.8—8.2
Sn	Sn	4.6—4.7
P* or Pb	—	6.6
S* or Sb	Li	3.7—3.8
Pg1	Πg	6.25
Sg1	Lg1	3.54
Pg2 or Pg or P̄	—	5.7
Sg2 or Sg or S̄	Lg2	3.37
Rg[2]	Rg[2]	3.02

[1] These may be taken as characteristic values, but minor regional variations occur.
[2] Rg is a short-period surface wave of Rayleigh type.

modes which eliminates the need for a low-velocity layer for their propagation. On the other hand, it is clear that the velocities of the channel waves exhibit such remarkably good agreement with the velocities of waves from near earthquakes that a pure coincidence is excluded. However, in both of these comparisons there still remains much to be investigated. For this purpose, it would be particularly useful to have records from densely spaced stations, from short distances out to great distances from seismic events, preferably along a few profiles. Only in such a way would it be possible to investigate details of the wave propagation, possible mode conversions, etc.

The wave symbols we have introduced in this chapter are summarized in Table 4. The principles for notation of phases are generally accepted and used by all seismologists. The notation has the advantage of giving immediate information of the way in which a wave has propagated through the interior of the earth. It is also easy to check that the system is fully unambiguous, i.e. a given notation cannot stand for more than one wave. In defining notation for new waves discovered in records or theoretically, it is also of importance to conform to the system already developed.

Table 4. Symbols for seismic waves.

a) *Body waves:* direct

Layer		Longitudinal	Transverse
Earth's crust:	granite	$Pg1$, $Pg2$ ($\bar{P}$)	$Sg1$, $Sg2$ ($\bar{S}$)
	basalt	P^* (Pb)	S^* (Sb)
	Moho	Pn	Sn
Mantle		P,p	S,s
Outer core		K	—
Inner core		I	J

b) *Body waves:* reflected or refracted

Discontinuity	Reflection from above	Reflection from below	Refraction
Earth's surface	—	PP, pP	—
Outer core	PcP	$PKKP$	$PKP1$, $PKP2$
Inner core	$PKiKP$	$PKIIKP$	$PKIKP$

In this table, P can everywhere be exchanged for S and I for J. P' and P'' are used as briefer notations for PKP and $PKIKP$, respectively: $P' = PKP$, $P'P' = PKPPKP$, $P'' = PKIKP$. $PKHKP$ = a wave with its deepest penetration in the transition zone between the outer and the inner core.

c) *Surface waves:*

Wave mode	Love	Rayleigh
Fundamental mode	$L(LQ, Q)$	$R(LR, Rg)$
Higher modes	$2^{nd}L$, $3^{rd}L$, ...	$2^{nd}R$, $3^{rd}R$, ...
Mantle waves	$Gn(n = 1, 2, 3, ...)$	Rn (Wn) $(n = 1, 2, 3, ...)$
Free vibrations	Torsional (toroidal) $_nT_l^m$	Spheroidal $_nS_l^m$
	l, m, $n = 0$, 1, 2, ... indicate the number of nodal surfaces: l = latitudinal, m = longitudinal, n = radial (cf. Fig. 34).	

d) *Special symbols:*

'Channel waves': Pa, Sa, PL, Li, $Lg1$, $Lg2$, $\varPi g$, T, etc.
Several of these can also be grouped under c).

3.6   Travel-time Diagrams

One of the most important problems in seismology since the turn of the century has been the determination of travel times (also called transit times) with the highest possible accuracy. The corresponding tables or graphs give the travel times of all seismic waves in dependence on epicentral distance and focal depth. Obviously, this problem can be solved only by successive approximations, at least when uncontrolled sources (earthquakes) are used. We start from an assumed source (with a certain latitude and longitude) and an assumed time for the occurrence of the earthquake, both obtained, for instance, by direct observations or otherwise. Then the onset times from seismograph records at different distances permit the construction of a preliminary travel-time curve which gives the dependence of the travel time on distance. This curve can then be used for a more accurate determination of the location of the source and the origin time which in turn gives another, more accurate travel-time curve, and so on, until the curve does not change. In all determinations of such diagrams it is assumed that all seismic waves are generated by one and the same motion at the source, i.e. all waves are assumed to have the same origin time. Also, the source is assumed to be a point source.

A number of travel-time tables have been constructed. Among those best known and most used are those of GUTENBERG–RICHTER and of JEFFREYS–BULLEN. The latter were first published in 1940, of which those for a surface focus are shown in graphical form in Figure 39. The ordinate t is the travel time and the abscissa Δ is the geocentric angular distance between the epicentre and the station. As the arrival time of the P-wave is the one which can be determined with the highest accuracy the same holds also for the corresponding travel times. Obviously, the travel times of different waves are not independent of each other. For instance, the travel times of PP and of PPP are related to those of P by the following relations:

$$\left. \begin{array}{l} t_{PP}(\Delta) = 2t_P(\Delta/2) \\ t_{PPP}(\Delta) = 3t_P(\Delta/3) \end{array} \right\} \tag{5}$$

provided that both focus and reflection points are located at the earth's surface. Such relations are used to calculate the travel times of PP and PPP from empirically determined times for P. Comparisons with seis-

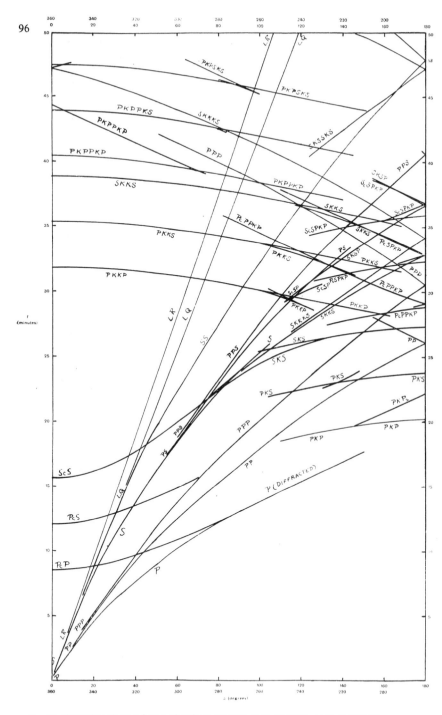

Fig. 39. Travel-time diagram for a surface focus according to H. Jeffreys and K. E. Bullen.

mograms have made it possible to identify *PP* and *PPP*. But the observed times of *PP* and *PPP* may show deviations from the calculated ones, because of deviations from the assumptions made in the calculations (focus and reflection points at the earth's surface). That foci are deeper is of course a general occurrence, but it has also been found, especially for the distance range around 90°–110°, that *PP* exhibits early arrivals, due to reflections at discontinuities within the earth's mantle. These early arrivals are generally weaker than the surface reflections which follow later in the same record. Even though the early *PP* has also got alternative explanations (Section 3.2), there seems to be no doubt that deeper reflections do occur. In an analogous way the travel time of *PS* can be deduced from the times of *P* and *S*:

$$t_{PS}(\Delta) = t_P(\Delta_1) + t_S(\Delta_2)$$
where $\Delta = \Delta_1 + \Delta_2$

(6)

Similar relations can be formulated also for other composite waves. In the Jeffreys–Bullen tables, empirical times are used for *P*, *S*, *PcP*, *ScS*, *PKP* and *SKS* as a base and all other times are calculated from these.

Determination of accurate travel times has naturally involved an enormous amount of work, including a comprehensive statistical treatment. On the other hand, the travel times have no doubt given the most important information about the earth's interior. Some conclusions may be mentioned here. We see from Figure 39 that the travel-time curves of *P*, *S* and all other body waves are curved in such a way that waves from greater distances have travelled during part of their path with a greater velocity than exists near the surface. The curvatures are such that they can only be explained by a velocity which increases with depth in the earth. See further Chapter 7. In comparison, the travel-time curves for *LQ* and *LR* in Figure 39 are rectilinear. This means that these waves have travelled along some layer of constant velocity, in this case the surface layers.

In general, the travel time of any body wave is a function both of the epicentral distance and of the focal depth. By means of considerations of the wave propagation in homogeneous spheres, made earlier in this chapter, it is relatively easy to demonstrate that the travel-time difference *S–P* depends much more on distance than on depth. Similarly, it can be shown that the difference *pP–P* depends much more on focal depth than on distance. These conditions prevail also in the real earth. See data in Table 5.

A remarkable result is that the travel times for body waves for a given distance and depth are very nearly the same, independently of the position of the source and the station, after eliminating the influence of the earth's ellipticity. This would mean that the earth has a symmetrical structure (inside the crust). In fact, this is only an approximation, but it is so good that even today the same travel-time tables are used as were constructed around 1940, without making any distinction between different parts of the earth. However, improvements have been made more

Table 5. Some travel-time data.

a) Travel times of *P*- and *S*-waves and the time differences *S–P* for different distances (Δ) and a surface focus. Calculated from JEFFREYS-BULLEN tables.

Δ°	P min sec		S min sec		S–P min sec	
10	2	28	4	22	1	54
20	4	37	8	17	3	40
30	6	13	11	10	4	57
40	7	38	13	45	6	07
50	8	58	16	09	7	11
60	10	11	18	23	8	12
70	11	15	20	26	9	11
80	12	13	22	17	10	04
90	13	03	23	55	10	52
100	13	48	25	20	11	32

b) Travel-time differences *S–P* (min sec) for different distances (Δ) and focal depths (*h*). Calculated from GUTENBERG-RICHTER tables.

Δ°/h km	100		300		500		700	
20	3	34	3	24	3	12	3	07
30	4	51	4	37	4	26	4	16
40	5	57	5	41	5	27	5	17
50	6	59	6	44	6	31	6	19
60	8	02	7	44	7	30	7	18
70	8	54	8	37	8	22	8	09
80	9	48	9	31	9	18	9	05
90	10	42	10	25	10	10	9	56
100	11	26	11	09	10	51	10	37

Table 5. Some travel-time data (continued).

c) Travel-time differences $pP-P$ (min sec) for different distances (Δ) and focal depths (h). Calculated from GUTENBERG-RICHTER tables.

$\Delta°/h$ km	100		300		500		700	
20	0	16						
30	0	19	0	57				
40	0	23	1	00	1	29		
50	0	25	1	05	1	36	2	03
60	0	25	1	07·	1	41	2	08
70	0	26	1	10	1	47	2	15
80	0	26	1	12	1	51	2	22
90	0	27	1	12	1	52	2	27
100	0	27	1	13	1	52	2	28

recently, especially by means of records of large explosions (nuclear tests) with well-known data on source location and origin time (thus eliminating the inaccuracies inherent in any earthquake study). Such investigations have demonstrated that the tables used up to now are in need of minor corrections, for example, that the Jeffreys–Bullen times need reduction by approximately 2 sec. Revised world-wide tables have recently been published by HERRIN et al. (1968) in the USA. Moreover, it appears as if regional tables would be the next step in the development. A few such tables have been worked out. See further in Chapter 11.

Travel-time diagrams (sometimes also called *hodographs*) or travel-time tables are the most important tools for the seismologist when he identifies the phases he has read on a seismogram. If an interpretation of a record should be accepted, then all measured phases (at least all clear and significant phases) should be explained and they should also agree mutually. In such work travel-time tables and graphs are needed not only for surface focus (as in Fig. 39) but also for greater focal depth. Indications of greater depth than normal are obtained partly from the existence of pP and similar waves, partly from underdeveloped surface waves.

3.7   Microseisms

A description of seismic waves would not be complete without mentioning also the seismic background noise or microseisms on the records.

These are such a characteristic feature of every seismic record that, if a record should exhibit perfectly straight lines with no microseisms, the most immediate conclusion is that the seismograph is out of operation. By microseisms or seismic noise we mean small elastic wave motions in the solid crust of the earth, which as a rule originate because of external influences from the atmosphere and the sea. Thus, they have in general nothing to do with earthquakes and their study is a border-line subject between seismology, meteorology and oceanography. These waves, which always exist on seismic records but with varying intensity, no doubt constitute a disturbance to the earthquake records. On the other hand, they have been the subject of numerous investigations themselves ever since the latter part of the nineteenth century. The most important problem is to explain their origin. Since the 1920's, two opposing theories have dominated the discussions: the coast theory, which explains the microseisms as originating from surf or other action at steep coasts, and the cyclone theory, which ascribes the microseism origin to the cyclones over deep water. Among the numerous microseism investigations made in different parts of the world, often with apparently contradictory results, I might mention an investigation by means of records at Uppsala. A combined study of seismograph records and the corresponding meteorological and oceanographic data suggest that in this case the most important source is some kind of coast effect, primarily at the Norwegian coast, but that in addition there is also some cyclone effect. As a consequence, it seems as if both theories are correct, and the result obtained at any particular station depends very much on its location in relation, for instance, to steep coasts, etc. During World War II, the Americans were able to trace tropical cyclones in the Caribbean Sea and the Pacific Ocean by means of the microseisms which originated from the cyclones. This became of great practical significance, while corresponding efforts in Europe failed.

The essential problem is to explain how the energy can be transmitted from the atmosphere through the sea to the ocean bottom. A theory developed by the British oceanographer LONGUET–HIGGINS in 1950 seems to unify the coast and the cyclone effects. According to his theory, the microseisms depend upon pressure variations from standing sea waves. Standing sea waves will arise by interference between two wave trains which propagate in opposite directions, as for example, around a storm centre at sea or by reflection against a steep coast. As distinct from propagating sea waves, the pressure variations under standing sea waves will be transmitted undiminished to the sea bottom. However, the problems are still far from solved and there are also other theories. Difficulties of dominating significance in the interpretation of microseisms as compared to earthquake records derive partly from simultaneous influence of a greater number of uncontrolled factors, partly from the distribution in time and space of the microseism source, as distinct from an earthquake. These factors render the analysis of the microseisms considerably more difficult. In spite of this, methods have been worked out for the determination of the direction of arrival of such waves. The nature of the waves is not quite clear, even though in most cases they seem to consist of Rayleigh and Love waves, including their higher modes. Recent observations from some array stations

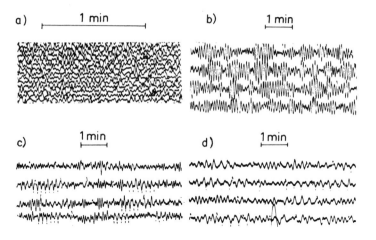

Fig. 40. Microseisms within different period ranges: a) short-period, $T < 2$ sec (Karlskrona Z, February 7, 1965), b) medium-period, $T \cong 6$ sec (Umeå N, April 5, 1969), c) medium-period, $T = 9$—10 sec (Umeå Z, March 11, 1969), d) long-period, $T = 17$—20 sec (Umeå N, April 24, 1965).

in the USA (especially at the large station in Montana, called LASA= = Large Aperture Seismic Array) have revealed that the microseisms at that place consist to about equal degrees of Rayleigh waves and P-waves.

Figure 40 shows examples of the most common types of microseisms at Swedish stations and which probably have quite a general validity as typical examples the world over (all cases shown are storm situations when microseisms are largest):

a) Short-period microseisms, T (period) <2 sec. Such microseisms depend mostly on near disturbances, such as wind over adjacent seas, also traffic, etc. They decrease significantly with increasing distance from nearby coasts, even those of more enclosed basins, like the Baltic Sea.

b) $T \simeq 6$ sec. These are due to a direct effect on the Norwegian coast and they are strongest when a cyclone centre is situated over northern Norway or northern Russia with winds and sea waves impinging almost perpendicularly along the whole west coast of Norway.

c) $T = 9-10$ sec. These microseisms exist when large low-pressure areas are situated at greater distance, especially over the North Atlantic between Greenland and the British Isles.

d) $T = 17-20$ sec. These are observed more seldom, and occur only a few times per year. They have been ascribed to coastal effects, due to ocean swell of the same period as the microseisms.

While microseisms of type a are most disturbing for short-period seismographs, those of types b, c and d are disturbing for medium-period and long-period seismographs. Earlier seismograph types (Wiechert, Galitzin, etc.) had their maximum magnification within the range of the microseisms of types b and c. See Chapter 2.

A still not completely solved problem is whether microseisms of type c are generated within the low-pressure area or if they are generated at the Norwegian coast from the sea waves which emanate from the storm centre. In this case as well, ocean-bottom seismographs could contribute efficiently to a solution of the problem. The Russians have already made some investigations of microseisms by means of such installations.

Quite distinct from the microseisms mentioned so far, which have external sources, are those which derive from internal sources, i.e. from earthquakes. This kind of microseisms is termed signal-generated noise. Theoretically, a seismic record should look much 'cleaner' and simpler than is really does. A theoretical record exhibits only a series of distinct phases, P, PP, S, etc., and between these the trace is undisturbed. Real seismograms, on the other hand, exhibit a continuous and rather irregular motion right through the whole record of an earthquake. Many efforts have been devoted to an explanation of this phenomenon. The motion appears to originate from the seismic waves in various ways, such as by scattering at irregularities in the crust or by reverberations (i.e. repeated reflections) between different layers in the crust. Another type of microseisms due to internal sources may be observed in the near vicinity of active volcanoes, due to motion of molten lava.

Chapter 4

Source Parameters and Their Determination

4.1 Parameters

A source of seismic waves (whether an earthquake or an explosion) is defined by the following parameters:

1. The latitude and longitude of the epicentre (i.e. the point on the earth's surface located vertically above the source).
2. The depth of the source, or the focal depth (the source itself is called the focus or hypocentre).
3. The time of the event, or the origin time of the seismic waves (simply called origin time).
4. The size of the event: magnitude or seismic wave energy.

In order to calculate parameters 1–3 only time measurements are needed (i.e. arrival times of the seismic waves at various seismograph stations), while parameter 4 requires measurements of amplitudes and periods. Therefore, we may call 1–3 kinematic parameters, while 4 is a dynamic parameter.

The calculation of an earthquake is thus concerned with the determination of a number of unknowns. It may be instructive to see what least number of given quantities are needed for such a calculation. For simplicity, we consider a plane (Fig. 41) and we assume the wave velocity v to be constant. Let us assume that we have measured the arrival times t_1, t_2, t_3 for the P-wave (or any other given wave) at three stations 1, 2, 3 and that $t_3 > t_2 > t_1$. Considering first the station pair 1, 2, then obviously the epicentre must be located on a curve for which $t_2 - t_1$ is constant and equal to the measured time difference. Such a curve is a hyperbola with 1 and 2 as foci. As we have already assumed that $t_2 > t_1$, the epicentre location is limited to one branch of the hyperbola, as shown in Figure 41a. In a similar way, we can proceed with the station pairs 2, 3 and 3, 1. The three hyperbola branches will intersect each other in one point or very nearly so, and this point is the epicentre. This method is called the *hyperbola method*.

a)   b)

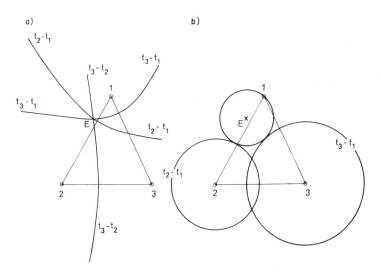

Fig. 41. Determination of the position of the epicentre (E) by means of the hyperbola method (a) and by means of the circle method (b). t_1, t_2, t_3 are the arrival times for the P-wave (or any other given wave) at the three stations 1, 2, 3, respectively.

In practice, however, the hyperbola method is not very convenient, unless hyperbolas are available, constructed in advance for given station pairs. Another method, we may call it the *circle method* (Fig. 41b), is considerably simpler. With the stations 2 and 3 as centres we construct circles with radii equal to $v(t_2-t_1)$ and $v(t_3-t_1)$, respectively, where v is the velocity of the particular wave read at the three stations. Then, the epicentre E is the centre of a circle which passes through the station 1 and is tangential to the two above-mentioned circles. In practical applications, it is possible to find the location of E after a few trials with no need to perform any calculations.

Thus, in order to determine the three parameters longitude, latitude and origin time, we need observations of the arrival time of the P-wave at least at three stations. This is also to be expected: for calculation of three unknowns, we need at least three given quantities.

If, in addition, we introduce the focal depth as unknown, then the minimum number of stations with given P-phases will be four. Figure 42 gives an analogous picture for the vertical plane (the problem is assumed two-dimensional). By means of only two P-observations (t_1 and t_2), it is not possible to distinguish between focal depth and epicentre location, because F is situated on the hyperbola $t_1 - t_2$. But if two more P-observations are added (t_3 and t_4), the solution becomes unambiguous. The four P-observations permit a determination both of epicentre location, focal

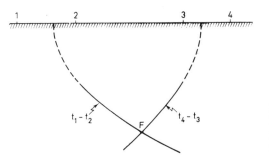

Fig. 42. Principle of the hypocentre (F) determination from four P-wave observations. For simplicity, the problem is assumed to be two-dimensional and the velocity constant.

depth and origin time. In this schematic two-dimensional case, where the epicentre is defined by only one coordinate, in fact three P-observations would suffice.

The arrival times of P are usually the primary data with which we have to work. But if, in addition, information is available on azimuth obtained from the P-wave or on arrival times for other waves (for instance S), then the number of stations can be diminished correspondingly. The following two cases give some typical examples:

1. Two stations: station 1 gives P-time and P-azimuth, station 2 gives P-time.

2. One station: P-time, S-time and P-azimuth given (see further below, this chapter).

In both the cases mentioned a unique solution is obtained under the assumptions of a surface focus.

Even though the seismological literature, especially that of older date, offers numerous examples of such geometrical games, these methods are generally of less significance today. The common method today is to base the determinations on as many different stations as possible. Local deviations exist from assumed travel-time tables depending, for instance, on local structures, and therefore a least-square calculation technique is applied. Then, the error limits of the results obtained can be indicated. It is customary to give the epicentral coordinates to 0.1°, the focal depth to 1 km and the origin time to 0.1 sec. The errors are usually 0.1°–0.4° in the epicentre location, 5–25 km in the focal depth and around a second in the origin time.

4.2   Coordinates and Origin Time

As a rule, source parameters are continuously calculated at a few world centres (see Section 8.1), where station readings are collected world-wide. These central determinations are based almost exclusively on the arrival times of the P-waves.

In principle, we can illustrate the method used by a simplified case (Fig. 43), in which we assume propagation over a plane surface with a

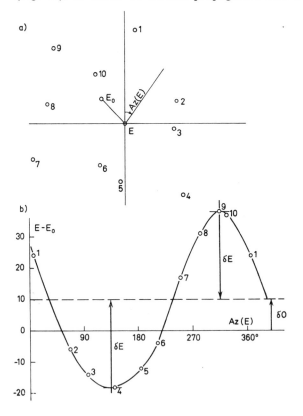

Fig. 43. Principles of determination of the epicentre by residual calculation using observations at a number of stations. 1, 2, ..., 10 = stations, E = assumed epicentre, E_0 = actual epicentre.

constant *P*-wave velocity. The following quantities are given or assumed known:

a) The arrival times of the *P*-wave are given at the ten stations in Figure 43;

b) a preliminary position of the epicentre (E) is assumed;

c) a preliminary origin time is assumed.

The problem is to adjust the assumed epicentre and the assumed origin time by means of the given observations at the ten stations, in such a way that the residuals become a minimum. In doing this, we proceed as follows:

a) Determine for each station the distance to the assumed epicentre E (this can be measured on a map or, better, calculated from given coordinates); we call these distances E.

b) Calculate for each station the epicentral distances from the *P*-wave travel times, using the assumed origin time and a travel-time table appropriate to our model; we call these calculated distances E_0.

c) Plot the differences $E - E_0$ against the azimuth $Az(E)$, measured from E to each station (Fig. 43b). As the assumed epicentre E is generally not exactly correct, finite differences $E-E_0$ arise. We find a clear variation with azimuth of a cycloid type. This demonstrates that the assumed epicentre E and the assumed origin time are in need of correction. The correct solution has not been obtained until all differences are zero or nearly so. With reference to Figure 43, this can be achieved by the following procedures:

d) Make the origin time earlier by an amount δO in Figure 43b, corresponding to the deviation from the true zero-line.

e) Shift the epicentre E towards 315° (northwest) to E_0 by an amount corresponding to the amplitude δE of the curve in Figure 43b.

In a renewed calculation of the residuals we would find that all differences were zero or nearly so and that they did not show any azimuthal variation. When this procedure is applied in practice, the distance $E-E_0$ is generally small in comparison with the distances to the different stations. In such a case the residual curve becomes almost exactly a sine curve.

This method evidently requires an approximate knowledge of the epicentre. In general, this can be easily obtained by a preliminary determination from a few stations at shorter distances. With the electronic computers, nowadays generally used for these calculations, there are no stringent requirements on the exactness of the preliminary epicentre. It would even be possible always to start from one and the same point on the earth (for instance, $0°$ longitude and $0°$ latitude), and the successive adjustments of this position which are needed are performed with high speed. The same holds for the preliminary origin time needed.

In order to illustrate the principle of the calculation, we have here used a simplified case: propagation in a plane with constant velocity. In practice, this case has to be replaced by the real earth which is represented by the corresponding travel-time table. Figure 44 gives an example of the determinations which are distributed from the U.S. Coast and Geodetic Survey (now the U.S. Geological Survey). Such determinations are clearly of very great help in the continued analysis of the records at the individual stations.

Frequently, it is necessary for individual stations to make preliminary determinations of the parameters shortly after the occurrence of an event. Then, only the records at one station may be available. As the method of such determinations provides a good example of the combined use of the different records and the different waves, we shall see how such a determination proceeds.

In order to determine the position of the epicentre from the records of only one station, we have to get its distance and azimuth from the station. The distance is relatively easy to calculate for well-recorded earthquakes and is obtained from the time difference between different phases, usually $S-P$. Table 5 gives a review of the values which may occur. The direction to the source is slightly more difficult to determine accurately and it puts higher requirements on the quality of the records. We have seen above (Fig. 19) that both P and R oscillate in the plane of propagation (i.e. the vertical plane through the epicentre and the station). By a combination of simultaneous amplitudes and directions of displacement on three components it is then possible to calculate the direction to the source. Figure 45 illustrates this procedure. In determinations of azimuth to the epicentre, as in Figure 45, it is possible to combine directly the *recorded* amplitudes (the so-called trace amplitudes) only in the event the three components have the same response curves. If this is not the case, then the trace amplitudes have first to be transformed into the

U.S. DEPARTMENT OF COMMERCE   COAST AND GEODETIC SURVEY   NO. 68-68
ENVIRONMENTAL SCIENCE   DON A. JONES   DIRECTOR   SEPTEMBER 24, 1968
SERVICES ADMINISTRATION   ROCKVILLE, MARYLAND 20852

PRELIMINARY DETERMINATION OF EPICENTERS   PAGE   1

AUG	GMT H M S	LAT	LONG	REGION AND COMMENTS	DEPTH KM	CGS MB	MAG MS	SD	N
2	07 07 02.5	39.5 N	111.0 W	UTAH	6			C.9	9
3	10 02 17.3	33.0 N	116.2 W	SOUTHERN CALIFORNIA MAG. 3.2 (PAS).	1	3.8		0.6	7
3	15 21 25.1*	37.8 N	112.3 W	UTAH	33 R			1.8	6
4	06 23 36.4	39.1 N	111.4 W	UTAH	15	4.0		0.8	8
5	23 07 04.7	39.5 N	111.0 W	UTAH	8			C.8	8
5	23 10 28.7	39.5 N	110.9 W	UTAH	5 K			1.0	9
11	22 07 56.9*	1.7 N	126.3 E	MOLUCCA PASSAGE	33 R			1.2	12
12	11 47 33.4*	1.7 N	126.7 E	MOLUCCA PASSAGE	33 R			1.1	7
12	23 58 33.2*	29.6 N	141.1 E	SOUTH OF HONSHU, JAPAN	48	4.4		0.4	9
13	11 59 16.1*	60.3 N	153.7 W	SOUTHERN ALASKA	127	4.3		0.6	14
14	10 13 46.8*	1.7 N	126.3 E	MOLUCCA PASSAGE	33 R	4.7		C.7	10
16	15 08 25.8*	21.2 S	179.3 W	FIJI ISLANDS REGION	667	4.2		1.0	15
16	21 33 46.7*	46.4 N	14.2 E	YUGOSLAVIA FELT AT RADOVLJICA AND GORJE AND TRZIC.	33 R	4.2		1.0	7
17	13 38 43.3*	1.8 N	126.4 E	MOLUCCA PASSAGE	33 R			1.5	10
21	06 08 39.5*	23.7 S	180.0 E	SOUTH OF FIJI ISLANDS	525	4.5		1.0	8
22	07 33 12.1	29.8 S	176.9 W	KERMADEC ISLANDS	39	4.4		1.4	15
23	11 03 24.8*	12.9 N	145.2 E	SOUTH OF MARIANA ISLANDS	45	4.5		1.3	8
24	12 21 28.7	56.2 S	143.5 W	SOUTH PACIFIC CORDILLERA	33 R	5.5		1.0	14
24	14 26 07.4*	30.0 N	95.1 E	TIBET	56	4.6		0.5	8
25	00 11 33.2*	1.2 N	126.1 E	MOLUCCA PASSAGE	62	5.3		1.1	17
26	05 59 09.7*	26.7 N	55.0 E	SOUTHERN IRAN	33 K	4.6		1.2	8
26	07 01 31.3	40.3 N	143.6 E	OFF EAST COAST OF HONSHU, JAPAN	24	4.4		1.1	13
26	14 35 08.5	0.2 S	121.8 E	NORTHERN CELEBES	295	4.5		1.2	26
26	17 24 35.C	17.6 S	69.6 W	PERU-BOLIVIA BORDER REGION	170	4.1		1.0	7
26	20 24 28.3*	21.3 S	179.5 W	FIJI ISLANDS REGION	671	4.2		C.4	11
27	06 53 41.4*	19.2 S	177.7 W	FIJI ISLANDS REGION	631	4.C		C.5	14
27	23 55 48.3	1.0 N	120.1 E	NORTHERN CELEBES	33 R	5.C		1.0	23
28	02 55 30.9	45.4 N	136.7 E	NEAR E. COAST OF EASTERN RUSSIA	248	4.2		0.7	11
28	15 05 31.5	14.7 S	167.3 E	NEW HEBRIDES ISLANDS FELT AT LAMAP.	123	4.8		1.1	27
29	14 26 22.9	5.4 S	145.5 E	EAST NEW GUINEA REGION	86	5.8		1.1	12
29	13 27 08.9*	15.3 N	122.6 E	PHILIPPINE ISLANDS REGION	33 R	5.5		1.0	5
29	21 15 44.1	6.9 N	73.0 W	NORTHERN COLUMBIA	151	4.6		1.1	15
30	06 18 30.4*	1.4 N	126.3 E	MOLUCCA PASSAGE	50	5.4		1.2	41
30	21 11 20.4	34.9 N	59.5 E	IRAN PROBABLE FORESHOCK OF IRANIAN QUAKE OF AUG. 31, HYPOCENTER POORLY DETERMINED.	33 R			4.3	6
31	10 47 37.4	34.0 N	59.0 E	IRAN MORE THAN 11000 KILLED AND 6000 INJURED. KAKHAK DESTROYED, HIGH DESTRUCTION IN SURROUNDING AREA. FELT THROUGHOUT KHURASSAN PROVINCE. MAG. 7-7 1/4 (PAS), 7.7 (BRK), 7 1/2 (GOL).	13 SEE COMMENTS AT BOTTOM PAGE 2.	6.0	7.3	1.5	56
31	11 34 32.9	33.9 N	59.2 E	IRAN	24	5.5		0.9	37
31	18 06 35.7	56.3 N	115.6 E	EAST OF LAKE BAIKAL	25 R	4.6		0.8	19
SEP									
1	00 39 54.8	1.6 N	126.3 E	MOLUCCA PASSAGE	33 R	4.9		C.9	18
1	01 42 50.5*	0.9 S	24.6 W	CENTRAL MID-ATLANTIC RIDGE	33 K	4.3		0.5	7
1	03 42 06.6*	0.9 S	24.7 W	CENTRAL MID-ATLANTIC RIDGE	33 R	4.4		C.8	8
1	05 18 07.1*	0.9 S	24.9 W	CENTRAL MID-ATLANTIC RIDGE	33 R	4.4		1.0	9
1	07 03 39.9*	1.1 S	24.9 W	CENTRAL MID-ATLANTIC RIDGE	33 R	4.4		0.8	9
1	07 27 30.2	34.0 N	58.2 E	IRAN MORE THAN 2000 KILLED AND EXTENSIVE PROPERTY DAMAGE AT FERDOWS MAG. 6.2 (PAS), 7.0 (BRK), 6 1/4-6 1/2 (GOL).	15	5.9	6.3	1.0	81

Fig. 44. Sample page of preliminary epicentre determinations from the U.S. Coast and Geodetic Survey, now the U.S. Geological Survey. MAG = magnitude, SD = standard deviation of the origin time, sec, N = number of stations used in the solution.

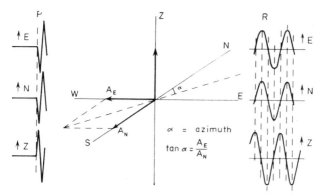

Fig. 45. Sketch showing how from given records (E, N, Z) of the *P*-wave it is possible to determine the direction (azimuth) to the epicentre. The corresponding Rayleigh waves are shown to the right, where one should note the 90° phase shift between Z and the horizontal components. Compare Figure 19.

corresponding *ground* amplitudes and, after this, combined to give the azimuth.

Obviously, three components are necessary and sufficient for such a determination. If we had only two horizontal components and no vertical component recorded, an unambiguous direction determination would not be possible. The Love waves (*L*) are recorded only by the horizontal components and do not permit an unambiguous determination of direction. The *S*-waves could be used as a check, but are less practical because of their composition of *SV* and *SH*. When, in practice, an epicentre determination is made from the records at only one station, it is advisable to use as many different waves as possible. If the determination has been made correctly all waves have clearly to agree with each other.

The two methods for epicentre determinations described above can be summarized as follows:

1. Many stations with times: can be used both for distant and near events.

2. One station with time, distance, azimuth: can be used both for distant and near events.

The difference between the two methods thus depends on the number of stations used and the character of the given data, but not on the location of the epicentre (near or distant).

While in general the epicentre location can be calculated with relatively good accuracy, it has proved more difficult to obtain the focal depth with the necessary precision. As indicated in the preceding section, the accuracy of the focal depth, as measured in kilometres, compares favourably with the accuracy in kilometres of the epicentre location. However, the requirements on focal-depth accuracy are more stringent. This concerns especially shallower depths, i.e. for events in the upper 100 km of the earth. The requirements on accurate depths stem from both practical and theoretical aspects of seismology, which can be summarized in a few points:

1. Nuclear explosion detection. Events for which the depth exceeds a few kilometres, say 5–10 km, are very likely natural (earthquakes), whereas shallower events may be artificial. See further Chapter 11.

2. Tsunami warning. Tsunamis or seismic sea waves may arise because of earthquakes near an ocean bottom (Chapter 1). It seems probable that only very shallow earthquakes give rise to tsunamis. Therefore, improved precision of the focal-depth determination would be of immense importance to warning services.

3. Pure seismology. In many seismological studies, a higher accuracy of focal depths is required, especially for the upper 100 km of the earth. This concerns every investigation where focal depth enters as an independent variable, for example in investigations of spectra as a function of focal depth or of energy content in higher-mode surface waves as a function of focal depth, etc.

At the same time as the requirements are especially stringent for focal depths less than 100 km, the difficulties for depth determinations are greater in this range than for deeper events. Table 5 gives information on the time differences $pP{-}P$ which correspond to different depths from 100 to 700 km. For depths less than 100 km the problems of accurate depth determination increase, partly because of the fact that pP follows closely after P and may be difficult to read, partly because of greater influence of local source conditions on the observed time difference $pP{-}P$.

The world centres mentioned above often do not consider it possible to guarantee a higher precision than ± 25 km in the calculated depths (even if these are given to 1 km). However, a higher accuracy can be achieved by the simultaneous inspection of the records from a whole network of stations. By direct visual comparison, it is possible to make more reliable wave identifications than just to go by reported times. Thus, experience from the Swedish network has shown that in this way it is generally possible to calculate focal depths to an accuracy of ±5 km. For most purposes, this is sufficiently accurate. This provides both an easy and an accurate method for depth calculations. More sophisticated techniques, especially array techniques, can lead to exact values in a more complicated way.

The correct identification of pP may offer certain problems. Considering the frequent occurrence of multiple shocks, it may often be questionable whether a later phase is really pP or if it is P of another shock in the same location. This is a frequently occurring problem in readings from networks of limited geographical extent. A better distinction can then be made by networks of greater extent.

4.3   Magnitude and Energy

In earlier earthquake statistics usually no attention was paid to the released energy. This naturally gave erroneous results, and it was a major step forward when the American seismologist RICHTER in the 1930's introduced the magnitude concept. In general, we can define the magnitude in the following way:

$$M = \log \frac{a}{T} + f(\Delta, h) + C_s + C_r \tag{1}$$

where M=magnitude, a=ground amplitude (expressed in microns; 1 micron=0.001 mm), T=wave period (sec), Δ=epicentral distance (degrees), h=focal depth (km), C_s=station correction, which corrects for special conditions at the station (local structure, etc.), C_r=regional correction, different for different earthquake areas, and depending on the earthquake mechanism (Chapter 6) and the wave propagation. All logarithms in this book refer to the base 10. The function f has been determined by a combination of theoretical and empirical results. It corrects for the effect of distance (which influences the amplitudes because of geometrical spreading and absorption) and for the effect of focal depth. f is different for different waves and also for different components of the same wave. The magnitude is a quantity which is characteristic of each earthquake (or explosion), and determinations made at different stations, or by means of different records or different waves at one and the same station, should agree within error limits. Even in the best cases, we have to expect errors of 0.2–0.3 units in a calculated magnitude. Equation (1) shows that the magnitude has no upper limit, theoretically speaking. It is thus completely erroneous to talk about the '12-degree Richter scale', which unfortunately is heard or seen in news media too often. From a practical point of view there is an upper limit for M, caused by the fact that the earth has a limited strength and cannot store energies beyond a certain upper limit. The largest earthquakes known have reached a magnitude of about 8.9. On the other hand, it is also clear from equation (1) that negative magnitudes are possible for small events.

So far, the magnitude concept appears simple and relatively free from complications. However, on closer examination, this is found to be far from the truth, and here I shall try to describe the present situation with regard to magnitude determinations. In fact, there is not one magnitude scale, but we can distinguish three different scales:

1. M_L is the original magnitude introduced by RICHTER in 1935. This was defined so as to be used for local shocks in southern California. M_L was defined by RICHTER as the logarithm of the maximum, *recorded* (trace) amplitude (expressed in microns) by a Wood–Anderson torsion seismograph with specified constants (free period = 0.8 sec, maximum magnification = 2800, damping factor = 0.8), when the seismograph was at an epicentral distance of 100 km. The magnitude for shocks at other distances can be calculated from a knowledge of the variation of the maximum amplitude with distance. Obviously, this definition is not in conflict with equation (1), even though it must be said that the definition was arbitrary and permitted only a limited application.

2. M is magnitude based on surface waves. In 1945, GUTENBERG developed the magnitude scale considerably, making it applicable to any epicentral distance and to any type of seismograph. This development required a better knowledge of the variation of wave amplitudes with distance, and in order to be able to use different seismograph types, it was henceforth necessary to use ground amplitudes instead of trace amplitudes. The first generalization was made for surface waves (R). By limiting the period range considered to 20 ± 2 sec, further by including only earthquakes of

normal depth (h constant) and finally by stating that the function f for surface waves is proportional to $\log \varDelta$, equation (1) is simplified to the following form:

$$M = \log a + c_1 \log \varDelta + c_2 \qquad (2)$$

where c_1 and c_2 are constants. In most formulas of this type which have been developed the amplitude a refers to the horizontal component of Rayleigh surface waves. Some similar formulas for the Rayleigh-wave vertical components have also been developed.

3. m is magnitude based on body waves and defined by the same equation (1). In 1945, GUTENBERG extended magnitude determinations also to body waves (P, PP, S) and to earthquakes of any depth. These generalizations have to be made on the basis of equation (1). Later, m was called the 'unified magnitude' by GUTENBERG.

However, the three scales do not agree with each other, which means that one and the same earthquake will have different magnitude values on the different scales. Relations between the scales have been deduced, which make relatively simple recalculations, from one scale to another, possible (Fig. 46):

$$\left. \begin{aligned} m &= 1.7 + 0.8 M_L - 0.01 M_L^2 \\ m &= 0.56 M + 2.9 \end{aligned} \right\} \qquad (3)$$

The relation between m and M expresses an interesting result, which has emanated from magnitude studies. The relation implies that for earthquakes at normal depth the surface waves become more and more important in relation to the body waves the bigger the earthquake is. As seen from the relation (3) and Figure 46 the slope is greater for the M-scale than for

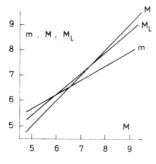

Fig. 46. Relations between different magnitudes.

the m-scale, and the two scales intersect at $m = M = 6.6$. This circumstance is explained by the fact that in larger earthquakes the focal mechanism is more extended both in space and time, and this has a favourable influence on the generation of longer-period surface waves.

M_L has only a rather limited application, but is still generally used in studies of regional events. The use of any well-calibrated short-period seismograph system and conversion into the amplitudes of the standard Wood-Anderson seismograph have enlarged the applicability of the M_L-scale. Best results are obtained if such measurements are combined with regionally determined attenuation values, preferably by spectral methods, instead of using those derived for California. An example of this procedure is presented by BÅTH et al. (1976), where an M_L-scale is developed for Fennoscandia. The other magnitudes (m and M) have a much wider application. One complication arises from the fact that some stations use only the M-scale, others only the m-scale, and it is not always obvious which scale has been applied.

But there are also a number of other complications, depending particularly on the way in which the measurements on the seismograms are made. One important reason for discrepancies is illustrated by a schematic case in Figure 47. While the magnitude scales were defined by GUTENBERG as applicable to the maximum amplitudes in each wave group (corresponding to A_m in Fig. 47), some seismologists are instead in the habit of measuring the very first swing (A_f in Fig. 47). Obviously, in the two cases shown in Figure 47, the magnitudes would erroneously be the same, if A_f is used, whereas the magnitude would be 0.7 units higher in the latter case if the maximum amplitudes are measured (assuming other parameters to be the same in the two cases).

Another cause of discrepancies arises when surface waves are also used for magnitude determination for focal depths in excess of normal. We have seen in Chapter 3 that the surface waves are best developed for a surface focus and that they decrease with increasing focal depth. However, consistent values of magnitudes can be obtained from surface waves for focal depths down to about 100 km, provided the magnitudes are corrected by $+0.008\,h$, where h is the depth in km. Even for such a relatively small depth as 60 km this magnitude correction amounts to no less than $+0.5$. However, there are quite large variations in this correction from

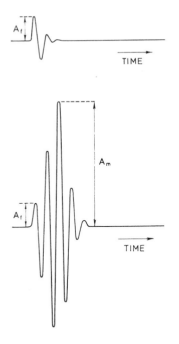

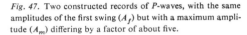

Fig. 47. Two constructed records of P-waves, with the same amplitudes of the first swing (A_f) but with a maximum amplitude (A_m) differing by a factor of about five.

earthquake to earthquake, and, exceptionally, even very deep earthquakes can produce remarkably well developed surface waves.

An important step towards greater unification of the magnitude determinations was taken at the International Geophysical Assembly in Zurich in 1967. There the following recommendations were adopted for magnitude determinations of distant events ($\Delta \geqq 20°$), here reproduced in full (with minor explanations):

'1. Magnitudes should be determined from $(a/T)_{\max}$ for all waves for which calibrating functions $f(\Delta, h)$ are available: PZ, PH, PPZ, PPH, SH, LH, (LZ). (Z=vertical component, H=horizontal component, L=surface waves).

2. Amplitudes and periods used ought to be published. Two magnitudes (m=body-wave magnitude, M=surface-wave magnitude) should be used. For statistical studies M is favoured. The conversion formula $m = 0.56M + 2.9$ is recommended.

3. For body waves the Q-values $\left(=f(\varDelta, h)\text{-values}\right)$ of GUTENBERG and RICHTER (1956) are used. For surface waves, the Moscow–Prague formula of 1962:

$$M = \log \frac{a}{T} + 1.66 \log \varDelta^0 + 3.3$$

is used. Determinations of station and epicentre corrections are encouraged. (a is the horizontal component of Rayleigh surface waves; T should be in the period range of 10–30 sec.)

4. If short-period records are used exclusively, too low magnitudes result. In order to eliminate this error, it is strongly recommended that for short-period readings either a/T or $f(\varDelta, h)$ be adjusted such that agreement with long-period instruments is achieved.'

Nowadays, magnitudes are regularly reported both by several individual stations as well as by networks of stations and by world centres. As an example, I may mention that the Seismological Institute in Uppsala from January 1968 reports both m (from body waves) and M (from surface waves) in agreement with the Zurich recommendations for all cases with well-measurable amplitudes. The reported values are averages of determinations obtained from records at Uppsala and Kiruna. As a rule, the following waves are used in these calculations: $P(\text{H}, \text{Z}, \text{Z}')$, $PP(\text{H}, \text{Z}, \text{Z}')$, $S(\text{H})$, $R(\text{H})$, where H=long-period horizontal component, Z=long-period vertical component, Z'=short-period vertical component. For larger earthquakes with a greater number of well-readable amplitudes, m may thus be an average of 10–15 individual determinations. For smaller shocks, on the other hand, m may be an average of just two determinations, like M. As we shall see later in Chapter 6, the combination of P and S implies to a certain extent an elimination of the azimuth-dependent radiation from the focus. Later, from January 1970, the procedure has been simplified such that m is always an average of only two determinations, i.e. $P(\text{Z}')$ from Uppsala and Kiruna, just as M is an average of $R(\text{H})$ from the two stations. The corresponding amplitudes and periods are reported in our bulletins when the amplitudes amount to 0.1 micron or more, but not otherwise.

In order to get an apprehension of representative magnitude values, a collection of these has been given in Table 6.

Table 6. Magnitude formulas.

a) Evaluation of $M = \log \dfrac{a}{T} + 1.66 \log \Delta° + 3.3$ for $T = 20$ sec.

$\Delta°/a$ micron	0.1	1	10	100	1000	10 000
20	3.2	4.2	5.2	6.2	7.2	8.2
40	3.7	4.7	5.7	6.7	7.7	8.7
60	4.0	5.0	6.0	7.0	8.0	(9.0)
80	4.2	5.2	6.2	7.2	8.2	(9.2)
100	4.3	5.3	6.3	7.3	8.3	(9.3)
120	4.5	5.5	6.5	7.5	8.5	(9.5)
140	4.6	5.6	6.6	7.6	8.6	(9.6)
160	4.7	5.7	6.7	7.7	8.7	(9.7)
180	4.8	5.8	6.8	7.8	8.8	(9.8)

b) Evaluation of $m = \log \dfrac{a}{T} + f(\Delta, h)$ for *PZ*, $T=1$ sec and h = normal.

$\Delta°/a$ micron	0.01	0.1	1	10	100
20	4.0	5.0	6.0	7.0	(8.0)
40	4.4	5.4	6.4	7.4	(8.4)
60	4.8	5.8	6.8	7.8	(8.8)
80	4.7	5.7	6.7	7.7	(8.7)
100	5.4	6.4	7.4	(8.4)	(9.4)

The main significance of the magnitude lies in the fact that it permits a classification of earthquakes based upon the energy released. The magnitude has a simple relation to the total energy of the seismic waves (E, ergs), which has been released in an earthquake:

$$\log E = 12.24 + 1.44M \tag{4}$$

A magnitude increase by one unit in the M-scale thus corresponds to an energy increase by 25–30 times. Equation (4) has been deduced for $M > 5$, and it should not be used for lower magnitudes. By means of equation (3) we can easily substitute m or M_L for M in equation (4). Expressed in m, the energy formula thus becomes:

$$\log E = 4.78 + 2.57m \tag{5}$$

Considering such energy-magnitude relations, we have to realize that magnitudes were originally defined in a more arbitrary way. It was not until a later stage that efforts were made to derive relations between M

and E. This has involved great problems, and ever since the 1930's a great number of different relations of this kind have been proposed, some of them deviating strongly from each other. Equation (4) seems to represent one of the more reliable solutions. In the derivation of an equation like (4) M is determined according to the usual method, while the energy E has to be computed, usually by an integration over the whole wave train under study, i.e. both in time and space. Such computations have been made both for surface waves and for body waves. Modern techniques (especially magnetic-tape recording combined with the use of electronic computers) have considerably simplified the extensive integration work. In comparison with energy calculations from seismograms, energy determinations from geodetic measurements in the epicentral area have only a limited application. In fact, only relatively few earthquakes permit reliable measurements in the field; moreover, it is unclear what the relation is between such measurements and the total seismic wave energy E released.

From Tables 7 and 8 it is clear that earthquakes cover an enormous energy range. This great range can be even better elucidated by the following comparison. An earthquake of the relatively modest magnitude $M = 6.8$ corresponds to an energy release of about 290 million kWh, i.e. the electric energy consumption in Uppsala during 1966 (then about 100 000 inhabitants). An earthquake of $M = 8\frac{3}{4}$ releases an amount of energy equivalent to the electric energy consumption in Uppsala for 670 years, assuming the same annual value as for 1966. And a very small earthquake, with a magnitude of 1.6, would correspond to the average amount of electric energy

Table 7. Corresponding values of seismic wave energy (E), magnitude (M and m), maximum intensity (I_0) and maximum acceleration (a_0).

E erg	M	m	I_0	$a_0{}^1$ cm/sec^2
10^{20}	5.4	5.9	6 1/2	50
10^{21}	6.1	6.3	7 1/2+	100
10^{22}	6.8	6.7	8 1/2+	250
10^{23}	7.5	7.1	10 −	550
10^{24}	8.2	7.5	11 −	1250
10^{25}	8.9	7.9	12	2800

[1] Refers to periods of 0.1—0.5 sec of the ground motion.

122 Source Parameters and Their Determination

Table 8. Review of some sample earthquakes.

Earthquake	M	I_0[1]	Energy[2]	Number of deaths
Agadir, February 29, 1960	5.8	7+	1	10 000—15 000
Skopje, July 26, 1963	6.0	7 1/2	2	1 100
Mediterranean Sea, July 19, 1963	6.4	8	7	0
Georgia, USSR, July 16, 1963	6 3/4—	9—	35	(0)
Kurile Islands, October 13, 1963	8 1/4	11	3400	0
Alaska, March 28, 1964	8.5	11+	7800	131

[1] I_0 has been calculated from M by means of equation (10), Chapter 4, and in single cases it may deviate significantly from values really observed.
[2] In order to facilitate comparison among the shocks listed, the energy of the Agadir earthquake has been chosen as unit.

used in Uppsala in just one second. However, hitherto it has not been possible to bring the enormous energy of earthquakes under control or to extract any useful work from it.

Table 8 also demonstrates the practically complete absence of correlation between released energy and casualties (number of deaths). In Chapter 5 we shall learn more about destructive earthquakes.

Thanks to the ease of its determination and its reliability, the magnitude has rightly come to stay as a dynamic source parameter of unparalleled significance. However, in recent time several other dynamic parameters have appeared, notably the *seismic moment* M_0. This is defined by the equation

$$M_0 = \mu DS \qquad (6)$$

where μ=modulus of rigidity (cf. Section 3.1), D=average dislocation on a fault plane and S=area of fault plane. The seismic moment can be determined from field data, from spectra of records of seismic waves or from empirical relations to magnitude. An example of such a relation is the following, deduced for earthquakes in the western United States (WYSS and BRUNE, 1968):

$$\log M_0 = 15.1 + 1.7 M_L \qquad (7)$$

where M_0 is expressed in dyne · cm and $3 < M_L < 6$.

4.4   *Intensity*

Intensity is not a source parameter, but it will be convenient to treat this concept in immediate connection with magnitude. We have to differ clearly between the magnitude and the intensity of an earthquake. The magnitude is calculated from instrumental records, while the intensity is based on direct effects of the shock, such as on buildings, topography, etc., i.e. so-called macroseismic effects. While the magnitude has a certain definite value for each earthquake, as we have seen, the intensity varies with the position of the observation point. The intensity is greatest within the epicentral area, and from there it usually decreases in all directions. As a rule, intensity is expressed on some scale, nowadays generally on a scale of 12 degrees.

The first more generally used intensity scale (ROSSI–FOREL, see Chapter 1) had 10 degrees. However, as it did not permit a sufficiently clear distinction among the strongest earthquakes (degree 10), it was replaced by a 12-degree scale. This is still usually referred to one or several of the names MERCALLI, CANCANI, SIEBERG. It was revised by WOOD and NEUMANN in the USA in 1931 and was then called the 'Modified Mercalli Scale' or simply the 'MM-scale'. Another revision was made by RICHTER, who called the result the 'Modified Mercalli Scale, 1956 version'. The latest modification is called the MSK-scale, after the seismologists MEDVE-DEV, SPONHEUER and KÁRNÍK. Appendix I to this chapter gives the MSK-scale of 1964. Table 9 gives corresponding intensity values in some different scales.

In order to determine which intensities have prevailed at different places during an earthquake, it is frequently necessary to send special expeditions to investigate the epicentral area. In addition, questionnaires may be distributed to the general public. A number of slightly different questionnaires are in use in different countries. Just to give an example, Appendix II of this chapter reproduces (in translation) the form used in Sweden. The forms may be sent to post offices or other agencies with close contact with the general public. When all data on the effects of an earthquake have been accumulated and the effects expressed in intensities, the result is usually represented by *isoseismals* on a map (Fig. 48). Isoseismals are nowadays usually defined as curves demarcating areas with different

Table 9. Corresponding intensity values on different scales.

MSK 1964[1]	Japanese Scale[2] 1950	Rossi-Forel 1874
I	0	I
II	1	II
III	2	III
IV	2.3	IV
V	3	V—VI
VI	4	VII
VII	4.5	VIII
VIII	5	IX
IX	6	X
X	6	X
XI	7	X
XII	7	X

[1] Intensity values according to MSK 1964 agree with those of the Mercalli-Cancani-Sieberg Scale (1917), the Modified Mercalli Scale (1931) and the Soviet Scale (1952).
[2] The seven-degree Japanese Scale is often used also in international communications issued from Japan. It is to be observed that this is an intensity scale and not a magnitude scale.

intensities from each other. The isoseismal maps are valuable complements to the instrumental records, and they are of special importance in engineering seismology. However, it has to be emphasized that the construction of isoseismal maps from macroseismic observations requires a critical attitude towards the observations and a good knowledge of the psychology of observations. Good knowledge about the geological conditions and the types of building construction used in the area is also of essential importance. As a rule, the intensity is greater on loose ground than on solid bedrock, an observation which has been confirmed by instrumental records. Whether the intensity is smaller below the surface (e.g. in mines) than on the surface, has not been established reliably. Some observations, which seem to indicate such a behaviour, may depend on different ground conditions (solid rock in the mine as compared to loose ground on the surface). Especially for use in engineering seismology, instrumental records of accelerations are of indipensable value and more reliable than direct observations alone. Such recording is also performed continuously in some earthquake areas, where damaging effects are expected.

Some difficulty in evaluating macroseismic observations may be

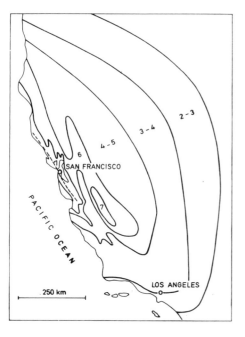

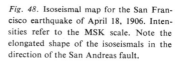

Fig. 48. Isoseismal map for the San Francisco earthquake of April 18, 1906. Intensities refer to the MSK scale. Note the elongated shape of the isoseismals in the direction of the San Andreas fault.

caused by the fact that several non-seismic phenomena give rise to effects similar to those of earthquakes. One example is provided by the sonic booms from airplanes. These booms are not recorded by seismographs except in the immediate vicinity of the phenomenon. Crack formation in loose material due to frost action is another example. This may occur during cold winters, especially in connection with rapid temperature variations. The shaking thus produced may remind one of a small earthquake, but since such events are limited to loose material, they are not recorded by seismographs. Landslides are another phenomenon restricted to the soil and other loose material and are generally not recorded by seismographs.

By means of an isoseismal map, it is possible to calculate characteristic quantities for an earthquake, such as the position of the epicentre, the depth of the source (hypocentre) below the earth's surface and the released energy. On the basis of the macroseismic observations, the epicentre is the central point where the maximum intensity has been observed.

It has been found quite often that epicentres determined macroseismically and instrumentally do not agree exactly. The deviations can depend on inhomogeneities in the geological structure. In addition, an instrumentally determined epicentre (from initial P-wave readings) is generally located at the point where the fracture starts. In extended fault systems this point can deviate considerably from the area with the highest intensity. For a source located at some depth, the decrease of intensity outwards from the centre is evidently much slower than for a source at or near the surface. This provides a possibility to estimate the focal depth from the rate of decrease of the intensity from the central area outwards. After this, the released energy can be estimated. It can be formulated as a function of

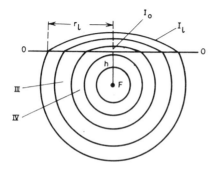

Fig. 49. Schematic picture of isoseismals, partly in a vertical section (below ground level OO), partly in a horizontal plane (coinciding with OO). F=focus, I_0=maximum intensity, h=focal depth, r_l=radius of the macroseismic area and I_l=the corresponding intensity.

any two of the following three variables: the maximum intensity (I_0), the average radius of the macroseismic area (r_l), the depth of the hypocentre (h). See Figure 49. It is of essential importance to notice that knowing only one of these quantities is not sufficient to define the released energy in an unambiguous way; instead two of these quantities are required in the general case. It is natural that an earthquake at a certain depth and with a certain macroseismic area has a certain energy. For instance, the maximum intensity alone is not sufficient to define the energy, because for a given maximum intensity the released energy has to be greater, the deeper the source is located.

The following equation gives at least approximate information about the relation between the three quantities mentioned:

$$I_0 - 2 = 3 \log \frac{r_l^2 + h^2}{h^2} \tag{8}$$

Equation (8) is a special case of the following more general formula, which informs us about the decrease of intensity with increasing distance. At horizontal distance $r=0$, the intensity is I_0, and at distance r, it is I. Then, the following approximate relation holds:

$$I_0 - I = 3 \log \frac{r^2 + h^2}{h^2} \tag{9}$$

A number of relations have been given between magnitude (M) and maximum intensity (I_0), of which the following, derived for earthquakes in southern California, can serve as an example (the focal depth h assumed constant = about 17 km):

$$M = 1 + 2I_0/3 \tag{10}$$

Similarly, the following approximate relation holds between I_0 and the corresponding acceleration a_0 (expressed in cm/sec^2):

$$I_0 = 3 \log a_0 + 1.5 \tag{11}$$

We have to observe that there may be strong variations in the numerical coefficients in equations (10) and (11) between different seismic areas. The formulas given here are nevertheless close to a world average and can therefore be used as a first approximation, when more accurate knowledge of an area is missing. Numerical values are given in Tables 7 and 8.

More general formulas, expressing ground acceleration a as a function of magnitude M, focal depth h and epicentral distance r, i.e. $a = a(M, h, r)$, can be found for example in BÅTH (1975). The approximations involved are about as severe as in the formulas above. More accurate results can be achieved by deduction of formulas for each specified region, preferably by spectral means. An example of such a study, referring to Fennoscandia, is given by BÅTH et al. (1976).

4.5   Seismic Effects on Structures

A special field of science, named engineering seismology or earthquake engineering, deals with the effects of seismic forces and vibrations on all kinds of structures. Field investigations after each destructive earthquake together with investigations on laboratory models and theory all play an important role in such research. Its practical aspects are clear and do not need any further explanation.

The ground motion during an earthquake, especially in the epicentral region, is a complicated phenomenon. A building or any other construction also represents a dynamically complex structure. We then understand that great complexities are involved in the interaction between the two. The motion of the ground will induce oscillatory stresses and strains in a structure. The characteristics of the vibratory motion of a structure will depend on the characteristics of the ground motion as well as on the properties of the structure, such as its size, shape, mass, rigidity, damping, etc. As a rule, it is the weakest parts in a structure that will be damaged first. Therefore, it is considered most suitable to avoid weak parts, i.e. to make the structure as a whole of similar strength throughout, sufficient to withstand expected seismic forces. In this way, concrete structures or reinforced buildings can be made to withstand even very large seismic forces. Flexible joints between walls, floors and roofs are of great help, and are exemplified in wooden structures, but are also a general principle followed in the design of any earthquake-proof structure. Resonance effects have to be avoided as much as possible. In case some choice for the location of a building is given, then experience from California has demonstrated that a location on as solid rock as possible should be selected (this being the primary request), then should in addition active faults be avoided. The requirements on earthquake-proof constructions have recently become much more severe than earlier. This is true especially in the design of nuclear power plants, for which earthquake effects are considered even in countries (like Sweden) where they were never before taken into account. Structures of considerable horizontal extension, e.g. larger bridges, dams, pipelines, are also sensitive to slow motions in the ground, such as in the vicinity of active faults, which has to be considered in their location and construction.

In the following we shall give a brief outline of calculations on the basis of which earthquake-resistant buildings may be constructed. The mathematics used is rather simple, and the main difficulty derives from lack of accurate knowledge of numerical values of all contributing factors to be applied in any special case. The forces that an earthquake may exert on a structure can be written as follows:

Horizontal force: $\quad F = s \cdot Q$

Vertical force: $\qquad V = \pm \gamma \cdot s \cdot Q$ $\qquad\qquad$ (12)

where s is the *seismic coefficient*, Q is the weight of the structure and γ is a correction factor. The seismic coefficient s is the product of several factors as shown in the following formula:

$$s = \alpha \cdot \eta \cdot \beta \cdot \delta \qquad\qquad (13)$$

where $\alpha=$ an intensity factor, depending both on ground acceleration (a_0/g, where $a_0=$ acceleration and $g=$ gravity) and on *seismic risk* R, i.e. the frequency of occurrence of earthquakes: $\alpha = \dfrac{a_0}{g} \cdot R$; $\eta=$ a distribution factor, i.e. a mechanical characteristic of the structure and its mass distribution; $\beta=$ a factor characterizing the motion induced in the structure by an earthquake; $\delta=$ a ground factor, depending on the ground characteristics and the foundation of the structure. The seismic risk R is defined by the following formula:

$$R = 1 - \left(1 - \frac{1}{\overline{T}}\right)^n \qquad\qquad (14)$$

in which $n=$ a specified number of years for which the risk is calculated and $\overline{T}=$ the recurrence period, i.e. the interval in years between shocks of a given intensity. The factor β can be approximately calculated from the formula

$$\beta = \frac{B}{T^{1/3}} \qquad\qquad (15)$$

where B assumes values between 0.6 and 0.8 and T is the period of the vibration.

Table 10. Numerical example of the calculation of the seismic coefficient s for $T=0.5$ sec, $\beta=0.75$ and $\eta=1$; compare equation (13).

Intensity MSK	Acceleration: Gravity a_0/g	Seismic risk R	$\alpha\beta$	Seismic coefficient s		Correction factor $\gamma = V/F$
				Rock	Soil	
7	0.08	1.00	0.06	0.02	0.13	1.2—1.5
8	0.15	0.90	0.11	0.03	0.24	0.9—1.2
9	0.30	0.73	0.17	0.05	0.37	0.8—1.0
10	0.60	0.52	0.23	0.07	0.51	

9 Båth: Introduction to Seismology

As an illustration, Table 10 summarizes some typical values of the factors, extracted from a Spanish publication (*Norma Sismorresistente*, 1968). The values of the seismic risk R naturally vary considerably from region to region, and those of Table 10 are only to be taken as examples. They are computed assuming $n=50$ years. For a period $T=0.5$ sec and $B=0.6$, we get $\beta=0.75$ which, in addition, may serve as an approximate average for most cases. The factor η can be calculated for any given structure. Here we assume for simplicity that $\eta=1$. The factor δ varies between 0.3 for hard rock (with longitudinal wave velocity $v_P>4.0$ km/sec) to about 2.2 for loose soil ($v_P\leqq0.3$ km/sec).

As an example, we then find from Table 10 that for an intensity of 10, the horizontal seismic force affecting a structure reaches the following amounts:

Hard rock: $F=0.07Q$

Loose soil: $F=0.51Q$

And the vertical force has approximately the same size as the horizontal. Such calculations may serve as a guide on how to construct a building or other structure to withstand the seismic forces. Obviously, there is quite a number of considerations which have to be taken and which are adequately summarized in the expression (13) for the seismic coefficient. Detailed and accurate evaluations no doubt require close cooperation between specialists within various branches, such as geology, seismology, structural engineering, etc. Developments similar to those given here apply to any type of mechanical vibrations and their influence on structures. For example, the modern problem of traffic vibrations can be treated in. an analogous way, then putting $R=1$ and considering that dominant periods T are generally short and that structural effects are of an integrated fatigue nature.

4.6   *Appendix I: The MSK Intensity Scale of 1964* (slightly simplified)

I. *Not noticeable:* The intensity of the vibration is below the limit of sensibility; the tremor is detected and recorded by seismographs only.
II. *Scarcely noticeable* (very slight): Vibration is felt only by individual people at rest in houses, especially on upper floors of buildings.

III. *Weak, partially observed only:* The earthquake is felt indoors by a few people, outdoors only in favourable circumstances. The vibration is like that due to the passing of a light truck. Attentive observers notice a slight swinging of hanging objects, somewhat more heavily on upper floors.

IV. *Largely observed:* The earthquake is felt indoors by many people, outdoors by few. Here and there people awake, but no one is frightened. The vibration is like that due to the passing of a heavily loaded truck. Windows, doors and dishes rattle. Floors and walls creak. Furniture begins to shake. Hanging objects swing slightly. Liquids in open vessels are slightly disturbed. In standing motor cars the shock is noticeable.

V. *Awakening:* The earthquake is felt indoors by all, outdoors by many. Many sleeping people awake. A few run outdoors. Animals become uneasy. Buildings tremble throughout. Hanging objects swing considerably. Paintings knock against walls or swing out of place. Occasionally pendulum clocks stop. Unstable objects may be overturned or shifted. Open doors and windows are thrust open and slam back again. Liquids spill in small amounts from well-filled open containers. The sensation of vibration is like that due to a heavy object falling inside the building.

Slight damages in buildings of poor construction are possible.

Sometimes change in flow of springs.

VI. *Frightening:* Felt by most indoors and outdoors. Many people in buildings are frightened and run outdoors. A few persons lose their balance. Domestic animals run out of their stalls. In few instances dishes and glassware may break, books fall down. Heavy furniture may move and small steeple bells may ring.

Slight damage is sustained in single buildings of medium construction. Moderate damage in a few buildings of poor construction.

In a few cases, cracks up to widths of 1 cm are possible in wet ground; in mountains occasional landslips; changes in flow of springs and in level of well-water are observed.

VII. *Damage to buildings:* Most people are frightened and run outdoors. Many find it difficult to stand. The vibration is noticed by persons driving motor cars. Large bells ring.

In many buildings of good construction slight damage is caused; in many buildings of medium construction damage is moderate. Many poor buildings suffer heavy damage, a few even destruction. In single instances landslips of roadway on steep slopes; cracks in roads; seams of pipelines damaged; cracks in stone walls.

Waves are formed on water, and water is made turbid by mud stirred up. Water levels in wells change and the flow of springs changes. In a few cases dry springs have their flow restored and existing springs stop flowing. In isolated instances parts of sandy or gravelly banks slip off.

VIII. *Destruction of buildings:* Fright and panic; also persons driving motor cars are disturbed. Here and there branches of trees break off. Even heavy furniture moves and partly overturns. Hanging lamps are in part damaged.

Many buildings of good construction suffer moderate damage and a few heavy damage. Many buildings of medium construction suffer heavy damage and a few destruction. Many poor buildings suffer destruction and a few total damage. Occasional breaking of pipe seams. Memorials and monuments move and twist. Tombstones overturn. Stone walls collapse.

Small landslips in hollows and on banked roads on steep slopes; cracks in ground up to widths of several centimetres. Water in lakes becomes turbid. New reservoirs come into existence. Dry wells refill and existing wells become dry. In many cases change in flow and level of water.

IX. *General damage to buildings:* General panic; considerable damage to furniture. Animals run to and fro in confusion and cry.

Many good buildings suffer heavy damage, a few destruction. Many medium buildings show destruction, a few total damage. Many poor buildings suffer total damage. Monuments and columns fall. Considerable damage to reservoirs; underground pipes partly broken. In individual cases railway lines are bent and roadways damaged.

On flat land overflow of water, sand and mud is often observed. Ground cracks to widths of up to 10 cm, on slopes and river banks more than 10 cm; furthermore a large number of slight cracks in ground; falls of rock, many landslides and earth flows; large waves on water. Dry wells renew their flow and existing wells dry up.

X. *General destruction of buildings:* Many good buildings suffer destruction, a few total damage. Many medium buildings show total damage; most of the poor buildings are totally damaged. Critical damage to dams and dykes and severe damage to bridges. Railway lines are bent slightly. Underground pipes are broken or bent. Road paving and asphalt show waves.

In ground, cracks up to widths of several decimetres, sometimes up to 1 meter. Parallel to water courses broad fissures occur. Loose ground slides from steep slopes. From river banks and steep coasts considerable landslides are possible. In coastal areas displacement of sand and mud; change of water level in wells; water from canals, lakes, rivers, etc. thrown on land. New lakes occur.

XI. *Catastrophe:* Severe damage even to well-built buildings, bridges, water dams and railway lines; highways become useless; underground pipes destroyed.

Ground considerably modified by broad cracks and fissures as well as by movement in horizontal and vertical directions; numerous landslips and falls of rock. The intensity of the earthquake has to be investigated specially.

XII. *Landscape changes:* Practically all structures above and below ground are greatly damaged or destroyed.

The surface of the ground is radically changed. Considerable ground cracks with extensive vertical and horizontal movements are observed. Falls of rock and slumping of river banks over wide areas; lakes are dammed; waterfalls appear and rivers are deflected. The intensity of the earthquake has to be investigated specially.

4.7   *Appendix II: An Example of a Macroseismic Questionnaire*

According to available information, an earthquake has occurred on.......
.... (date and hour) in ....
(geographical area).

You are kindly asked to answer the following questions as completely as possible, and return the form to .... (address of institute collecting the data).

As every piece of information is of value, the questionnaire ought to be returned even if only a few of the questions can be answered. If the earthquake was not noticed at your place, this is also significant information.
Was the above-mentioned earthquake noticed at your place?.............
If the earthquake was observed, will you please answer the following questions.

Where were the observations made (county, city, village, farm, etc.)?.....
....

When did the earthquake occur?............................

Were the observations made indoors (which floor? what kind of building?)

....

or in the open air or at sea?...

What are the ground conditions at the place where the observations were made (solid rock or loose ground, sand, clay, or similar)?................

Describe below the observations as accurately as possible

How did the earthquake start?.....................................

By rumbling sound or by a shock?..................................

Was any shock noticed?..

How many shocks?...

How much time elapsed between the shocks?.........................

What was the duration of each shock?..............................

Was any rumbling noticed?...

Before, during or after the shock?.................................

What was the duration of the rumbling?............................

How was the earthquake motion felt (as a shock, as shaking motion or as oscillatory motion)?..

From which direction did the motion appear to come?................

What effects did the earthquake produce?...........................

Was it noticed generally by the population or only by individual persons?

....

Give the approximate number of persons you have interviewed..........

Were any easily movable objects shaken or set rattling, such as lamp-shades, windows, doors or furniture?.................................

Were creakings of floors or walls heard?............................

Were smaller objects moved or did they fall down?...................

Were heavier objects, such as furniture, moved?......................

Did bells ring?...

Did hanging lamps swing?...

Were sleeping persons awakened (in general or only in individual cases)?

....

Did pendulum clocks stop?..

Were trees and bushes clearly swaying?.............................

Did general uneasiness and fright arise?............................

Were heavier objects overturned?..................................

Were buildings damaged? ....

Did pieces of plaster fall down from buildings? ....

Did cracks arise in walls? ....

Were chimneys overthrown? ....

Did fissures appear in the ground? ....

In which direction did swinging objects move? ....

In which direction were chimneys and other objects overthrown? ....

Was anything else of interest observed in connection with the earthquake (change of river flow, unusual motion of water in lakes or on the sea, etc.)? ....

Is there any other reliable information available about the earthquake? Was it felt equally strong at all places in your area or was it felt stronger at some places? Where were the strongest effects observed (on hills or in valleys)? ....

Chapter 5

Earthquake Statistics and Earthquake Geography

5.1 Earthquake Catastrophes

It has been estimated that during historical time 50 to 80 (according to one estimate 74) million people have lost their lives in earthquakes or their immediate aftereffects, as fires, landslides, tsunamis, etc. The figure is rather modest compared to some other kinds of catastrophes, not to mention modern traffic. The distribution of population takes generally no account of earthquake risk, at least not on a large scale. In fact, some of the seismically most active areas are among those most populated; this is especially true for Japan. The distribution of population is dictated primarily from economic viewpoints, and it is considered far more practical to find protective measures against earthquake disasters. Active research is going on in this field at present. This concerns partly the problem of earthquake prediction (Chapter 10), which would make evacuations possible, partly the introduction of building codes, which are still lacking in many earthquake countries.

There is usually no clear correlation between the magnitude of a shock and the number of killed people or other destruction (cf. Table 8). An earthquake may be large but not destructive, on the other hand, an earthquake may be destructive but not large. The absence of correlation is due to the fact that a great number of other factors enter into the problem when considering casualties: first of all, the location of the earthquake in relation to populated areas, but also soil conditions, building constructions, etc. The Mediterranean and Middle East area has provided numerous examples of relatively small, but destructive earthquakes (for example Agadir, Morocco, February 29, 1960; Lar, Iran, April 24, 1960; Skopje, Yugoslavia, July 26, 1963; also Managua, Nicaragua, December 23, 1972). On the other hand, earthquakes of large magnitude occur from time to time at sea or in unpopulated areas, with no damage to people or structures, and they often pass unnoticed by the general public. Earthquake statistics based on news reports can at most be used to get an apprehension of destruction and casualties, but is not useful from a purely scientific viewpoint. There are numerous descriptions of earthquake effects to be found in the literature, also there are a number of summaries of such descriptions. Older literature becomes more and more unreliable the farther back one goes. There are frequently gross overestimates of the number of deaths, which also happens from time to time in the very first news from an earthquake disaster. Official statements from governments and rescue organizations are generally reliable. In addition, older literature often dwells upon secondary effects, while primary effects – such as observations of fault displacements – do not attract the same attention. A thorough knowledge of earthquake effects and a sound criticism is necessary in using this literature. On the other hand, it is often questionable to what extent surface observations of fault displacements are representative of focal mechanism, considering that for most shocks with observable effects the hypocentre is 20–30 km deep.

The effects of earthquakes on buildings and other constructions, such as bridges, dams, etc., have been studied extensively for many earthquakes, e.g. San Francisco in 1906, Tokyo and Yokohama in 1923, Kern County, California, in 1952, Chile in 1960, Skopje in 1963, and others. Nowadays, usually every earthquake which has caused damage is carefully investigated on the spot, to a great extent thanks to special missions sent out on the initiative of UNESCO. The experience gained by such studies is naturally of great importance in formulating any building codes. From the engineering side it is frequently and correctly emphasized that a greatly expanded network of strong-motion accelerographs is needed throughout the seismic regions of the world.

Experience has shown that it is practicable to build in an earthquake-proof way. Buildings and other constructions, specially built according to building codes and located on good ground (preferably solid rock), have been able to resist the stresses which arise even in the strongest earthquakes. The examples are still few but nevertheless of the greatest significance, as they point to the possibility that earthquake disasters to buildings can be avoided. The additional costs required in earthquake-proof building are in general considerably lower than the costs for rebuilding. However, hitherto some form of building code has been applied only in about one-third of the approximately 70 countries, where earthquake risk exists.

Table 11. Destructive earthquakes.

Date	Region	Magnitude (M)	Number of deaths	Remark
1556 January 23	China: Shansi		830 000	This is usually referred to as the earthquake with the greatest number of persons killed. Although the figure may of course be uncertain, it is by no means unrealistic.
1693 January 9	Italy: Sicily		(60 000)	
1703 December 31	Japan: Odowara, Tokyo		5 230	Divergent information on the number of people killed: SIEBERG (1932) gives 150 000, RICHTER (1958) 5233, the latter figure probably more correct.
1730 December 30	Japan: Hokkaido		137 000	Uncertain; may be confusion with preceding earthquake.
1737	India: Calcutta		300 000	Uncertain, as reliable confirmation is lacking.
1755 November 1	Portugal: Lisbon	(8 3/4)	60 000	One of the best-known earthquakes. The epicentre was probably located about 100 km west of Lisbon. Unusually large macroseismic extent, the radius of perceptibility reaching about 2000 km and the seiche limit about 3500 km, i.e. to Scandinavia and Finland.
1759 October 30	Syria		30 000	
1783 February 5	Italy: Calabria		30 000	
1797 February 4	Ecuador, Peru		40 000	
1812 March 26	Venezuela: Caracas		20 000	
1857 December 16	Italy: Naples	(6 1/2)	12 000	Studied extensively by R. MALLET, especially regarding direction of displacements.
1868 August 13	Peru, Ecuador		40 000	

Table 11. Destructive earthquakes (continued).

Date	Region	Magnitude (M)	Number of deaths	Remark
1883 July 28	Italy: Casamicciola (Ischia)		2 300	Strictly local destruction, the area of destruction not being more than 3 km wide, suggesting very shallow focus. Probably volcanic earthquake.
1891 October 28	Japan: Mino-Owari		7 270	Displacements up to 4 m horizontally, 7 m vertically. Great damage.
1896 June 15	Japan: Riku-Ugo		27 120	
1897 June 12	India	8.7		Extensively studied by OLDHAM. Two faults, Samin and Chedrang, the latter with displacements reaching 11 m. Several secondary effects: flooding in the Brahmaputra plain, slides near the Assam hills. Shillong destroyed.
1899 September 10	Alaska: Yakutat Bay	8.6		Vertical uplifts of 14.5 m. Effects on glacier flow. Several major shocks in this area in September 1899, the next largest on September 4, $M=8.2$.
1902 December 16	Turkestan		4 500	
1905 April 4	India: Kangra	8.6	19 000	Epicentral area, including Kangra, on Tertiary rocks in the Himalaya foothills.
September 8	Italy: Calabria	7.9	(2 500)	
1906 January 31	Colombia	8.9	1 000	
March 16	Formosa: Kagi	7.1	1 300	Investigated in detail by OMORI. Strike-slip displacement reaching 2.5 m, vertical displacement half as much.

Table 11. Destructive earthquakes (continued).

Date	Region	Magnitude (*M*)	Number of deaths	Remark
1906 April 18	California: San Francisco	8.3	700	Studied in greater detail, especially macroseismically, than most other earthquakes. Strike-slip dextral displacement on San Andreas fault extending over 330 km and amounting to maximum 6.4 m; maximum vertical displacement not quite 1 m. Great fires.
August 17	Chile: Santiago, Valparaiso	8.6	20 000	
1907 January 14	Jamaica: Kingston	6.0	1 600	
October 21	Central Asia	8.1	12 000	
1908 December 28	Italy: Messina, Reggio	7.5	83 000	
1911 January 3	North Tien-Shan	8.7	450	Extensive ground displacements.
1912 August 9	Marmara Sea (west coast)	7.8	1 950	
1915 January 13	Italy: Avezzano	7.5	29 980	
October 3	Nevada: Pleasant Valley	7.8	(0)	Vertical displacements of 3—4 m.
1920 December 16	China: Kansu, Shansi	8.6	100 000	Macroseismic area included the whole of China. Largest destruction around Tsinning. Major fractures, landslides.
1922 November 11	Peru: Atacama	8.4	600	
1923 September 1	Japan: Tokyo, Yokohama	8.3	99 330	Usually called the Kwanto earthquake. Maximum displacements of 4.5 m. Destructive fires.
1925 March 16	China: Yunnan	7.1	5 000	Talifu almost completely destroyed.

Table 11. Destructive earthquakes (continued).

Date	Region	Magnitude (M)	Number of deaths	Remark
1927 March 7	Japan: Tango	7.9	3 020	A very well-investigated earthquake.
May 22	China: Nan-Shan	8.3	200 000	Large fractures. Felt as far as Peking.
1929 May 1	Iran: Shirwan, Kutshan, Budshnurt	7.1	3 300	
June 16	New Zealand: Buller, Murchison	7.6	17	Vertical displacements about 5 m on White Creek fault (South Island). Considerable landslides. Severe damage.
1930 July 23	Italy: Ariano, Melfi, Calitri	6.5	1 430	
1931 February 2	New Zealand: Hawke's Bay, Napier, Hastings	7.9	225	The first great earthquake disaster in New Zealand. Displacements reached 1—2 m. Isoseismals elongated in the direction of the two islands; felt all over New Zealand. Fires and landslides.
1933 March 2	Japan: off Pacific coast	8.9	2 990	
1934 January 15	India: Bihar-Nepal	8.4	10 700	Effects typical for large thicknesses of sediments and alluvium. Fissures opened up, one being 5 m deep, 10 m wide and nearly 300 m long. Almost all houses either tilting or sinking into the ground, up to 1 m.
1935 April 20	Formosa	7.1	3 280	
May 30	Pakistan: Quetta	7.5	30 000	Quetta almost completely destroyed.

Table 11. Destructive earthquakes (continued).

Date	Region	Magnitude (*M*)	Number of deaths	Remark
1939 January 25	Chile: Concepción	8.3	28 000	
December 26	Turkey: Erzincan	7.9	30 000	Roughly elliptical macroseismic area, 1300 km long (east-west) and 600 km wide. Strike-slip displacements up to 3.7 m with smaller vertical displacements. Tectonics similar to San Andreas in California. Large activity continued during 1942—4.
1943 September 10	Japan: Tottori	7.4	1 190	
1944 December 7	Japan: Tonankai, Nankaido	8.3	1 000	
1945 January 12	Japan: Mikawa	7.1	1 900	
1946 November 10	Peru: Ancash	7.3	1 400	Purely vertical displacements, up to 3.5 m, along a 5 km fault scarp trending NW (eastern block moving up): first time clear faulting was demonstrated for an earthquake in South America. Great destruction, landslides.
December 20	Japan: Tonankai, Nankaido	8.4	1 330	
1948 June 28	Japan: Fukui	7.3	5 390	Larger intensities on alluvium (where large fissures opened up) than on surrounding bedrock.
October 5	Turkestan: SE of Ashkhabad	7.3	(400)	Considerable damage in an area which had not been affected by strong earthquakes for several centuries.
1949 August 5	Ecuador: Ambato	6.8	6 000	Large landslides and topographic changes in the Andes.

Table 11. Destructive earthquakes (continued).

Date	Region	Magni-tude (M)	Number of deaths	Remark
1950 August 15	India: Assam, Tibet	8.7	1 530	Radius of macroseismic area reaching 1400 km. Great topographical changes, landslides, floods. More damaging in Assam than the shock of 1897. Numerous aftershocks in an area of more than 700 km in east-west extent.
1952 March 4	Japan: Tokachi	8.3	(28)	Several hundred people killed, of which 28 on Hokkaido. Considerable damage, tsunami, significant variations in the geomagnetic field following the earthquake.
July 21	California: Kern County	7.7	14	The largest earthquake in California since 1906, with considerable damage. Main shock on White Wolf fault, mainly dip-slip (vertical displacements over 1 m), aftershocks mainly strike-slip. Both main shock and aftershocks very well covered by seismograph recordings in the area.
1953 February 12	Iran: Mazandaran	6.5	970	Ground fracturing.
March 18	Turkey: Marmara Sea, Bandirma	7.4	240	
August 12	Ionian Islands	7.4	460	
1954 September 9	Algeria: Orléansville	6.8	1 250	Strongest earthquake in North Africa in the last hundred years. Occurred on the Dahra range along the coast. Vertical soil displacements of 0.6—1 m

Table 11. Destructive earthquakes (continued).

Date	Region	Magnitude (M)	Number of deaths	Remark
				over more than 20 km in length. Focal depth 8—9 km. A 'relay shock', $M=$ 6.2, occurred on September 10, 1954, about 40 km north of this epicentre.
1955 March 31	Mindanao	7.9	430	
1956 June 9	Afghanistan: Kabul	7.7	220	Varying information on number of deaths: 224, 600 or 10 000, the lowest figure probably most reliable.
July 9	Aegean Sea: Santorin	7.4	57	Tsunami, volcanic eruption.
1957 July 2	Iran: Tehran to Caspian Sea, Damovand, Laridjan, Bi Bol	7.2	130	Extensive property damage.
July 28	Mexico	7.8	55	Intensity distribution found to depend heavily on local geology and soil conditions. Damage at greater distance often due to longer-period waves (resonance effects with some structures).
December 4	Mongolia: Altai-Gobi	7.8	30	Great topographical changes, landslides. In the Altai Mountains a rift was found, 250 km long. One mountain range is reported to have increased in height by 6 m.
December 13	Iran: Farsinaj, Hamadan, Kermanshah	7.1	1 130	
1958 July 10	S. Alaska, British Columbia, Yukon	7.8	5	Felt over an area 200 km in extent along the Fairweather fault. Horizontal

Table 11. Destructive earthquakes (continued).

Date	Region	Magni-tude (M)	Number of deaths	Remark
				displacements 7 m, vertical 1 m, especially at Lituya Bay and Yakutat Bay, motions conforming to San Andreas' system. Landslides, sand blows, fissures, extensive minor faulting.
1960 February 29	Morocco: Agadir	5.8	10 000 to 15 000	The greatest number of deaths ever for such a small shock. Maximum observed intensity was 11.
April 24	Iran: Lar (Girash)	5.9	450	Damage almost total along a very narrow band right through Lar. Larger destruction to buildings on the alluvial plain (where Lar is located) than to those on surrounding rock.
May 22	Chile: 36° S–48° S	8.3	5 700	Active area 1600 km long and 160 km wide. Fault movement began in the north and progressed southwards. Both horizontal and vertical displacements. Floods, tsunami, volcanic activity. This is a repetition of events in 1835—7 in the same area.
1962 September 1	Iran: Qazvin	7.1	12 230	Probably no earthquake as strong as this one has occurred in this area since 1630. Progressive faulting, extending over 100 km, on the Ipak fault. Landslides, rock falls. About 1800 aftershocks recorded in Iran up to the middle of November 1962.

Table 11. Destructive earthquakes (continued).

Date	Region	Magnitude (*M*)	Number of deaths	Remark
1963 July 26	Yugoslavia: Skopje	6.0	1 100	Like Agadir 1960, this shock occurred at shallow depth right under a city; 80% of the buildings in Skopje destroyed; 295 aftershocks felt up to August 15, 1963. Skopje had earlier been destroyed by earthquakes in 518 and 1555.
1964 March 28	Alaska: Anchorage, Seward	8.5	131	Most destructive earthquake in this area since 1899. Fissures in Anchorage, 10 m deep and 15 m wide. Horizontal and vertical displacements of 6 m. Aftershocks within an area of 900 km$\times$300 km. Fires, tsunamis (Yakutat Bay 8 m high, Kodiak Island 3 m, Easter Island 30 cm), minor volcanic activity. Seiches, 2 m high, on the coasts of Texas and Louisiana.
June 16	Japan: Niigata	7.4	36	Great material damage. Tsunami. Great topographical changes.
1965 March 28	Chile: Valparaiso	7.5	600	
1966 August 19	Turkey: Varto	6.9	2 520	
1968 January 15	Sicily	6.1	740	A series of shocks, of which $M = 6.1$ was the largest.
May 16	Japan: SE of Hokkaido	8.6	48	
August 1	Luzon: Manila	7.7	300	
August 31	Iran: 280 km south of Mashad	7.4	12 000 to 20 000	Horizontal displacements of maximum 4 m over a distance of 27 km.
September 1	Iran: Ferdows	6.7	> 2 000	Extensive property damage.
1969 February 28	Atlantic Ocean, off Portugal	7.9	13	Damage in Portugal, Spain and Morocco. Tsunami.

Table 11. Destructive earthquakes (continued).

Date	Region	Magni- tude (*M*)	Number of deaths	Remark
1969 July 25	Eastern China	6.1	(3 000)	
1970 March 28	Turkey: Gediz	7.3	1 086	Dislocations of 3.5 m in rock.
April 7	Luzon: Manila	7.7	15	Major damage in Manila.
May 14	Caucasus	6.9	?	Unknown number of deaths. Great damage.
May 31	Peru: Huarás, Chimbote, Trujillo, Yungay	7.7	66 794	Avalanches, landslides, floods. Largest catastrophe in Peru this century.
December 10	Peru-Ecuador	7.6	81	
1971 February 9	California: near Los Angeles	6.8	65	Extensive property damage.
May 12	Turkey: Burdur	6.3	72	
May 22	Turkey: Bingöl, Genc	6.7	863	
July 9	Chile: Illapel, Valparaiso	7.7	90	Great fires. Tsunami. Landslides.
1972 April 10	Iran: Zagros	7.1	5 374	Major damage. Landslides.
December 23	Nicaragua: Managua	6.0	10 000	Heavy damage. Fire. 80% of Managua destroyed.
1973 January 30	Mexico: Colima, Jalisco, Michoacan	7.8	52	Damage.
February 6	China: Szechwan	7.8	?	Many casualties. Considerable damage.
August 28	Mexico: Vera Cruz, Ciudad Cerdan, Orizaba, Cordoba	6.8	> 600	Fires, considerable damage.
1974 October 3	Peru: Lima, Callao	7.8	78	Extensive damage. Tsunami.
December 28	Pakistan: Polas	6.4	5 300	Pattan destroyed.
1975 February 4	China: Liaoning	7.6	?	Considerable damage.
September 6	Turkey: Lice	6.8	2 700	Lice destroyed.
November 29	Hawaii	6.9	2	Tsunami. Volcanic eruption. Ground cracks 1 m wide.
1976 February 4	Guatemala	7.3	22 836	Guatemala City almost completely destroyed. Landslides.

Table 11. Destructive earthquakes (continued).

Date	Region	Magni-tude (*M*)	Number of deaths	Remark
1976 May 6	Italy: Friuli	6.9	918	Considerable damage, numerous aftershocks. Rockfalls, landslides. Strongest earthquake in this region for the last 2000 years.
May 17	Soviet: Uzbek	7.3	?	Considerable damage at Gazli and Bukhara. April 8, 1976, same region, had *M* = 7.3 also.
May 29	China: Yunnan	7.0 6.7	?	Two larger shocks.
June 25	Irian	7.3	443	Landslides, flooding, damage.
July 14	Indonesia: Bali	6.6	466	
July 27	China: Hopeh Prov., Tangshan, Tientsin, Peking	8.2	700 000	Exact number of deaths unknown (Hong Kong reported 655 237). Severe damage. Numerous aftershocks, largest *M* = 7.8 on July 28 and *M* = 6.8 on November 15.
August 16	China: Szechwan	7.2	?	Aftershocks August 21 and 23, with *M* = 6.5 and 6.7, respectively.
August 16	Mindanao, Moro Gulf, Sulu	8.1	8 000	Tsunami. Considerable damage. Aftershock August 17, *M* = 7.1.
November 24	Turkey: NE of Lake Van	7.3	3 626	Considerable damage.
1977 March 4	Rumania	7.2	1 541	Extensive damage in Bucharest.
April 21	Guadalcanal	7.6	12	
August 19	Sumbawa	7.8	90	
1978 September 16	Iran	7.7	16 000	Extensive damage.
November 29	Mexico	7.8	(9)	

Table 11 summarizes information on the most destructive earthquakes. The material has been limited in the following way:

1. Earthquakes with magnitude $M > 7.5$ with reported casualties, even if small, are included, in particular those which have been studied in considerable detail.

2. Earthquakes with $M \leqq 7.5$ are included, if the number of deaths exceeded 1000. In addition, several others with a less number of casualties are included, that occupied much space in newspapers, especially from the last quarter century.

Even though we have aimed at some completeness in Table 11 for the present century, we cannot naturally claim this. The obvious increase towards later years clearly reflects the much better communication systems in recent time. Before 1900 only major earthquake disasters are included, usually only where the number of deaths has exceeded 20 000. Many larger shocks, with magnitude around 8 or over, do not appear in Table 11. This is the case for earthquakes at sea, especially around the Pacific Ocean, which may cause tsunamis and some damage, but remarkably few casualties, if any at all. No focal depths have been given in Table 11, as it is obvious that nearly all destructive earthquakes have been at shallow depth and usually not deeper than 60 km. Rumania is an exception with destructive earthquakes with $h \cong 150$ km.

The total number of deaths listed in Table 11 for the period 1900–1976 amounts to more than 1.6 million, i.e. over 22 000 per year. However, these figures are heavily influenced by the China earthquake on July 27, 1976. Excluding this earthquake with 700 000 deaths, the total for 1900–1976 amounts to over 0.9 million and the annual figure to nearly 12 000.

There is no relation between the number of deaths N and the magnitude M which could serve as a guide in individual cases. Only a statistical relation exists such that $\log N$ is approximately proportional to $0.8\,M$, which means an increase of N by a factor of about 6 for a unit upward step in M. But it has to be strongly emphasized that this is only a statistical result from Table 11, derived from the data for the years 1905–1976, and it cannot be used in individual cases. The reason is that a number of other factors has a strong or even dominating influence on the number of casualties.

Geographically, the number of destructive earthquakes exhibits a striking distribution. Of the 100 earthquakes in the 80-year period 1897–1976, 64% fall within the belt $35° \pm 10°$ N. This belt contains Japan, Central Asia, the Middle East (Iran) and the Mediterranean area. If instead of just counting the number of destructive earthquakes, we consider the number of deaths, we find that this belt accounts for nearly 90% of the total loss of lives in earthquakes. The corresponding percentage figures for the belt $35° \pm 5°$ N are 45 for the number of destructive shocks and over 80 for the number of deaths. The existence of such a pronounced 'destruction belt' is due to a number of factors:

1. High seismicity.
2. Densely populated areas (at least some of them).
3. Old cultures, where most buildings are not earthquake-proof, but in rather bad condition.

The year 1976 deserves a special comment. Numerous earthquake disasters occurred throughout this year, with maximum number of casualties in China. In fact, we have to go back to the year 1556 to find a similar death toll, i.e. not less than 420 years, and then it was also China that was hit. On the other hand, looking at 1976 from a purely seismological point of view, there were not more or stronger earthquakes this year than in average. Figure 50 shows the annual energy release and the annual number of all earthquakes with $M \geqq 7.0$ since 1968, with corresponding averages marked. As far as energy release is concerned, it is seen that we need to go back only to 1972 to find a year with a higher value. The explanation of this combination of extraordinary catastrophes with just average seismological behaviour lies in the anomalous distribution of earthquakes in 1976. They were hitting sensitive, densely populated areas. On the other hand, several well-known seismic areas, as for example the Aleutians, Kamchatka, Kurile Islands, were remarkably passive in 1976. In other words, a seismicity map based exclusively on 1976 would be a rather good map of catastrophes but a very poor map of the earth's true seismicity.

We have been using the number of people killed as an indication of the destruction. It is difficult to summarize in one figure such a complicated thing as the destruction during an earthquake, and even the number of deaths is not without shortcomings. But I have found it to be much more useful than, for instance, information on the number of destroyed houses or the monetary value of destroyed property. The latter information in

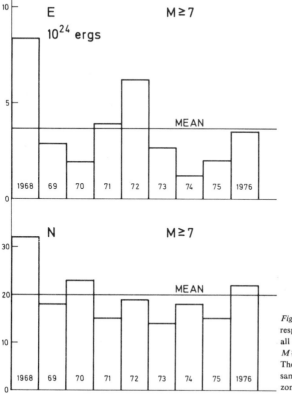

Fig. 50. Total number N and corresponding energy release E for all earthquakes with magnitude $M \geqq 7$ for the years 1968 to 1976. The respective mean values for the same period are indicated by horizontal lines.

particular does not mean very much, even though it will be of the greatest interest to companies issuing earthquake insurances. The maximum intensity (I_0) is naturally a useful quantity, but frequently not enough. A complete description has naturally to take due account of all details and as far as possible express the effects numerically.

Concerning the strength of an earthquake, this is judged by the general public from the damage that the earthquake may have caused. This is quite clear, and only the seismologist, who also has the opportunity of examining his records, may arrive at another opinion. If we look upon earthquakes as expressions of the dynamics of the earth then the magnitude or the released energy are the only reliable measures of the strength of an earthquake. As already emphasized the damages depend on a large number of other factors, which have nothing to do with the dynamics of the earth. The Agadir earthquake in Morocco in 1960 is considered by the general public as a large earthquake; however, it was in fact relatively small, magnitude = 5.8, whereas the destruction was very great. Similarly, the Tashkent earthquakes in 1966 appeared as large to the newspaper-reading public. However, these were also very small shocks; the biggest reached a magnitude not more than about 5. For the seismologist, who could see both sides (both his records and the news reports), it appeared as a problem why such small shocks could have such large effects. The probable explanation, later confirmed by Russian seismologists, was that the shocks occurred at shallow depth (a few kilometres) right under the city. The conditions were similar in Agadir in 1960, in Skopje in 1963 and in Managua in 1972. A small local shock of shallow focus right under a city can produce considerable damage to the buildings just above (by displacements, crack formation in walls, and slides), but it does not produce much in the way of seismic waves which can be observed at a greater distance.

5.2   Frequency and Energy of Earthquakes

As already emphasized, no reliable earthquake statistics can be founded on news reports alone. This is naturally no criticism of our news agencies. On the contrary, these perform a highly recommendable service by transmitting information on earthquake effects and then primarily those which have caused destruction and death tolls. And that our news agencies should give reports much beyond this is hardly their duty; this is instead a duty of the seismologists. It should be emphasized that it is the magnitude concept which has made it possible to measure the seismicity exactly in a quantitative way. Seismological statistics and seismicity maps from the time before the use of magnitude scales were very incomplete and unreliable.

The number of earthquakes is considerably greater than most people realize. Over magnitude 8 there are on the average one to two earthquakes per year for the earth as a whole. Proceeding towards lower magnitudes, the number increases exponentially. The total number of earthquakes on the earth is estimated to be about 1 million per year. This corresponds to

Table 12. Number and energy of earthquakes for the earth as a whole.

Magnitude M	Number of earthquakes per 10 years	Energy release 10^{23} ergs/10 years
8.5—8.9	3	156
8.0—8.4	11	113
7.5—7.9	31	80
7.0—7.4	149	58
6.5—6.9	560	41
6.0—6.4	2100	30

about two shocks every minute. Of these only the stronger ones, numbering a few thousand per year, are recorded world-wide by seismograph networks (see Chapter 2).

The number of earthquakes of magnitude 6.0 and over occurring on the earth as a whole is summarized in Table 12, using 10 years as a reference time unit. For magnitudes $M \geqq 7.0$, the numbers given are those actually observed for an interval of 47 years (1918–64), based upon a compilation by S. J. DUDA at the Seismological Institute, Uppsala. For magnitudes $M < 7.0$, the numbers in Table 12, have been obtained by extrapolation, and therefore they have to be taken with some reservation.

The number N of earthquakes with the magnitude M can be expressed by the following simple formula:

$$\log N = a - bM \tag{1}$$

where a and b are positive constants. Referring N to *one* year and a magnitude interval of 0.1 unit ($M \pm 0.05$), we find for the above-mentioned material for 1918–64 that $a = 8.73$ and $b = 1.15$ (Fig. 51):

$$\log N = 8.73 - 1.15M \tag{2}$$

Thus, the number of earthquakes increases rapidly as we proceed towards lower magnitudes. The value obtained for b ($= 1.15$) is an average for the whole material. However, b exhibits significant differences between different earthquake regions as well as between different depth ranges. Nearly all b-values fall in the range from 0.5 to 1.5. The b-value has served as a kind of tectonic parameter in many discussions.

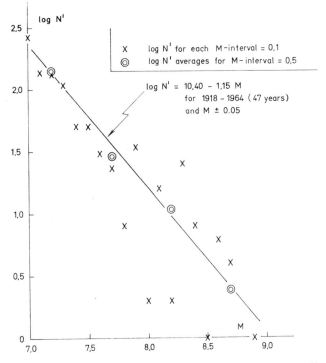

Fig. 51. Earthquake frequency for the earth as a whole for the interval 1918—64.

We have seen (Chapter 4) that the energy E (expressed in ergs) of the seismic waves depends on the magnitude according to the following formula:

$$\log E = 12.24 + 1.44M \qquad (3)$$

In a certain way the two effects, expressed by (2) and (3), counteract each other. As we proceed towards lower and lower magnitudes, the energy of each shock decreases according to (3), but at the same time the number of shocks increases according to (2). In order to see what the net result will be, we combine (2) and (3) into the following formula:

$$\log EN = 20.97 + 0.29M \qquad (4)$$

where EN is the energy release, expressed in ergs per year, corresponding to the magnitude M. The right-hand column of Table 12 has been calculated by means of equation (4).

Obviously, the greatest energy release is due to the largest earthquakes, in spite of their smaller number. With decreasing magnitude the energy release decreases according to equation (4), but clearly much slower than for individual shocks according to (3). From equation (4) we see that a step upward on the magnitude scale of one unit corresponds to a doubling of the total energy released. These calculations show that earthquakes with magnitudes $M \geqq 8.0$ are responsible for about 50% of the total energy released, those with magnitudes $M \geqq 7.0$ correspond to about 75% and those with $M \geqq 6.0$ to nearly 90% of the total energy.

In summary, from statistical viewpoints we can divide earthquakes into three groups, considering the earth as a whole and the era of instrumental seismology:

1. $M > 8$: the material is complete but insufficient for reliable statistical conclusions.

2. $M = 7 - 8$: the data are both completely known and statistically sufficient; however, this is a valid statement only for one magnitude unit.

3. $M < 7$: the data are not completely known or catalogued in an easily accessible way; on the other hand, it is clearly statistically sufficient as soon as a homogeneous series is available. For some more restricted areas attempts have been made to eliminate this deficiency by studies of microearthquakes (see below).

For continued statistical studies of the earth's seismicity one of the main tasks is obviously to get homogeneous catalogues for the magnitude range $M = 6 - 7$ in particular, extending as far back in time as possible. The handicap under item 1, however, can be remedied only by continued observations in the future.

Frequently, it is of great importance to know the seismic activity not only for the earth as a whole, but even more so for limited areas. From what has already been said, it would in general be necessary to wait a considerable number of years before reliable statistics could be gathered. In recent years seismologists have tried to eliminate this undesirable circumstance by observations of very small shocks, so-called *microearthquakes* ($M < 3$). Within the area to be investigated highly sensitive seismographs are installed at many places. In seismic areas the microearthquakes are usually frequent and there is no need to wait a long time until statistically sufficient material has been accumulated. When this is available, it is possible to calculate the constants a and b in equation (1) and then to extrapolate this result to earthquakes of higher magnitudes. In so doing, it is generally assumed that a and b are the same for large as for small earthquakes. In some cases the method has been successful, for example in Nevada, where it has been possible to reproduce the secular statistics within a short time. But certain other results suggest that the coefficient b is not the same for all magnitude ranges. Thus, it has been found in some cases that b increases considerably with increasing magnitude. In such cases a simple extrapolation would lead to an overestimate of the number of large earthquakes.

In concluding this section, I have summarized in Table 13 complete data for all the largest earthquakes (magnitude $M \geqq 8.5$) which have occurred in the interval 1897–1976. The data are based on the most reliable sources available. It should also be noted that all data in Table 13 are deduced from instrumental records.

5.3   Geographical Distribution

A seismicity map can be constructed according to different principles, for instance:

1. In a purely qualitative way, as in Figure 52.
2. More exactly physically defined, especially by indicating the energy released per unit time and unit area.
3. By indicating the maximum observed intensity at different points; such maps, called *seismic zoning maps*, are of importance primarily in engineering seismology.

Figure 52 shows the geographical distribution of earthquakes, based upon homogeneous data for the interval 1904–52 (GUTENBERG and RICHTER). About 80% of all seismic energy is released within the marginal areas of the Pacific Ocean, particularly along the coasts of eastern Asia down to New Zealand. In comparison with this, the contributions from other seismic zones are relatively modest, even though great catastrophes occur from time to time.

The map in Figure 52 gives a correct picture of the main features of the earth's seismicity. However, we have to note that the observation

Table 13. Earthquakes with magnitude $M \gtreqqless 8.5$ during the interval 1897—1976.

Date	Origin time Greenwich h m s	Latitude deg	Longitude deg	Focal depth km	Region	Magnitude M
1897 June 12	11 06	26.0N	91.0E	n^1	India	8.7
August 5	00 12	38.0N	143.0E	n	Japan	8.7
September 20	19 06	6.0N	122.0E	n	Mindanao	8.6
September 21	05 12	6.0N	122.0E	n	Mindanao	8.7
1899 September 10	21 41	60.0N	140.0W	n	Alaska	8.6
1902 August 22	03 00	40.0N	77.0E	n	Sinkiang	8.6
1905 April 4	00 50 00	33.0N	76.0E	n	India	8.6
July 23	02 46 12	49.0N	98.0E	n	Mongolia	8.7
1906 January 31	15 36 00	1.0N	81.5W	n	Colombia	8.9
August 17	00 40 00	33.0S	72.0W	n	Chile	8.6
1910 June 16	06 30 42	19.0S	169.5E	100	New Hebrides	8.6
1911 January 3	23 25 45	43.5N	77.5E	n	Tien-Shan	8.7
June 15	14 26 00	29.0N	129.0E	160	Ryukyu	8.7
1914 November 24	11 53 30	22.0N	143.0E	110	Mariana Islands	8.7
1917 May 1	18 26 30	29.0S	177.0W	60	Kermadec	8.6
June 26	05 49 42	15.5S	173.0W	n	Samoa	8.7
1920 December 16	12 05 48	36.0N	105.0E	n	China	8.6
1929 March 7	01 34 39	51.0N	170.0W	60	Aleutians	8.6
1933 March 2	17 30 54	39.3N	144.5E	n	Japan	8.9
1938 February 1	19 04 18	5.3S	130.5E	n	Banda Sea	8.6
November 10	20 18 43	55.5N	158.0W	n	Alaska	8.7
1939 December 21	21 00 40	0.0	123.0E	150	Celebes	8.6
1941 June 26	11 52 03	12.5N	92.5E	60	Andaman Islands	8.7
1942 August 24	22 50 27	15.0S	76.0W	60	Peru	8.6
1950 August 15	14 09 30	28.5N	96.5E	n	India, Assam, Tibet	8.7
1958 November 6	22 58 06	44.5N	148.5E	75	Kurile Islands	8.7
1964 March 28	03 36 13	61.1N	147.6W	20	Alaska	8.5
1968 May 16	00 49 54	40.8N	143.2E	7	Japan	8.6
1972 January 25	02 06 23	22.5N	122.3E	n	Formosa	8.5

^{1} n = normal depth, i.e. around 20—30 km below the earth's surface.

interval is too short to guarantee against surprises. The map is based exclusively on instrumental records obtained at seismograph stations around the world, and these do not extend further back in time than to around the turn of the century. Some seismic phenomena occur at far greater intervals. For instance, it was believed up to 1954 that earthquakes with focal depths exceeding 300 km occur only in the Pacific marginal areas.

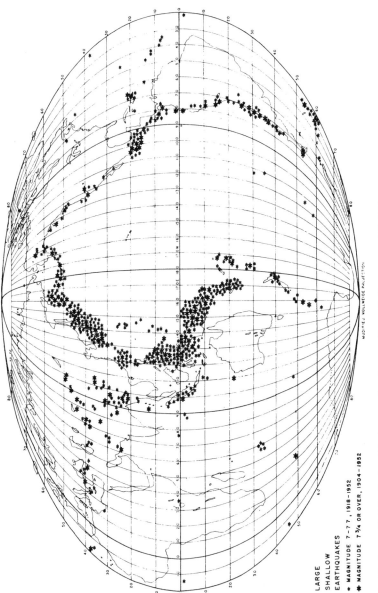

LARGE
SHALLOW
EARTHQUAKES

✳ MAGNITUDE 7 – 7.7 , 1918 – 1952

✱ MAGNITUDE 7 ¾ OR OVER, 1904 – 1952

Fig. 52. Geographical distribution of the larger earthquakes according to B. GUTENBERG and C. F. RICHTER.

However, on March 29, 1954, an earthquake occurred in southern Spain at a depth of over 600 km – then perhaps the greatest seismological sensation within the previous 25 years. The catastrophe at Agadir on February 29, 1960, had a precedent, but then we have to go back to the year 1731. Skopje in southern Yugoslavia was partially destroyed on July 26, 1963, and there we have to go back to 1555 to find a parallel case. Still another example is furnished by an earthquake at the northern end of the Red Sea on March 31, 1969 (origin time=07 15 54.4 GMT, epicentre=27.7 °N, 34.0 °E, magnitude $M = 6.6$). This shock has no correspondence within the approximately 70-year-old statistics, instrumentally based, nor have we been able to find any historical indication of such strong earthquakes earlier at this place. The examples of 'surprises' can easily be multiplied.

For easier review we distinguish three main belts of earthquake activity:

1. The circum-Pacific belt.
2. The mid-Atlantic to East African and Easter Island belt.
3. The Asiatic-European belt (the Alpide belt).

Belt 1 is no doubt the most important by far, considering the released energy. It should be emphasized that seismic activity is limited to the marginal areas of the Pacific Ocean, at the transition between oceanic and continental structure. In contrast, the Pacific Ocean itself is really 'peaceful' even from a seismic point of view, and it is in fact one of the greatest stable areas on the earth. Exceptions have to be made for the volcanic Hawaii Islands and the Easter Island rift structure. Belt 2 is a global structure, which is clear from Figure 53. This belt starts at the northern coast of Siberia (near the estuary of the river Lena), it passes across the Arctic and continues over Spitsbergen and Iceland into the Atlantic. It follows the centre of the Atlantic Ocean along its whole extent, proceeding into the Indian Ocean. There it branches into the East African rift system, and another belt passing over the Easter Island ridge in the Pacific up to the Rocky Mountains in North America. Belt 3 can be considered as a link between belts 1 and 2. It starts partly in the Kamchatka area, partly in Indonesia. The first belt passes diagonally across Central Asia; the other passes along southern Asia (the Himalayas). These belts unite in Pamir, from where the belt continues westwards through Iran, Turkey and the Mediterranean area out to the Azores, where it joins the

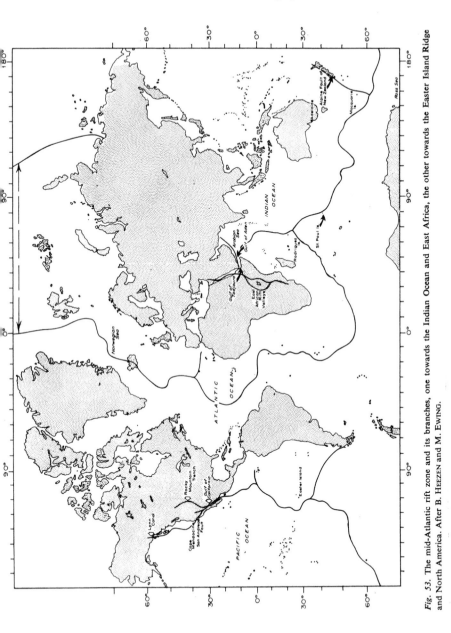

Fig. 53. The mid-Atlantic rift zone and its branches, one towards the Indian Ocean and East Africa, the other towards the Easter Island Ridge and North America. After B. HEEZEN and M. EWING.

mid-Atlantic belt. Concerning released energy, belt 1 takes the first place, as already mentioned, followed by belt 3, while belt 2 comes in the third place. Outside the narrow belts mentioned the seismic activity is considerably lower, but it is hardly completely absent anywhere. Relatively stable areas are to be found in Russia-Siberia (to the north and northwest of the mountain ranges in Asia), in Africa (excepting the East African rift zone and northern Africa), and in parts of North and South America. Shield areas are exceptionally stable. Examples of such areas are to be found in Canada, Brazil, Fennoscandia, northern Asia, parts of Africa, Arabia, western Australia. All seismic belts are also obvious on the frontispiece map (opposite the title page of this book), which shows more recent results, obtained by computer calculation.

However, even within the three belts mentioned, the activity is not evenly distributed. In some parts it may be very high while intervening parts exhibit a lower activity. A numerical comparison of different regions is presented in Table 14, based on all earthquakes with $M \geq 7.9$ in the interval 1904–64. The total energy E released in the stated interval of time and magnitude amounted to 2.4×10^{26} ergs. Its percentage distribution over the most important areas is clear from Table 14. All circum-Pacific areas together correspond to 77% of the total energy release. As the different regions in Table 14 are of different size or of different length, I

Table 14. Geographical distribution of the seismic energy release for $M \geq 7.9$ for the interval 1904—64.

Region	Percentage energy release	Energy per degree along the belt 10^{23} ergs
Alaska	4.3	6.1
Western North America	1.0	0.8
Mexico–Central America	4.2	2.3
South America	16.4	6.4
Southwest Pacific Ocean–Philippine Islands	26.5	7.0
Ryukyu–Japan	15.8	13.5
Kurile Islands–Kamchatka	5.8	7.0
Aleutian Islands	3.0	2.9
Central Asia–Turkey	16.9	5.6
Indian Ocean	4.5	—
Atlantic Ocean	1.6	—·

have also expressed the energy concentration by dividing the total energy (ergs) in each belt by the length (degrees) of each belt. The figures thus obtained are clearly to some extent dependent upon the division into regions, but still they furnish a reliable picture of the relative distribution of the activity. The highest energy concentration is found in the Ryukyu Islands-Japan area. The energy concentration is about half as big as this in the Kurile Islands-Kamchatka and in the southwest Pacific Ocean-Philippine Islands. Only slightly lower values are found for South America, Alaska and Central Asia to Turkey. Smaller values are obtained for the Aleutian Islands and Mexico-Central America, these two areas being comparable. For all other areas the energy concentration is considerably lower and it has not been possible to calculate it. For the Pacific area we can distinguish two 'poles' with maximum activity, one located in the Japan area, the other in Chile.

On the whole, earthquakes are located in areas with active tectonics, formation of mountain ranges, folding, etc. The relation between earthquakes and volcanic eruptions is a question which is frequently raised. In general, it can be said that earthquakes which have a direct connection with volcanic eruptions (so-called volcanic earthquakes) are small, whereas all important earthquakes are tectonic. Nevertheless, there is often a certain parallelism in the occurrence of tectonic earthquakes and volcanoes, but presumably with no immediate genetic connection between the two.

For the tectonic earthquakes there are essentially two types of tectonics which are of significance: arc tectonics and block tectonics. Arc tectonics is illustrated by Figure 54; block tectonics implies that blocks of the solid earth move in relation to each other, often along rectilinear, vertical fault surfaces. Arc structures dominate in most of the circum-Pacific earthquakes as well as in the Alpide belt. Along the western margin of the Pacific Ocean, the arc structure consists mostly of island arcs, while in the Alpide belt it appears mostly as mountain ranges (as the Himalayas, the Alps, the Carpathians). Block tectonics dominates in certain parts of the circum-Pacific area, especially in California and the central part of New Zealand as well as in the Pamir-Baikal zone and along the mid-Atlantic belt, i.e. especially where earthquakes of normal depth occur. In some areas the two types overlap, as in Japan, Peru, the Philippines. Earthquakes of arc tectonics are in the majority, but information about them derives mostly from in-

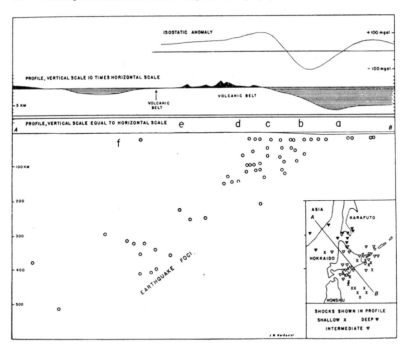

Fig. 54. Vertical section of a typical so-called Pacific arc in northern Japan. After B. GUTENBERG and C. F. RICHTER.

strumental records. In contrast, most macroseismic information derives from earthquakes of the block type.

The most important structural elements of a typical Pacific arc are the following (see Figure 54 for the different elements a–f):

a) Deep-sea trench.

b) Most important tectonic line with shallow-focus earthquakes, negative gravity anomaly and a non-volcanic anticline (which may appear as a ridge on the sea bottom or as an island chain).

c) A belt of positive gravity anomaly with earthquakes at about 60 km depth.

d) Most important arc structure from late Cretaceous or Tertiary period, often consisting of large islands. Active or recently extinct volcanoes. Earthquakes at a depth of about 100 km.

e) An inner structural arc. The volcanism is usually older, either in an advanced stage or extinct. Earthquakes at depths of 200–300 km.

f) A belt of earthquakes at depths of 300–700 km.

At the west coast of Chile there are elevation differences of up to 14 km between the bottom of the deep-sea trench and the summits of the Andes. Such height differences seem to exclude a static equilibrium but rather suggest a dynamic equilibrium. The regular arrangement of different elements a) to f) indicates that all phenomena are different expressions of one and the same process in the earth, namely a relative motion of the sea bottom and the continental structure. Moreover, it is this relative motion that causes the stresses which are necessary for the occurrence of earthquakes. As we shall see in the next chapter, it is possible to carry the generalizations much further and to consider arc structures and rift structures as special expressions of one world-wide dynamic system.

Arc structures have been found in many places around the Pacific Ocean, even though one or another of the components a)–f) may be missing in some places. While a)–d) are regularly found in the Pacific arcs, the element e) and particularly f) are missing in several of them. Arc structures outside of the Pacific area, as, for example, in Burma, the Himalaya Mountains, Baluchistan, the eastern Mediterranean, the Tyrrhenian Sea, the Carpathian Mountains, are in general more rudimentary.

One of the most important results from a closer study of deep earthquakes was the finding that in arc structures the foci are located on sloping planes. These planes exhibit a slope angle of about 45° in under the continents. Later, it was found that the shallower shocks are located along a plane with a smaller slope (34° in the Kurile Islands; see Fig. 55), while the deeper ones are found along a plane with a greater slope (58° in the Kurile Islands). The break in the plane occurs between earthquakes of intermediate depth and the deep ones (see next paragraph).

The variation of the seismic activity with depth is shown in Figure 56. We see that the largest activity is located in the uppermost 75 km. From here the activity decreases down to about 400 km depth, exhibiting a pronounced minimum between 425 and 475 km (at the arrows in Fig. 56a). Below this level, renewed activity starts between 475 km and about 700 km depth. On the basis of these results it appears most natural to divide earthquakes with regard to focal depth into normal or shallow

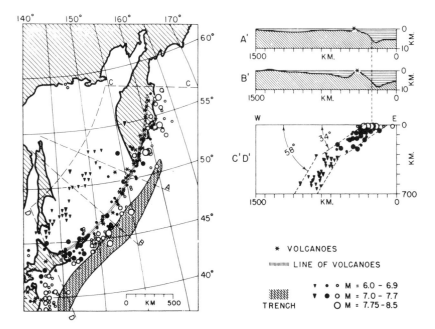

Fig. 55. A typical arc structure: Kamchatka-Kurile Islands. After H. BENIOFF.

earthquakes ($h < 60$ km), intermediate-depth earthquakes ($60 \text{ km} < h < 450$ km) and deep earthquakes ($h > 450$ km). The depth variation is no doubt closely connected with the mechanism of earthquakes, but still much remains to be investigated before a complete explanation can be given.

The curves in Figure 56 have been calculated from data given by S. J. DUDA (1965). We have to observe that the data cannot yet be considered statistically significant, particularly not for the deeper levels. It is easy to show that even one larger earthquake at greater depth can modify the curves. This fact is reflected as a certain fluctuation of the values in Figure 56a, especially at deeper levels. In order to eliminate this deficiency to some extent, I have also calculated values for each interval of 100 km (Fig. 56b). Then, the curves become clearly smoother, but at the same time we may run the risk of losing significant details in the curves.

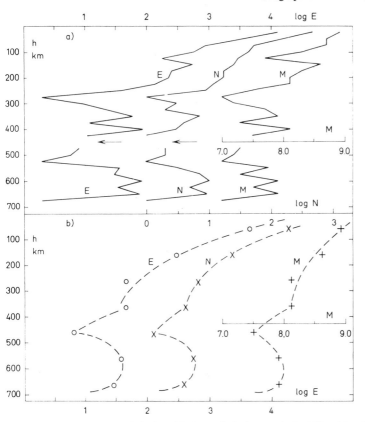

Fig. 56. The depth dependence of seismic activity. E = total seismic energy (unit = 10^{22} ergs) for all earthquakes with $M \geqq 7$ for the interval 1918—64, N = corresponding total number of earthquakes, M = maximal magnitude: a) for depth intervals of 25 km (the points are connected by straight lines), b) for depth intervals of 100 km (the dashed curves suggest a plausible generalization).

Chapter 6

Earthquake Mechanism

6.1 *Solutions for Individual Earthquakes*

Figure 57 shows a photograph of part of the San Andreas fault in California – one of the best investigated fault zones in the world. The displacements are in this case mainly horizontal, in which the eastern side (towards the mainland of the USA) is moving towards the south in relation to the western side (towards the Pacific Ocean).

Figure 58 shows a schematic picture of the mechanism, that is responsible for so-called tectonic earthquakes, i.e. all earthquakes of any significance. The heavy full line marks a fissure or fault in the solid earth. By slow motions in the earth, proceeding over geologic epochs, one side (a) of the fault is displaced in relation to the other (b). This is shown by a deformation of the straight lines drawn across the fault, from the first situation (A) to the situation at a later time (B). This process continues until the stresses thus generated in the fault zone are large enough to overcome the friction between the two sides (a) and (b). Then, a rupture occurs, i.e. a sudden displacement, after which the configuration is as shown in (C). It is this sudden rupture which constitutes the earthquake. The slow motions at the fault continue unimpeded, the whole process is repeated and a new shock occurs at some later time, and so on. This picture of the earthquake-generating process is ascribed to the American seismologist REID. He arrived at this result after careful studies of the 1906 San Francisco earthquake, which incidentally occurred on the San Andreas fault. The theory is generally called the *elastic rebound theory*.

Similar relative motions can exist along planes of any orientation, e.g. sloping planes. This is illustrated in Figure 59, showing a schematic vertical section perpendicular to the coast zones of the Pacific Ocean. In this case the sea bottom moves in relation to neighbouring continents both in vertical and horizontal directions. This motion pattern gives rise to the seismically most active areas in the world. The relative motions shown in Figure 59 can extend to depths of 600–700 km, different at dif-

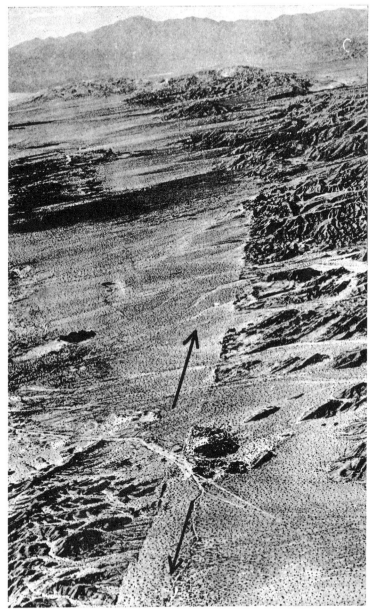

Fig. 57. Part of the San Andreas fault in California. After H. BENIOFF, modified.

ferent places, and the fault zone slopes in under the continent in the way that the figure shows (compare Fig. 54). Later in this chapter, in the section on the new global tectonics, we shall see that the sloping fault zone in Figure 59 has a modified explanation. In fact, both Figure 58 and 59 are nowadays more of a historical significance in the development of our knowledge, but their importance for present-day seismology is still so great that they certainly deserve to be described and known.

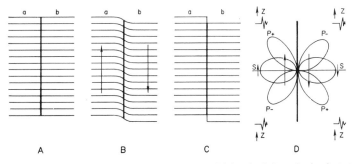

Fig. 58. A simple model (in the horizontal plane) explaining the fault mechanism in tectonic earthquakes. $P+$ =compression (anaseism), i.e. the first motion of the P-wave is directed *away from* the source (NW and SE quadrants above), $P-$ =dilatation (kataseism), i. e. the first motion of the P-wave is directed *towards* the source (NE and SW quadrants above).

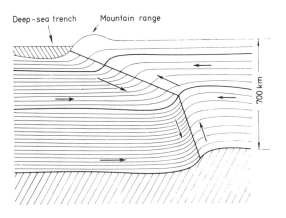

Fig. 59. A schematic vertical section perpendicular to the Pacific coast, showing orientation of fault surfaces and relative motions.

From these simple models we understand that three conditions are required to produce earthquakes: relative motions in the earth, stress generation and stress release.

Motions of different parts of the earth in relation to each other [such as between the sides (a) and (b) of the fault in Fig. 58] constitute an old and much discussed problem, including such questions as continental drift, mountain formation, etc. That such motions really do exist, is beyond any doubt. However, it is still an unsolved problem what the reasons are for such motions. Here, there is still room for hypotheses of various kinds. Among the different hypotheses which have been proposed, the following ought to be mentioned:

1. Contraction of the earth by cooling. This was the old hypothesis, which has nowadays been mostly abandoned, partly because the solar origin of the earth is no longer considered reliable. Moreover, it is uncertain, whether the earth's internal temperature is at present decreasing or increasing.

2. Expansion of the earth. This is a relatively new hypothesis, which has gained more and more adherents. As a reason for the expansion one has referred to phase changes at the boundary of the outer core at 2900 km depth.

3. Convection currents (circulation cells) within the earth. These imply an unchanged volume of the earth. These currents are attributed to the temperature structure in the earth's interior and correspond approximately to the atmospheric circulation in thunderstorms.

Seismological observations alone are not able to give any final solution to the problem concerning the driving mechanism of the slow motions in the earth. On the other hand, they no doubt provide the most complete information on the present dynamics of the earth.

Stresses have to be accumulated for earthquake generation, which means that earthquakes cannot arise in liquid matter. In such a case stresses are released at the same time as they are generated. No earthquakes have been observed at depths exceeding 720 km, which has been interpreted as due to a greater mobility of the matter at greater depth. An increase of the electric conductivity at about the same depth also indicates a greater mobility of the matter.

The release of the stresses will happen sooner or later, depending on the amount of friction. The friction at a fault surface certainly has an upper limit and therefore this also puts a certain upper limit on the earthquake strength.

The degree of cooperation between the three factors mentioned determines the seismic conditions on the earth: the geographical distribution is determined by the location of the relative motions (which are quite unevenly distributed over the earth), whereas the speed of the motions and the size of the friction determine the frequency and strength of the earthquakes.

The case shown in Figure 58 corresponds to a *dextral* horizontal displacement along a vertical fault surface. By dextral we mean that, in looking across the fault, the other side is displaced towards the right. If instead this displacement is directed towards the left, it is called *sinistral*. In nature, the fault surface can have any orientation and, similarly, the displacement can have any direction along this surface. Figure 60 shows

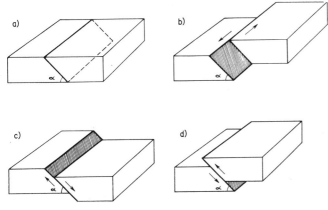

Fig. 60. Fundamental types of fault motion.
a) A block with a fault before any motion has taken place.
b) Strike-slip fault, transcurrent fault, lateral fault, tear fault, wrench fault, α is often near 90°.
c) Dip-slip fault, normal-slip fault, normal fault, tensional fault, gravity fault. Generally $45° < \alpha < 90°$.
d) Dip-slip fault, reverse-slip fault, reverse fault, compressional fault, thrust fault. Often $0° < \alpha < 45°$.

the fundamental types of faults, including the various alternative names used for different faults. Obviously, the radiation of seismic waves will vary considerably with the direction from the focus. Thus, we may expect that transverse waves will reach maximal amplitudes in a direction perpendicular to the fault but will be small or vanishing in the same direction as the fault itself. Longitudinal waves will have their maximum amplitudes in a direction of about 45° with the fault direction and will disappear both along and perpendicular to the fault surface. In Figure 58 the radiation diagrams are shown for P and S for the simple model illustrated in that figure.

While earthquake parameters can be calculated, at least approximately, from the records at only one station (Chapter 4), the determination of the source mechanism evidently requires a combination of records obtained at a great number of stations in different directions from the source. Earlier studies of this kind were almost exclusively based on the direction of first motion of the P-wave (compression or dilatation). In more recent studies, both P- and S-waves as well as reflected body waves and surface waves have been used and in addition spectral methods are applied.

By including as many different wave types as possible in the calculation of focal mechanisms, we eliminate to a certain extent the lack of observations in critical directions. In addition, such a combination provides a much deeper insight into the properties of different waves, their propagation, etc., as all waves have to agree with one mechanism solution, when correctly interpreted. Under the assumption that all waves have originated by one and the same motion at the hypocentre, then the different phases (their amplitudes, direction of displacement in the records, etc.) are evidently not independent of each other. If we start from the record and, so to speak, follow the waves backwards to the hypocentre with due regard to all changes in amplitudes, phase angles, etc., which have occurred at the different discontinuity surfaces or within continuously varying media, then all phases should agree. The partitioning into two reflected waves (P and S) and two refracted waves (P and S), which occurs when an elastic wave (P or S) strikes a given discontinuity surface (in density or elastic properties or both), is unambiguously determined. In principle, it is easy to calculate, even though algebraically rather cumbersome expressions are met with in the general case.

However, as always when a problem is inverted we have to be careful and check if the solution is unique. For a given focal mechanism it is a straightforward task to calculate the radiation of different waves and their distribution over the earth's surface. This is what we did in Figure 58 for a simple case. But in practice, we have the inverted problem: to draw conclusions about the mechanism from given distributions. In addition, we could not exclude the possibility of curved fault surfaces or of curvilinear fault displacements, even if such cases are presumably of subordinate significance.

For the earth the problem is not as simple as in the plane (Fig. 58) since we have to consider the shape of the earth and the curvature of the wave paths. One method which has proved to be useful in practice is to imagine the focus surrounded by a small sphere, the *focal sphere*, and to project the focal motions onto this sphere. The *P*-waves are depicted as

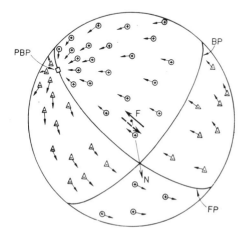

Fig. 61. Focal sphere for a single force couple F with moment (single dipole with moment). Radial displacements (*P*): plus outwards, minus inwards. Tangential displacements (*S*): indicated by arrows. F = focus, FP = trace of the fault plane, BP = trace of the auxiliary plane, perpendicular to the fault plane, PBP = pole of the auxiliary plane, N = null vector.

radial displacements on the surface of the sphere and the *S*-waves as tangential displacements. These are in turn split into *SV* and *SH*. Figure 61 shows a typical case.

It is quite clear that we need observations with a good coverage of the focal sphere in order to be able to make any reliable conclusions about the mechanism. Unfortunately, in many practical applications only a small part of this sphere is covered by observations (Fig. 62). For example, if *P*-observations are available only for distances exceeding 10°, then only about ⅓ of the surface of the focal sphere is covered by observations (for 20° and 30° the corresponding fractions are only about ⅙ and ¹/₁₀, respectively). This is important to keep in mind in any focal mechanism solution. This deficiency can be remedied in two ways:

1. By including observations as near to the epicentre as possible (i.e. for distances less than 10°).

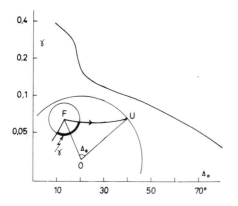

Fig. 62. The fraction (γ) of the surface of the focal sphere that is covered by P-wave observations when these are available only for distances larger than Δ_0. The figure has been derived with an assumed focal depth of 100 km.

2. By using also other waves in addition to P; in particular, pP would be suitable, as it leaves the focus in an upward direction.

But a focal sphere is not sufficient to demonstrate the focal mechanism. It is also necessary to be able to project the focal sphere onto a plane, and for this purpose a number of projection methods have been applied. The details will not be described here. This is mainly a technical problem of interest only in immediate connection with the work on a mechanism solution.

As said, the number of observation points can be increased by using, besides P, also the recorded initial displacements of other P-waves, such as PP, PcP, pP, with due regard to possible phase changes upon the reflections. These waves will give information about the radial component in other directions than shown by the direct P-wave, for one and the same station. Similarly, S-wave readings could be supplemented by readings of SS, ScS, sS. Unfortunately, it is frequently difficult to read the direction of motion for all waves after P, depending on the already existing motion in the seismogram. For the S-waves, one frequently uses the whole oscillation figure, which can yield results of higher reliability than a reading of just the direction of the first onset.

In addition to such kinematical properties, observations of a more dynamic character are also of great importance, e.g. of the amplitude ratios between different waves. Such methods have been particularly developed by Russian seismologists. In recent years, studies of seismic wave spectra have also been added as a further useful contribution. However, there are still certain difficulties inherent in a complete explanation of an observed spectrum, as it depends both on the focal mechanism and on changes during the wave propagation.

From Figure 58 it is clear that *P*-waves alone are insufficient to decide which plane is the fault plane. As we see from this figure, the plane perpendicular to the fault plane, i.e. the so-called *auxiliary plane*, is oriented in exactly the same way to the distribution of compression and dilatation of *P*. In order to decide which plane is the fault plane, one has sometimes combined the *P*-readings with geological or tectonic information. Even better, however, would be to combine the *P*-readings with *S*-readings, as these would be able to discriminate between the two planes, in case we are justified in assuming the model in Figure 58 to be correct.

As the *P*-waves do not permit an unambiguous solution and as many seismologists have been working only with this wave, a new concept, the so-called *null vector* (Fig. 61), was introduced. This is defined as the intersection between the two planes, and it is obviously unambiguous. It is called null vector, because it is perpendicular to the displacement and is thus not displaced. The null vectors have been mapped from solutions of

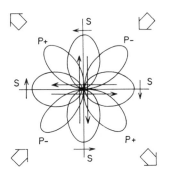

Fig. 63. Radiation of *P* and *S* from a double couple (double dipole with zero net moment). *P* + = compression, *P* − = dilatation. Heavy arrows suggest a possible stress distribution.

many earthquakes within certain areas, in efforts to draw more general conclusions from their distribution. For instance, in the southwest Pacific Ocean it has been found that the null vectors are parallel with vertical planes along the geographical structures.

The conditions in Figure 58 correspond to the action of a single force couple. However, in many solutions it has been shown that the *S*-waves do not exhibit the simple distribution scheme of Figure 58, but instead show four lobes, just like the *P*-waves but rotated 45° in relation to the *P*-lobes (Fig. 63). Such a radiation pattern can be explained by assuming two force couples at the source instead of just one couple. In such cases, obviously not even the *S*-waves are able to discriminate between the fault plane and the auxiliary plane. In the seismological literature much discussion can be found about the nature of mechanisms, especially if there has been a single-couple source or a double-couple source. BENIOFF has shown that both hypotheses are in agreement with REID's elastic rebound theory. Modern investigations are much more in favour of the double-couple explanation (Fig. 63) than the single-couple one (Fig. 58).

In recent years, surface waves have also been included in mechanism studies. By and large, it is true to say that the radiation of Rayleigh waves is related to the radiation of *P* and that Love-wave radiation is related to *S*. In most major earthquakes the active fault has a considerable extent in the horizontal direction, sometimes up to around 1000 km. Observations of surface waves have been particularly informative in such cases. We then have to imagine that the fault motion starts in some part of the fracture zone, usually near one end of this zone, and then propagates along the whole fracture. Obviously, by such a phenomenon the major earthquakes will be extended, both in space and in time with certain ensuing complications in the seismic records. It has been found theoretically and also confirmed by observations that surface waves which propagate in the same direction as the fault motion will have shorter periods and larger amplitudes than those surface waves which propagate in the opposite direction (compare the DOPPLER principle). In an application of this theory to observations of Rayleigh waves from the Chile earthquake of May 22, 1960, a velocity of fault propagation of 4.5 km/sec was found and a fault length of 750–800 km of the active segment. The seismologist A. BEN-MENAHEM (Israel) has performed pioneering investigations of great importance in the application of surface waves to the study of focal mechanisms. Hitherto, the studies have mostly concerned fundamental-mode surface waves, and the relations between mechanism and higher-mode surface waves largely remain to be investigated.

Another method to calculate the propagation of a fault depends on phase relations between vertical and horizontal components of spheroidal free vibrations of the earth. The phase difference can have any value between 0° and 180° depending upon the wave-length, the fault length and the velocity of fault propagation. For the same Chile earthquake as just mentioned it was found in this way that the fault propagation velocity was 3–4 km/sec and the active fault length was 960–1280 km. Taken together, this means that the total duration of the source phenomenon was not less than about 5 minutes.

These theories are of great significance, especially as they permit a calculation of fault length and fault propagation velocity also in cases which are not accessible to direct observation on the spot. Direct observations have revealed displacements of the order of a few centimetres in small shocks up to about 14.5 m (Yakutat Bay, Alaska, 1899). The horizontal extent of the active fault zone also varies from a few metres in small shocks to around 1000 km in the largest. The greatest extent of directly observed displacements was encountered in the San Francisco earthquake of 1906 which could be followed over a length of 330 km. Most of the large fault zones are located around the Pacific Ocean, but with the exception of the San Andreas fault zone in California they are mostly located off the coasts and thus in general evade direct observations.

In interpreting mechanism solutions, we have to remember that our immediate results concern the orientation of the fault surface and the direction of motion along this surface. The next step is to try, by means of these results, to learn something about the stress system, which has caused the rupture. These stresses must be considered to have a more primary importance than just the fault motion.

However, the relations between fault motions and earthquakes have in recent time been subjected to renewed scrutiny. Some seismologists prefer to consider the primary phenomenon in the earthquake source as a sudden volume change (for instance, caused by some phase change of the matter), and then the formation of fracture should be of a more secondary nature. To date, this view has not gained many adherents, and it may be true that this picture would fit better for deep earthquakes ($h >$ >450 km) than for the shallower ones. Particularly for the deep earthquakes there still remains much to be investigated in this respect.

BENIOFF reports that records on strain seismographs in South America of two deep South-American earthquakes suggest that the elastic rebound theory is in need of revision. BENIOFF maintains that his records cannot be interpreted as anything else but a mechanism consisting of a sudden collapse, caused by some change of state of the matter. It has also been shown by others that the dimensions of such an implosive source can be considerably smaller than according to the fault theory for the same energy release. This hypothesis is also in accord with the similarity found between P-wave spectra of deep-focus earthquakes and explosions. It ought to be observed that collapse due to change of state can provide for a large energy release, whereas collapse of cavities (Chapter 1) is only of very small significance.

6.2   Geographical Combination of Earthquake Mechanisms

At present, a large number of focal mechanism solutions are available for individual earthquakes. These solutions can be considered as point information about present tectonic motions, which need to be coordinated and also extended both in space and time. Even if hitherto not too many attempts have been made to combine the point solutions into larger pictures of continental or global extent, it must be emphasized that such works will give the most reliable information on present tectonics and the dynamics of the earth's interior. In particular, tectonic motions in the solid earth giving rise to mountain formation, *orogenesis*, are in the centre of interest.

However, further complication arises in such a combination of mechanism solutions. For instance, the majority of the mechanism solutions, especially around the Pacific Ocean, have given evidence of horizontal displacements. This means that the displacements in most of the investigated cases are not directed up or down sloping planes (Fig. 59), which would be expected whatever the driving force may be. Several expedients to avoid this dilemma have been envisaged, such as that the differential displacements at the rupture do not coincide with the rupture surface. There has also been expressed some criticism towards the focal mechanisms, partly concerning methods, partly concerning the selection of earthquakes for investigation, even if nobody seems to have been able to point to any particular errors.

The focal mechanisms of earthquakes around the Pacific Ocean exhibit to a great extent dextral horizontal displacements. By a combination of such circum-Pacific solutions, BENIOFF concluded that the bottom of the entire Pacific Ocean is rotating counter-clockwise in relation to surrounding continents and that a complete revolution would take about 3×10^9 years to complete. Later, BENIOFF modified his hypothesis to some extent. We shall be looking a little closer at this idea. In addition to the apparently dominating displacement system (dextral horizontal displacements, tangential to the coasts), as we encounter in the Pacific marginal areas, there are also radial, horizontal displacements as well as displacements along sloping planes. The San Andreas fault system in California is among the best investigated ones, with clear dextral horizontal displacements (6.4 m in the San Francisco earthquake in 1906, 5.8 m in the Imperial Valley earthquake in 1940). Geodetic measurements have shown that the relative motion of the two sides of the fault amounts to as much as 5 cm/year. The total length of the fault is estimated to be about 3400 km. Earthquakes in Nevada indicate that the San Andreas' stress system extends at least 500 km into the continent. In addition, there is a multitude of other faults in California, for instance, the Garlock fault, which deviates sharply from the San Andreas direction, but which can be included in the same tectonic system.

Field observations after the Chile earthquake of May 22, 1960, revealed a system of dextral horizontal displacements, parallel to the coast of Chile and extending along most of Chile. The active fault segment extended about 1000 km to the south of the epicentre of the large earthquake on May 22, 1960. Records of G- and R-waves on strain seismographs in California gave the same result. The Peru–Ecuador segment of the earthquake belt is apparently closely related to the Chilean segment and constitutes a direct northward extension of the latter. For Central America and Mexico the character of the displacements has not been fully clarified from instrumental information, but the similarity to adjacent segments is a strong indication of related or similar motions.

The earthquake belt from British Columbia to Alaska exhibits clear dextral horizontal displacements. This has been clearly shown for three large earthquakes in the area, whereas the large vertical displacements at Yakutat Bay in 1899 are considered as a secondary phenomenon. The Alaska earthquake of March 28, 1964, also showed great vertical displacements of several metres, which are also considered to be secondary. In the Aleutian Islands a major earthquake occurred on March 9, 1957, which has been studied in considerable detail. The active fault extended over a length of 1100 km, to judge by the extent of the aftershock area. The width was only about 160 km. The great length of the area is a strong indication that a horizontal displacement has dominated. The Kurile Islands–Kamchatka area has been the site of many large earthquakes. The Kamchatka earthquake of November 4, 1952, has been carefully studied. It extended over an area about 1000 km long and 220 km wide. A focal mechanism solution for this earthquake demonstrated a dextral horizontal displacement, parallel to the southeast coast of Kamchatka. For the area from Japan to New Zealand there is much evidence for dextral horizontal displacements, but also for the opposite, notably in the Philippine Islands.

The preponderance of dextral horizontal displacements for earthquakes around the Pacific Ocean led to the hypothesis of the ocean bottom rotating counter-clockwise in relation to adjacent continents. The fracture systems and the earthquake belts are located at the contact zone between the two blocks moving in relation to each other. However, this imaginative summary picture suffers from a number of difficulties and problems. Most of the arcs (as in the Aleutians) are convex towards the Pacific Ocean. Between Samoa and New Zealand the belt is not tangential but rather radial to the Pacific Ocean. Some belts form sharp angles with each other (for instance, Aleutians–Kamchatka, Aleutians–Alaska). All these facts do not agree with the picture of the Pacific Ocean bottom rotating like a solid body (a disk) relative to surrounding continents. A possibility would be that we instead have to deal with plastic flow, but proof is lacking for this. Even if the picture of the rotating Pacific Ocean seems to be difficult to defend, still the hypothesis represents a commendable effort to integrate different point solutions. That some form of connection exists between the motions in one and the same earthquake belt (as the circum-Pacific) seems to be difficult to dismiss. But the connections are probably more complicated than generally envisaged hitherto.

BENIOFF's notion of large-scale rotations has recently been revived, but in quite a different form (see next section). It is of considerable interest to compare BENIOFF's results with the picture which LENSEN (New Zealand) has deduced for the direction of the principal horizontal stress for the circum-Pacific belt (Fig. 64).

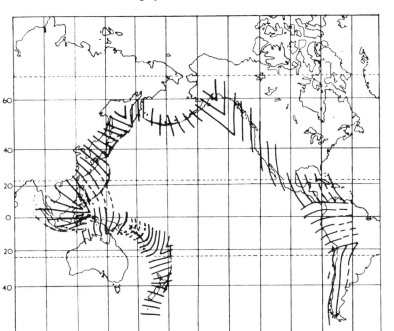

Fig. 64. Directions of the principal horizontal stress around the Pacific Ocean. After G. J. Lensen (1960).

By a similar comparison of local mechanisms for the Asiatic continent, Scheidegger has demonstrated the possibility of combinations over large areas. He uses the null vector and applies a statistical treatment of the individual solutions. A southward motion of the Asiatic continent is proposed as a possible solution for the observed mechanisms. India is considered to have a northward motion, relative to central Asia. A similar, somewhat modified picture of the motion in Asia has been suggested by Båth (Fig. 65). This picture is not based on focal mechanism solutions but on the geographical distribution of the earthquake belts. Many more observations from this area are needed, both of slow motions and of earthquake mechanisms, before any of these hypotheses can be considered as

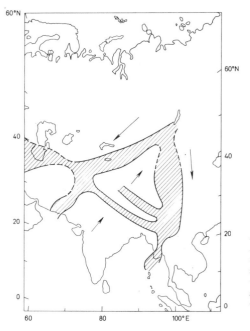

Fig. 65. Earthquake belts in continental Asia and a hypothetical explanation (the arrows are supposed to indicate possible slow motions in the earth). Earthquakes with $M \gtrsim 7$ for the interval 1897—1964 according to S. J. DUDA's table are located within the shaded areas.

well founded. Moreover, it is necessary to clarify the conditions in eastern Asia, on the transition to the Pacific Ocean.

Another global structure, which has been studied particularly by scientists of Columbia University in New York, consists of the mid-Atlantic ridge (Fig. 53). Along the whole extent of this ridge, there is a seismically active fissure or rift. However, the seismicity is considerably lower than in the circum-Pacific belt or in the Asiatic–European belt. It is generally believed that this rift system depends on tensile stresses, due to which the earth's crust has cracked. But there are several indications that also other factors may be of importance. These derive partly from solutions of focal mechanisms within this system, partly from direct stress measurements on Iceland (conducted by N. HAST). The latter indicate large horizontal compressional stresses instead of tensile stresses. A considerable extension of observations of all kinds within this rift system no doubt appears to be an important geophysical task. In the next section, we shall see how modern theories of global tectonics integrate both the rift system and the arc structures into one world-wide structural pattern.

In this connection, it should be mentioned that the planes, along which deeper earthquakes ($h > 300$ km) are located, exhibit dip angles of about 55°–60° (Fig. 55). Two American scientists found in 1968 that these planes are tangential to the outer core boundary, provided they are sufficiently extended. According to their suggestion, relative motions of core and mantle generate tangential stresses which in some way are transmitted towards the earth's surface and released there in the form of earthquakes. However, it appears improbable that tangential stresses could arise at this boundary between a liquid and a solid medium, especially as it concerns stress accumulation over long periods of time. In contrast, normal stresses could be envisaged at the boundary between core and mantle, especially if one adheres to the hypothesis of an expanding core. Such pressures could cause the mantle to rupture in its outer part along planes which would be approximately tangential to the core boundary. This has been demonstrated by laboratory experiments.

In this section, we have tried to take the step from mechanisms of individual earthquakes to systems of continental size. We have seen that some generalizations in this way are possible. The next step would be to extend the system to comprise the whole earth, i.e. to try to explain all the different expressions of the earth's activity, in the form of earthquakes and otherwise, as due to one and the same underlying reason. Efforts in this direction have not been lacking. In the most recent years a theory of ocean-floor spreading has played a very great role in the discussions, but still much remains to be investigated.

6.3   *The New Global Tectonics*

Seismology as well as other branches of geoscience provide us with numerous pieces of information. Our goal is to combine all these pieces into some coordinated system which will inform us better about underlying mechanisms of the earth of which all our observations may come out as special cases. Coordination of different observations into bigger systems has always been appealing to the imagination, and history can report many such attempts. A most significant attempt goes under the name of *continental drift* and was proposed by the German scientist A. WEGENER around 1915. There is no doubt that horizontal displacements of large extent occur, such as observed along the San Andreas fault in California. Likewise, large vertical motions are known to have occurred. But WEGENER's theory went further and considered that the continents have drifted apart from an original block. It was stimulated by among other things, the similarity of the coast lines of Africa and South America, also by similarities in rocks on the two sides. WEGENER's hypothesis gave rise to lively discussions, presented in several hundred papers, involving much of scientific controversy. It appeared as if the negative criticism gained and for some time the hypothesis was almost completely abandoned.

However, during the 1960's the hypothesis has again been proposed and now it has gained a significant number of adherents, in fact so many that a more detailed account of the recent findings is justified here. The situation is naturally much better now as we have a considerably larger amount of various geophysical observations to base a theory upon. The new theory appears under various names, such as the *ocean-floor spreading hypothesis* or the *new global tectonics* or *plate tectonics*. In its essential parts, it implies that the crust is drifting apart in opposite directions at the ocean ridges (Fig. 66). Roughly perpendicular to the ridges, there are faults

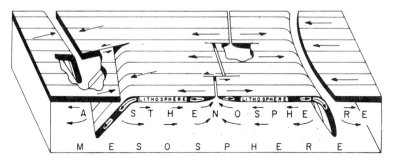

Fig. 66. Relative motions envisaged in the ocean-floor spreading hypothesis. Ridge structure in the centre, arc structures to the left and right. After B. ISACKS et al. (1968).

(called transform faults, rather than simple strike-slip faults), which mark the lines of breakage. New material is supplied from below. The ocean floor, which drifts away from a ridge, will proceed until it hits a continental edge. Here it bends downwards, and creates the dipping planes well known from the distribution of deep earthquakes in island arc structures. This system of movements is shown in Figure 66. The top layer, the so-called *lithosphere*, is a layer of finite strength about 100 km thick. This moves on the *asthenosphere* layer which has practically no strength and extends to several hundred kilometres depth. Below this we have the *mesosphere* which extends to the core and in which no tectonic processes take place. The lithosphere consists of blocks moving over the asthenosphere layer. The major tectonic features are the results of relative movement and inter-action of these blocks, which spread apart at the rifts (ocean ridges) leaving a pattern of ridges and transform faults, slide past one another at the large strike-slip faults and are underthrust at the arc structures. This means a modification to our earlier picture, Figure 59, according to which there is a sloping fault zone extending in under the continent. In the new hypo-thesis this fault zone is replaced by a slab of the oceanic lithosphere under-thrusting the continent *(subduction)*. The immediate causes of earthquakes occurring along the sloping zone, i.e. the existence of stresses between adja-cent parts of the solid earth, are still there.

A world-wide combination of fault mechanisms, both on rift structures and in island arcs, has been possible. It would explain the present motions as rotations around a number of poles on the earth. Table 15 summarizes some such results presented by the French scientist LE PICHON (1968). He distinguishes a few rotation patterns, as indicated in Table 15, each

Table 15. Centres and rates of rotation after X. LE PICHON (1968).

Zone	Rotation centre		Rotation rate 10^{-7} degree/year
	Latitude	Longitude	
South Pacific (Antarctica–Pacific)	70°S	118°E	10.8
Atlantic (America–Africa)	58 N	37 W	3.7
North Pacific (America–Pacific)	53 N	47 W	6.0
Indian Ocean (Africa–India)	26 N	21 E	4.0
Arctic Ocean (America–Eurasia)	78 N	102 E	2.8

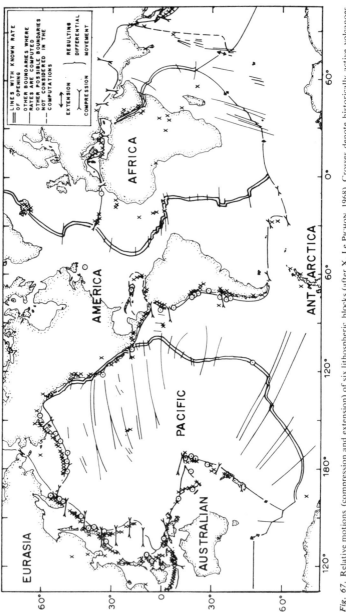

Fig. 67. Relative motions (compression and extension) of six lithospheric blocks (after X. Le Pichon, 1968). Crosses denote historically active volcanoes, open circles denote earthquakes which have generated major tsunamis.

with its rotation centre. The location of the rotation centres were determined by least-square procedures, both from the orientation of transform faults and from the known rates of displacement. The centres given in Table 15 are based upon the orientation of the fracture zones. We note, for instance, that the centres of rotation both for the Atlantic and for the North Pacific are relatively near to the southern tip of Greenland. These results can be interpreted in terms of a number of blocks which move relatively to each other. As a first approximation, Le Pichon discriminates between six major blocks, as indicated in Figure 67. Their relative motions are given in Table 16, which is a summary of a more detailed tabulation by Le Pichon. The system of motions forms a world-wide pattern in which all motions are interrelated. The pattern obtained permits extrapolation back in time, which reveals the nature and extent of continental drift in

Table 16. Differential movements between blocks (Fig. 67) after X. Le Pichon (1968).

Block	Relative motion cm/year	Number of observations averaged
Eurasia–Pacific	−8.7[1]	6
Australia–Pacific	−5.0	5
America–Antarctica	−4.3	8
America–Pacific	−5.9	3
America–Eurasia	−1.5	2
Africa–Antarctica	+2.1	2
Australia–Antarctica	+6.1	2
Africa–Eurasia	−2.1	4
Australia–Eurasia	−5.1	5

[1] Positive values indicate extension, negative compression.

past geological epochs. Of a detailed account of this aspect, given by Le Pichon, we may mention just one example. If we imagine that Antarctica is kept still, the displacements of all other continents can be shown as a counter-clockwise motion around Antarctica by amounts of the order of 5 cm/year, i.e. 500 km during the last 10 million years. Another interesting result can be obtained by considering Figure 54. The length of the dipping plane (in the direction of downdip) amounts to 800–850 km. With a relative compressional motion of 8.7 cm/year (Table 16), the time required for the slab to penetrate so deep is of the order of 10 million years. This time interval seems to have a world-wide validity. Related to such historical aspects are also investigations of polar wanderings and their possible interrelation with seismic phenomena.

Like any new theory of such a comprehensive nature, also this approach has been met with very much interest and much debate. However, it appears that such a large amount of observational data favour this explanation, that as a consequence the new ideas have been widely accepted. It is no doubt of great importance to test the idea with all possible geophysical and other means. Much information derives from magnetic measurements, which among others have revealed a striking alignment parallel to the rift structures and have permitted a more accurate dating. Similarly, seismic exploration of ocean-bottom sediment thickness and distribution has contributed substantially. From the purely seismological side there is naturally an abundance of relevant facts. It is maintained by several seismologists that within the entire field of seismology there appears to be no serious obstacle to the new global tectonics. Earthquake mechanism studies no doubt provide us with very significant information in this connection. Figure 68 shows a summary of about 100 such mechanisms. The relative motions resulting from this compilation agree remarkably well with LE PICHON's simplified model with relative motions of six large and rigid lithospheric blocks. Also the seismicity distribution over the earth confirms the new ideas. The earthquakes are confined to narrow continuous belts that bound large stable areas. In the zones of divergence and strike-slip motion, the activity is moderate and shallow (transform-fault motion), whereas in the zones of convergence the activity reaches much higher energy levels and extends also to greater depths (underthrusting lithosphere slabs). The presence of volcanism, the generation of many tsunamis (involving vertical motions at the source) and the frequency of occurrence of large earthquakes also seem to be related to underthrusting in island arcs. Another aspect is presented by anisotropy in the upper mantle. There is quite a lot of evidence for this, implying differences in wave velocities in different directions, and which could be explained by the new theories.

The new theory entails quite a different view on the existence of deep earthquakes. Earlier it was considered that the earth had sufficient strength to store stresses for considerable time down to depths around 700 km. Below, the material was considered as much more mobile, evidenced

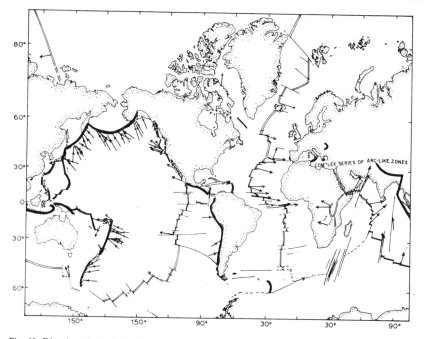

Fig. 68. Direction of relative horizontal motion, as derived from earthquake mechanism studies. Double lines mark the world rift system, heavy lines mark arc structures, thin, single lines mark major transform faults. After B. ISACKS et al. (1968).

by, among other things, an increase of the electric conductivity around this depth. BENIOFF also preferred to include the whole layer down to about 700 km depth in the crust of the earth. In the new theory, however, the layer able to hold stresses is limited to the lithosphere, which is at most 100 km thick. Its thickness agrees approximately with the crystalline layer of GUTENBERG, which he estimated to extend to about 80 km depth. The layer below, the asthenosphere, with practically no strength, can then be identified with the upper-mantle low-velocity layer. This layer is unable to store any stress to produce earthquakes. The deeper earthquakes (depths exceeding about 100 km) are localized to narrow zones and they are bound to the lithosphere. Where the lithosphere bends down under continental margins, it still preserves its rigidity and therefore also its ability to store stresses. And then it also provides a possible location for earthquakes to occur. The fact that earthquakes are not found to occur below about 700 km is therefore not explained by any world-wide property of the earth's interior, but is just a more localized feature, bound to island arc structures. But how do stresses arise in the descending lithospheric slab at the island arcs? If the surrounding medium at the continental margin is rigid enough, the relative motions can generate stresses which can be stored for some time. But with a non-rigid surrounding medium and with a uniformly moving lithospheric slab, any stresses would hardly be expected, except where the slab bends or meets resistance. Alternatively, it has been speculated that slow phase changes in the slab may generate stresses that ultimately produce fracture. The problem of the mechanism of deep earthquakes as compared to shallow ones is no doubt related to the present picture of a down-going slab. Moreover, it is to be noticed that in the motion of the lithosphere from the ridges towards the island arcs not enough stresses are generated to produce earthquakes. The oceanic areas outside the ridge systems are remarkably free from earthquakes, as we have seen.

There are still several problems connected with the new theories. For instance, we need explanations of the seismic differences between the coastal areas of the Pacific Ocean on the one hand and the Atlantic and Indian Oceans on the other. Also the high compressional stresses found in absolute stress measurements in Iceland must be considered in relation to the rift tension required in the new theories.

One major question concerns the nature of the driving force which can produce all these motions. In earlier discussions of possible continental drift, GUTENBERG maintained that no force is known which would produce such world-wide motions. Nowadays, the situation seems to have changed, as well-established motions do exist, and they cannot be dismissed for the reason that no large enough driving force is known. Instead, we have to find this force, whatever it is. There are two main ideas which dominate these discussions, i.e. the hypotheses of an expanding earth and of convection currents in the earth's interior. It has been considered unlikely that convection currents could be the reason, as these would then have to be oriented along the ridge structures. Rather it is assumed that the breaking apart of the lithosphere slabs and their motion are due to some underlying state of stress. It has been argued that a possible reason for such a state of stress is an expansion of the earth. However, with the known numerical values of the movements it is possible to demonstrate that, if the earth should keep its nearly spherical form, it is necessary to combine the expanding-earth hypothesis with compensating motions by compression or thrust. The problem of the driving force cannot yet be considered solved.

Even though the new theories are to a great extent only qualitative, lacking detailed quantitative information on the earth's interior, we can state in summary that they have had an extremely stimulating effect on discussions and investigations of the earth's dynamical properties. Also, they seem to be able to incorporate nearly all observations into their scheme and to provide acceptable explanations for a variety of phenomena. In contrast, BELOUSSOV (1970) maintains that 'not a single aspect of the ocean-floor spreading hypothesis can stand up to criticism'. Even though there are still problems to be clarified, there is no doubt that the new theories will have repercussions within many fields of the earth's study, for instance, earthquake prediction (Chapter 10), just to mention one case.

6.4   Time Series of Earthquakes

Besides the combination in space the combination in time of earthquakes is also of great significance. The normal process is that a bigger earthquake is followed by a long sequence of *aftershocks*. This may continue for several years after the main shock. Thus, it has been found for the earthquakes in Kamchatka in 1952 and in the Aleutians in 1957, that 3–4 years elapsed before the activity was back to normal again. However, most of the after-shocks are small, and the biggest of them as a rule reaches a magnitude about 1.2 units lower than for the main shock. This aftershock usually occurs a few hours to a few days after the main earthquake (Table 17). In addition, the main shock may be preceded by small earthquakes, so-called *foreshocks*. These are mostly so small that, as distinct from the aftershocks, they are usually not recorded at distant stations.

In several cases, deviations from this normal scheme are observed. It should be especially emphasized that frequently there is not just one main shock, but there may be several about equally strong or even increasing in strength. The latter case happened, for instance, for the earthquakes on the Ionian Islands in August 1953 (the series comprised one shock on

Table 17. Main shock and largest aftershock–a few examples.

Earthquake	Richter magnitude (M)		Time of occurrence of the largest aftershock
	Main shock	Largest aftershock	
Alaska, March 28, 1964	8.5	6.8	About 17 hours after the main shock
Aleutians, February 4, 1965	8.1	7.2	About 3.5 hours after the main shock
Japan, May 16, 1968	8.6	8.0	About 10 hours after the main shock
Kuriles, August 11, 1969	8.1	6.9	About 7.5 hours after the main shock

August 9, 1953, magnitude $= 6^{1}/_{4}$, then one on August 11, magnitude $= 6^{3}/_{4}$–7, and one on August 12, magnitude $7^{1}/_{4}$–$7^{1}/_{2}$, and in addition a large number of smaller shocks). The largest of these shocks could be regarded as the main earthquake, but then the foreshocks are exceptionally strong.

More correctly, the three shocks mentioned are to be looked upon as representing a three-stage release of what otherwise could have been one main earthquake, corresponding to the total energy in the three shocks.

Quite another time sequence is represented by the so-called *earthquake swarms*. In these there is no main earthquake, but all shocks are rather similar in magnitude and in general relatively small. The swarm starts with a few shocks, then their number gradually increases, until a maximum is reached, and then the whole phenomenon dies out again gradually. There are several examples of such swarms. Jan Mayen in the North Atlantic is a source of such phenomena from time to time. In the autumn of 1965, a swarm started at Matsushiro in Japan which lasted over one year. This swarm attracted much attention and was carefully observed by Japanese seismologists and others. At its maximum activity, the number of shocks reached several thousands per day. The magnitudes were low and do not seem to have exceeded 5; moreover, the focal depths were small, as a rule 2–5 km. Similarly, the series of shocks at Tashkent in 1966 has to be considered as a swarm. Swarms are especially apt to occur in regions with present or earlier volcanic activity. These shocks are most likely to be classified as volcanic, as distinct from tectonic ones.

Aftershocks of tectonic earthquakes have been studied for many years, by investigating among other things the variation of the number of aftershocks with time. Table 18 gives a review of the number of aftershocks

Table 18. Number of aftershocks per day (24 hours) counted from the time of the main shock, according to records firstly at Uppsala, secondly over the Swedish network, for some of the most important earthquakes during the 1960's. The minimum magnitudes are around 5—5.5 on the Richter scale.

Day[1]	A. Uppsala		B. Swedish network		B/A	
	True number	Reduced number	True number	Reduced number		
a) Alaska, March 28, 1964						
1.	145	100	389	100	2.7	
2.	51	35	121	31	2.4	2.5
3.	28	19	65	17	2.3	
b) Aleutians, February 4, 1965						
1.	170	100	227	100	1.3	
2.	48	28	77	34	1.6	1.4
3.	45	26	55	24	1.2	
c) Japan, May 16, 1968						
1.	45	100	171	100	3.8	
2.	10	22	47	27	4.7	4.3
3.	1	2	11	6	(11)	

[1] The days are counted from the time of the main shock: 1. = first day, i.e. 0—24 hours after the main shock; 2. = second day, i.e. 24—48 hours after the main shock; 3. = third day, i.e. 48—72 hours after the main shock.

recorded from some of the most important series during the 1960's. In addition to the numbers actually observed I have also given reduced numbers in order to facilitate comparisons. The column B/A gives the ratio of the number of shocks recorded over the whole Swedish network to the number of shocks recorded only at Uppsala. From Table 18 we can learn the following facts:

1. The ratio between the number of shocks recorded by the whole network (Table 2) and only at Uppsala (B/A) exhibits a strong dependence on the location of the shocks. Most likely, the mechanism of the earthquakes is the reason for such behaviour, along with different sensitivity of our different stations.

2. The decrease in the number of shocks from the first to the second and third days shows remarkably smaller variations. Disregarding minor differences, this decrease is about the same for all earthquake areas investigated and also it is about the same at all our stations as it is at Uppsala alone.

Studies of the number of shocks in an aftershock sequence are mainly of a statistical nature. An effort to make a more physical interpretation of the aftershock problem was made around 1950 by BENIOFF in Pasadena. Figure 69 demonstrates the behaviour of materials with elastic creep properties under an applied load. The stress variation with time is shown in Figure 69a and the corresponding strains in Figures 69b and 69c. When a stress OA is applied momentarily, there is first a purely elastic strain OA (also momentarily), but after this an elastic creep or aftereffect takes over. Depending on the properties of the material, either the strain can attain a certain limiting value after a sufficiently long time of loading (Fig. 69b) or elastic flow will occur (Fig. 69c) until the material breaks. Similarly, the conditions on removing the stress BC first correspond to purely elastic strain release BC and EF, respectively, followed by elastic creep. This unloading is considered to correspond to the conditions in an earthquake. BC in Figure 69a corresponds to the stress change in the earthquake, BC in Figure 69b to the main shock, while the creep curve corresponds to the aftershocks. A material which behaves as in Figure 69b is said to have *elastic creep* characteristics, whereas a material as in Figure 69c has *elastic flow creep* characteristics. By the introduction of such time effects, we have clearly abandoned the simple HOOKE's law. While this is well suited to treat such short-period phenomena as in ordinary seismic waves, it is not sufficiently accurate in dealing with the longer-period phenomena leading to earthquakes.

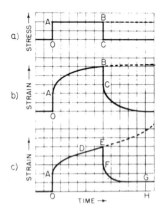

Fig. 69. Relations between the time variation of stress and of strain: a) The time variation of the applied stress, b) and c) resulting strains for different materials. After H. BENIOFF.

The behaviour of materials, such as we have just learnt from Figure 69, has an important general application to the earth's interior, in particular to its behaviour as solid or liquid. In a strict physical sense solid is equivalent to crystalline. Such a condition does probably not prevail within the earth to depths greater than about 80 km. From a seismological point of view it is more convenient to define the behaviour of a body (as solid or as liquid) in relation to the period of the applied force. One and the same non-crystalline material can thus behave as solid in relation to short-period influences (as in the usual seismic waves), but as liquid in relation to more long-period actions (as in secular motions, tide effects, etc.). In seismology the material is considered as solid if it can transmit S-waves (such as the crust, mantle and probably the inner core), but it is considered as liquid if S-waves are not transmitted (as in the outer core).

In his efforts to apply such considerations to aftershock sequences, BENIOFF had to introduce a number of simplifying assumptions. These led to the result that the strain (i.e. the deformation per unit volume) could be obtained as the square root of the seismic wave energy E. And E in turn was obtained from the magnitude M. Thus, a possibility was offered to calculate the strain in a simple way for each earthquake, both for the main shock and for the aftershocks. The strains were plotted in a cumulative way against origin times of the shocks, which led to diagrams corresponding to the latter part of Figure 69b.

BENIOFF's method was widely adopted and applied by many seismologists. In 1963, BÅTH and DUDA suggested a modified technique, which paid more attention to some factors for which BENIOFF had just made assumptions. Above all, it was obvious that the volume of an earthquake, i.e. the space within which the stress has been stored, is not constant, as BENIOFF assumed, but increases with the magnitude. By identifying the earthquake volume with the volume occupied by the aftershocks, the following relation was obtained between this volume V (cm³) and the magnitude M:

$$\log V = 9.58 + 1.47M \tag{1}$$

Moreover, it was clear that the essential difference between large and small earthquakes is not to be found in the strain (deformation per unit volume), but primarily in the size of the volume which is in a state of stress and in

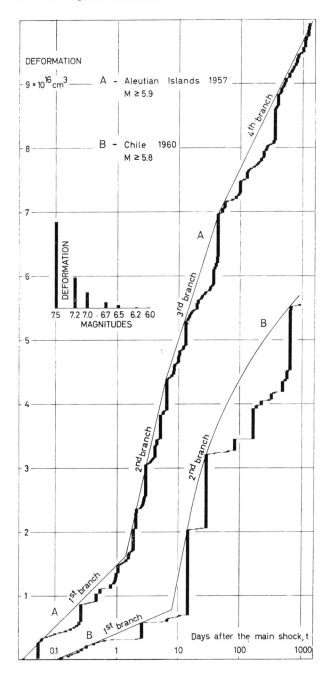

DEFORMATION

$9 \times 10^{16} cm^3$

A - Aleutian Islands 1957
M ≥ 5.9

B - Chile 1960
M ≥ 5.8

DEFORMATION

7.5 7.2 7.0 · 6.7 6.5 6.2 6.0
MAGNITUDES

A

B

4th branch

3rd branch

2nd branch

2nd branch

1st branch

1st branch

A

B

Days after the main shock, t

0.1   1   10   100   1000

which a simultaneous release occurs. As a consequence it was found more correct to investigate the time variation of the total deformation instead of the deformation per unit volume (the strain). The total deformation D (cm³) has the following relation to M:

$$\log D = 5.17 + 1.46M \qquad (2)$$

Instead of investigating the time variation of D, one can investigate the time variation of E. It is of interest to note that all three quantities $\log V$, $\log D$, $\log E$ vary in almost exactly the same way with M [compare equation (4) in Chapter 4]. This implies that D/V ($=$ strain) and E/V are constant, i.e. independent of magnitude.

Figure 70 shows the cumulative deformation, calculated according to equation (2), for two aftershock sequences, one in the Aleutians after March 9, 1957, and one in Chile after May 22, 1960. In both cases we see a marked increase of the activity, about 1.36 and 7.9 days, respectively, after the main shock. Such a marked increase appears to be a general characteristic of aftershock sequences, but its explanation is still not clear. Earlier it was assumed that the first part corresponds to the release of compressional stresses and the later part to the release of shear stresses. However, this idea has not been confirmed by laboratory tests. The possibility does not appear excluded that the improved version of BENIOFF's method will lead to more unified shapes of the release curves than resulted from the original technique. Such studies, suitably supplemented by investigations of laboratory models and theoretical calculations, are of the greatest significance to our understanding of the behaviour of the solid earth in relation to stresses, in other words, its rheologic behaviour.

Figure 71 shows a schematic diagram of the energy accumulation and release. It is a well-known fact that aftershocks of deep earthquakes are smaller and fewer than of shallow-focus earthquakes. Figure 71 gives a plausible explanation for this behaviour. For deep earthquakes a major part of the stored energy is released in the main shock, leaving only a smaller part for the aftershocks. For shallow-focus earthquakes the partitioning of the stored energy would favour the aftershocks more.

Fig. 70. Deformation characteristics for two aftershock sequences: Aleutian Islands 1957 and Chile 1960. After M. BÅTH and S. J. DUDA.

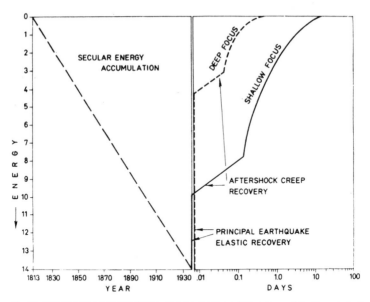

Fig. 71. Energy accumulation (arbitrary scale) and release. The purpose of the figure is to demonstrate the differences both between the main shock and the aftershocks as well as between shallow and deep earthquakes.

Studies of time sequences of earthquakes has also been extended to so-called secular series, i.e. usually comprising the time from the beginning of instrumental records around the turn of the century up to the present time. In this way, deformation and energy release curves have been constructed both for the earth as a whole and for more restricted areas. Figure 72 shows the accumulated energy release for all earthquakes with magnitude $M \geqq 7$ during the interval 1897–1976. The most pronounced feature in this figure is the change of slope around the year 1917. According to the straight lines, which have been put into Figure 72, the average energy release was 8.5×10^{24} ergs/year for the interval 1897–1917, but only about half as large or 4.0×10^{24} ergs/year for 1917–76. In terms of energy, this difference corresponds to one earthquake with $M = 8.6$ per year. However, it has to be remembered that such observations do not give more than a momentary picture, geologically speaking. Therefore, the results cannot be taken to represent more than the investigated interval, and they do not permit extrapolations in either direction.

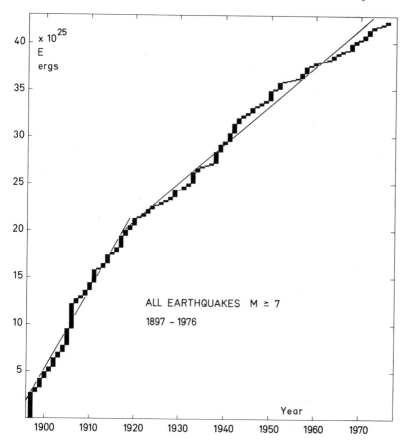

Fig. 72. Accumulated energy release for all earthquakes with $M \geqq 7$ for the interval 1897—1976. The curve has been calculated by means of equation (4) in Chapter 4, using S. J. DUDA's tables for 1897—1964 and the monthly bulletins of the Seismological Institute, Uppsala, for the following years.

Many laymen have the impression that the seismic activity on the earth has increased in recent years. The reason that many people have such an impression is to be found partly in an intensification of the seismological reporting activity, such that newspapers contain more frequent and more detailed information on earthquakes than earlier. Another reason is to be found in the expansion of the population to earlier uninhabited areas which is presently going on in many earthquake countries (for instance, Assam). But, as we have seen, reliable seismological statistics (Fig. 72) do not confirm the impression of increased seismic activity; on the contrary, there has been some decrease during this century.

Since the headline of this section reads *Time series of earthquakes*, one might expect something to be said about periodicities of earthquakes. These have, in fact, been the subject of much investigation, especially in earlier years. However, any results of general validity or of physical significance have hardly been produced. This whole field has also a rather poor reputation among seismologists.

Seismology has no doubt furnished the quantitatively most reliable information about the present tectonic conditions in the earth, partly by energy determinations (Chapter 4), partly by mechanism studies. In spite of this, the best results are still attained by a combination with other geophysical and geological methods of research. The knowledge we have accumulated about the dynamics of earthquakes is still incomplete and it is valid only for the time interval for which we have instrumental records. Extrapolations are not possible, neither over longer (geological) epochs back in time nor for even short periods of time ahead of us (see further Chapter 10). Nor do the seismological observations alone give any definite answer to the question about the driving mechanism behind the observed phenomena. In addition, there is another limitation whose scope it is still too early to estimate, i.e. stress release by slow motions (Chapter 10). If this should be responsible for a considerable portion of the stress release, this would mean that the seismic records will give information on only a part of the total stress release.

Chapter 7

Internal Structure of the Earth

7.1 The Main Features of the Earth's Physical Properties

It is suitable to divide the earth's interior into the following three parts:
1. The earth's crust.
2. The earth's mantle: the upper mantle and the lower mantle.
3. The earth's cores: the outer core and the inner core.

Table 19 gives a review of the extent of the different parts, both their depth extent as well as their volumes, masses, and average densities. See also Figure 73.

Table 19. Different structural components of the earth's interior.

Layer	Depth extent km	Volume True 10^9 km^3	Volume Percentage	Mass True 10^{12} megaton^2	Mass Percentage	Average density g/cm^3
Earth's crust	Surface—Moho[1]	5.1	0.5	15	0.3	2.94
Upper mantle	Moho—1000	429.1	39.6	1673	28.0	3.90
Lower mantle	1000—2900	473.8	43.7	2415	40.4	5.10
Outer core	2900—5100	166.4	15.4	1747	29.2	10.50
Inner core	5100—6370	8.6	0.8	125	2.1	14.53
Sum		1083.0	100.0	5975	100.0	5.52

[1] The thickness of the earth's crust averaged over the whole earth is about 10 km.
[2] 1 megaton = 10^6 tons.

Body-wave velocities. By means of travel times of seismic waves (Chapter 3) it is possible to calculate the velocities of longitudinal and transverse waves in the earth's interior. Just as in Chapter 3, it will be of advantage first to study the conditions in a homogeneous sphere (Fig. 74). If R = the radius of the sphere and v_P = the velocity of P-waves, then we find immediately from Figure 74 the following relation between travel time t and distance Δ:

$$t = \frac{2R}{v_P} \sin \frac{\Delta}{2} \qquad (1)$$

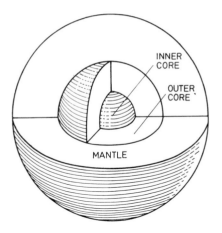

Fig. 73. Schematic picture of the earth's interior, showing the right proportions between mantle, outer and inner core.

Equation (1) is shown graphically in Figure 74, assuming $R=6370$ km ($=$the earth's radius) and $v_P=6.0$ km/sec. As we see, the travel-time graph $t(\varDelta)$ is curved, in spite of the fact that we have assumed that the velocity v_P is everywhere the same. The reason for this is purely geometrical: if we had measured the distance along the real wave path FU, then obviously the travel-time graph would have been rectilinear. But in seismology we usually measure distances $\varDelta$ as indicated in Figure 74 (either in degrees or

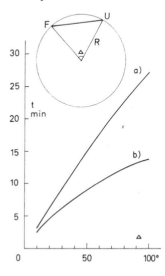

Fig. 74. Travel-time curves for P for a surface focus: a) for a homogeneous sphere ($v_P=6.0$ km/sec), b) for the real earth (according to JEFFREYS and BULLEN).

in kilometres along the earth's surface). Then the travel-time graph has to be curved even when v_P is constant.

Figure 74 also shows the *observed* travel-time graph for P for a surface focus. We find that the observed graph gives lower times than the one corresponding to a homogeneous sphere and moreover, the observed graph is more curved than in the homogeneous case. These circumstances can be interpreted in only one way: that the waves have propagated with higher velocities during some part of their path, or, in other words, that the wave velocities in general increase with depth in the earth. However, a consequence of this is that the wave paths are no longer rectilinear, but curved and in fact concave towards to earth's surface. And in turn this naturally means that the wave paths are longer than in the case of a homogeneous sphere. Thus, we have to determine such a velocity-depth distribution that the net result (of a longer wave path and higher velocities) is a decreased travel time.

It is possible to solve this problem exactly and it has also been subjected to extensive mathematical treatments, including solutions of integral equations. Symbolically, we can express the problem in the following simple way: $t(\Delta) \rightarrow v(r)$, i.e. from observed travel times t we calculate the velocities v at different distances r from the earth's centre. Conversely, a number of theoretical models have been investigated, where one starts from an assumed velocity distribution and deduces the corresponding travel-time curve, i.e. a problem which can be expressed as $v(r) \rightarrow t(\Delta)$. This curve can then be compared with observations and the theoretical model can be successively modified until agreement is achieved. The transformation $v \rightarrow t$ is a direct problem with a unique solution, whereas $t \rightarrow v$ represents the inverse problem, usually non-unique. Here, we shall be content with reporting the results of these investigations.

Figure 75 shows the velocities of longitudinal and transverse waves as a function of depth below the earth's surface. Obviously, the results according to the two geophysicists JEFFREYS and GUTENBERG are in general in very good agreement. Of particular interest is the pronounced decrease of the P-wave velocity from 13.6 to 8.1 km/sec at 2900 km depth, i.e. at the outer-core boundary. In addition, the transverse waves cease to exist at this depth. Waves which have traversed the outer core as shear waves have still not been found with reliability. This has led to the conclusion

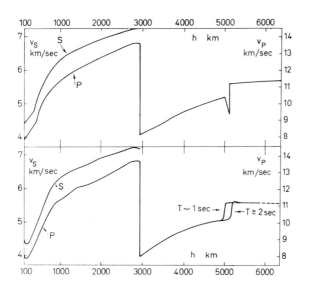

Fig. 75. The velocities (*v*) of longitudinal waves (*P*, *K*, *I*) and transverse waves (*S*) in the earth's interior according to H. JEFFREYS, 1939 (upper figure) and according to B. GUTENBERG, 1958 (lower figure). *T* = wave period, *h* = depth below earth's surface.

that the outer core behaves as a liquid – at least in relation to seismic waves. This conclusion is confirmed by observations of the tidal effect in the solid earth (corresponding to the ocean tides), which is too large to be explained by an entirely solid earth, but fits well with the assumption of a liquid core. The irregularities of the velocity around 5000 km depth (Fig. 75) mark the boundary of the inner core. Whether the latter is solid or liquid, is a problem which is not yet quite settled, but some facts seem to indicate that it is solid and thus should be able to transmit transverse waves. For his velocity values, JEFFREYS has given error limits of at most ±0.5%, except for the depth intervals 413–1000 km and 4982–5121 km, where the velocities are considered somewhat less reliable.

The good agreement between JEFFREYS' and GUTENBERG's velocity curves is an expression of the fact that their travel-time tables agree closely. The remaining discrepancies just illustrate the uncertainties which in spite of this may still remain. It is particularly at two levels where interesting deviations are found:

1. In the transition zone to the inner core. JEFFREYS has here introduced a low-velocity layer, while GUTENBERG does not consider this to have sufficient basis in the observations. The difficulties become large at these levels,

which is clear when we realize that a wave spends only a small fraction of its total travel time in any such layer. The problem about the transition to the inner core cannot yet be regarded as completely solved. Most likely, the transition zone consists of several layers, to some extent reminiscent of the layering in the earth's crust, even though the transition extends over a few hundred kilometres in depth.

2. In the upper mantle. At a depth of 413 km, JEFFREYS has a second-order discontinuity (i.e. a discontinuity in the velocity gradient with regard to depth), which would explain the behaviour of the travel-time curve around 20° distance (the so-called 20° discontinuity). GUTENBERG, on the other hand, has completely eliminated the discontinuity at 413 km depth and replaced it by a low-velocity layer in the upper mantle (Fig. 75). It is still not quite settled which one of these interpretations can be regarded as the correct one, even though the idea of a low-velocity layer has got strong supporting evidence from a number of different observations.

The close agreement between the two velocity curves for some time led to the idea that future research would only need to improve details in these curves, while all the main features were well known. However, in the last few years a number of velocity curves have been proposed which show more significant deviations from those just discussed. Above all, this concerns the depths around 400 and 600 km where relatively strong velocity increases with depth have been introduced (Fig. 76). These results are based on observations at so-called array stations, where frequently not the travel times themselves but the derivative $dt/d\Delta$ has been used as a basis for the conclusions. With a dense net of stations, this quantity can be observed in much greater detail than has been possible hitherto. In addition, early arrivals of PP and $P'P'$ have been observed which may seem to require sharp discontinuities around 400 and 650 km depth. However, recent alternative explanations of these early arrivals in terms of scattering (Section 3.2) have somewhat weakened this suggestion.

It should be emphasized that in the deduction of the velocities (Fig. 75), a spherically symmetric structure of the earth has been assumed. This assumption is made also in the calculation of other properties described below. Lateral variations of a more regional nature have been neglected, although it is known that such variations may have a considerable depth extent, including a major part of the lower mantle. From this viewpoint we may

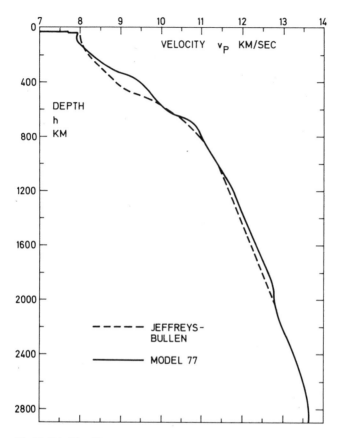

Fig. 76. Velocities of *P*-waves in the mantle according to array data ('model 77') compared with the Jeffreys-Bullen velocities. After M. N. Toksöz, M. A. Chinnery and D. L. Anderson (1967), modified.

regard the structure of the earth we are deducing in this chapter as a first approximation, considering the depth variation as far more significant than lateral variations.

It may be instructive to compare the seismic wave velocities with some other well-known velocities. Let us take as examples, on the one hand, the velocity of light ($=300\,000$ km/sec), on the other hand, the velocity of sound in air ($=0.3$ km/sec). A simple calculation shows that if the *P*-wave would propagate with the velocity of light it would need only 0.042 sec to traverse the entire earth. This means that with present-day accuracy of time measurements at the seismograph stations (about 0.1 sec), all stations would get a simultaneous recording and it would be impossible to locate the event. On the other hand, if the *P*-wave travelled with the velocity of sound in air, then it would need 11.8 hours to penetrate the whole earth. Even in this case it would be impossible to locate most events, because with present frequency of earthquakes it would be impossible to coordinate observations at different stations. The true time for a *PKP*-wave to travel 180° from a surface focus is 20 min 12 sec. It is evident that the seismic wave velocities fall within a range which is favourable from the viewpoint of measuring techniques.

Density. The next step in our study of the earth's internal constitution is to deduce the variation of density with depth. A number of density curves has been proposed through the course of time, the presently most accurate one by the Australian seismologist BULLEN. In deducing the density in the earth, we base our calculations on the following considerations:

1. In a complete expression, the density of matter depends on pressure, temperature and chemical composition. In the calculations, the effects of the latter two are neglected. Therefore, the method cannot be applied to any layer where there is reason to suspect a change in the chemical composition or a phase change. Furthermore, hydrostatic equilibrium is assumed.

2. The calculated density curve has to yield the correct value of the earth's total mass (Table 19) and of its moment of inertia with regard to the axis of rotation ($=0.3336MR^2$, with $M=$mass and $R=$radius of the earth). These quantities are well known from astronomical observations. In addition, the density curve has to be compatible with the velocity curves.

Let us introduce the following symbols:

$p=$pressure and $\varrho=$density at the distance r from the earth's centre,

$m=$the mass of the earth inside the radius r,

$g=$acceleration of gravity and $\gamma=$gravitational constant.

Then we arrive at the following four simple relations:

1. The condition for hydrostatic equilibrium:

$$\frac{\mathrm{d}p}{\mathrm{d}r} = -g\varrho \qquad (2)$$

14   Båth: Introduction to Seismology

2. The law of gravitation:

$$g = \frac{\gamma m}{r^2} \tag{3}$$

3. The expression for the incompressibility (at constant chemical composition):

$$k = \varrho \frac{dp}{d\varrho} \tag{4}$$

4. From equation (1) in Chapter 3:

$$\frac{k}{\varrho} = v_P^2 - \frac{4}{3} v_S^2 \tag{5}$$

Combining equations (2) to (5) we arrive at the following expression for the radial density gradient, which has been applied to the calculations of the density in the earth:

$$\frac{d\varrho}{dr} = - \frac{\gamma m \varrho}{r^2 \left(v_P^2 - \frac{4}{3} v_S^2 \right)} \tag{6}$$

While the density of the earth's crust (about 33 km thick in continents) is about 3 g/cm³, the mean density of the earth is = 5.517 g/cm³ with an error of about 0.08%, as obtained by JEFFREYS from determinations of the gravitational constant. This shows that the density in the interior of the earth is greater than in the surface layers. In applications of the formula (6) for the radial density gradient the mass of the surface layers is first subtracted. This mass is somewhat different in different places, but calculating with an average crust instead has only a very slight influence on the density values at greater depths.

In his density calculation BULLEN applied the velocity curves which JEFFREYS had determined (Fig. 75). Starting from a probable value of the density = 3.32 g/cm³ at 33 km depth, the density is first calculated for the depth range 33–413 km. In this range the velocity variations of P and S are smooth (constant gradient) and the assumption of a constant chemical composition is assumed to be fulfilled. In the following layer (413–984 km), where the velocity gradients are large, it is doubtful whether the composition is constant. In addition, at each discontinuity surface the possible density jump enters as a further unknown.

However, the density distribution has to be such that it gives the correct value of the earth's moment of inertia. For a homogeneous sphere the moment of inertia is $0.4MR^2$. It is easy to show, that for constant values of M and R, the moment of inertia is $> 0.4MR^2$, if the density decreases inwards, and it is $< 0.4MR^2$, if the density increases inwards. If we should continue the density calculation, mentioned in the preceding paragraph, right through the whole mantle down to 2900 km depth, we would find that the moment of inertia of the core would be $0.57M_cR_c^2$ (M_c and R_c are the core mass and radius), which is obviously too high. In order to avoid this difficulty and to be able to proceed, we make some additional assumptions. As v_P and v_S themselves are continuous at 413 km depth, then this is probably also the case with the density but not with its gradient. The depth ranges 984–2898 km, 2898–4982 km and 5121–6371 km are each assumed to be of constant composition just like the range 33–413 km. The coefficient for $M_cR_c^2$ in the expression for the moment of inertia of the core is probably between 0.375 and 0.395, which determines the density at 984 km depth to an accuracy of 0.1 g/cm³. For the range 413–984 km the density is assumed to be a quadratic function of the radius r. The three constants in this formula are determined by the densities at 413 km and at 984 km and from the density gradient at 984 km. BULLEN gives two extreme series of densities below 500 km, one of them for an hypothesis of a continuously varying density in the core, the other for an hypothesis of a density at the earth's centre 10 g/cm³ higher than according to the first hypothesis. The true density curve is probably to be found somewhere between his two curves. The densities shown in Figure 77 are considered to have errors of at most 1% down to 2700 km and at most 3% below this level. The most pronounced feature of the density curve is the jump from 5.7 to 9.4 g/cm³ at the crossing of the outer-core boundary.

If instead we should make GUTENBERG's velocity curve the basis for our density calculation, we encounter certain difficulties, especially for the low-velocity layer in the upper mantle. BULLEN assumed constant composition in this layer, but this is no longer certain, if we have to deal with a low-velocity layer. Later, BULLEN also made calculations for other models, leading to somewhat different results (see below).

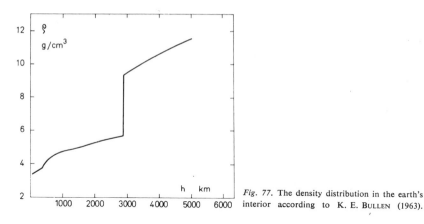

Fig. 77. The density distribution in the earth's interior according to K. E. Bullen (1963).

In this connection it is appropriate to correct some usual misunderstandings:

1. It is not correct to maintain that the densities are deduced from the velocities. The velocity curves serve as a guide, but for the densities other information and assumptions are needed as well. They are also less reliable than the velocities.

2. It is not correct to maintain that the velocities increase inwards in the earth because the density increases. As we see from equation (1) in Chapter 3, increased density alone means decreased velocity, not increased.

Just as the velocity curves have recently been altered in their foundations, something similar has to a certain extent happened also with the density curve. Some observations of amplitudes of *PcP* (i.e. the *P*-wave reflected at the outer core) indicate that the density jump at this surface would not be as large as in Bullen's model. The observations have been interpreted as indicating the same or nearly the same density on the two sides of the boundary. But then the question arises as to what extent the density curve will have to be changed elsewhere, in order still to satisfy the mass and the moment of inertia of the earth. Another possibility is that the core boundary is a transition zone with a thickness of several tens of kilometres and that this is the reason for the *PcP*-observations. See further Section 7.3 on the earth's core.

Elastic parameters, pressure, gravity. When the variations of density

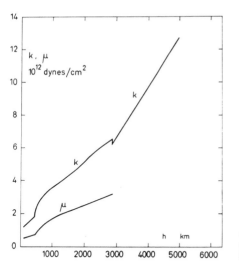

Fig. 78. Incompressibility (k) and modulus of rigidity (μ) at different depths (h) in the earth's interior. According to K. E. BULLEN.

and wave velocities with depth are known, it is a simple matter of calculation, using equation (1) in Chapter 3, to get the corresponding variations of the incompressibility k and the modulus of rigidity μ (Fig. 78). As we know, two elastic parameters are enough to describe the behaviour of the material in the earth in relation to seismic waves. There are other elastic parameters than k and μ, but they are simply related to these. From the density variation it is also relatively simple to calculate the depth variation of pressure and gravity (according to equations (2) and (3); see Fig. 79). The pressure (=the pressure of overlying layers) increases monotonically with depth, as expected, whereas the acceleration of gravity exhibits both maxima and minima. Thus, the acceleration has a major maximum at the outer-core boundary, where it attains the value 1037 cm/sec². The reason for this apparently irregular depth variation of gravity is to be found in the fact that g depends partly on the mass (m) inside the respective levels in the earth, partly on the distance (r) to the centre of the earth from the different levels. Proceeding towards the centre, both m and r decrease. As the acceleration is proportional to m/r^2, we could expect the curve shown in Figure 79.

Recent modifications of both the velocity and the density curves,

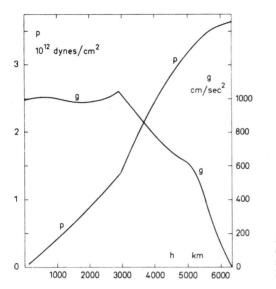

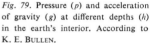

Fig. 79. Pressure (p) and acceleration of gravity (g) at different depths (h) in the earth's interior. According to K. E. BULLEN.

mentioned above, will entail some modifications of all curves which have been deduced from them.

As we have seen, the physical properties we have studied so far are intimately connected with each other and form a consistent system. The trend of the calculations is defined by the following points:

1. Travel times of P- and S-waves: deduced from seismograms.
2. Velocities of P- and S-waves: calculated from 1 by extensive operations.
3. Density: calculated by means of 2 and other assumptions and conditions.
4. Elastic parameters: calculated from 2 and 3.
5. Acceleration of gravity and pressure: calculated from 3.

Revised models. In recent years, improved information on a number of points has called for a revision of the model presented so far for the earth's interior. The newer development has been summarized by BULLEN and HADDON (1970), who have included the following new results in their revised models:

1. A new value of the earth's moment of inertia ($0.3309MR^2$ instead of $0.3336MR^2$), based on observations of artificial satellite orbits.

2. Revised *P*- and *S*-wave velocity distributions in the earth, both in the upper mantle and in the inner core.

3. More reliable travel-time data from nuclear-explosion and array-station results.

4. A generalization of equation (6) for the radial velocity gradient to the following form:

$$\frac{d\varrho}{dr} = - \frac{\eta\gamma m\varrho}{r^2\left(v_P^2 - \frac{4}{3}v_S^2\right)} \tag{7}$$

where the factor η is different from unity in the earth, except for the depth range 3600–4500 km.

5. Studies of compressibility of materials at high pressure.

6. Shock-wave experiments.

7. Observations of the earth's free oscillations.

The revised data have led to several new earth models of which one of the most recent ones is demonstrated in Figure 80. It is especially to be noticed that the *S*-wave velocity exhibits a minimum in the upper mantle in this model. According to the newest model of BULLEN and HADDON the density increases from 5.62 g/cm³ to 9.89 g/cm³ at the outer-core boundary at 2878 km depth and the gravity attains the value 1080 cm/sec² at this level.

A recent modification concerns the radius of the earth's core. A number of different observations indicates that this radius is somewhat larger than believed hitherto. The proposed increases vary depending upon the kind of observations, for instance, an increase of 10–30 km according to *PcP*-observations, an increase of about 15 km from observations of the earth's free vibrations and an increase of 64 km derived from some observations of *P*-waves diffracted around the core boundary.

Absorption. The physical properties discussed so far are based on purely elastic conditions in the earth's interior. However, it is a well-known fact that seismic waves are absorbed during their propagation. This indicates that also non-elastic effects are operative. Investigations of absorption must be based on amplitudes of seismic waves, as distinct from travel times used above. The studies have not yet been able to yield a completely unique picture of the absorption, but at least some

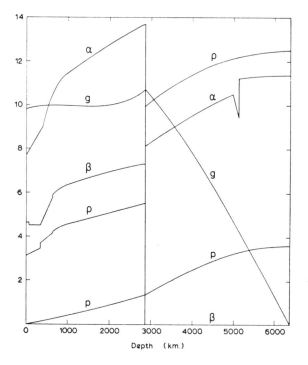

Fig. 80. Earth model according to R. A. W. HADDON and K. E. BULLEN (1969). p = pressure in units of 10^{12} dynes/cm², ϱ = density in g/cm³, $\alpha(=v_P)$ and $\beta(=v_S)$ = P- and S-velocities in km/sec, g = acceleration of gravity in 10^2 cm/sec².

preliminary results can be included here. The amplitude absorption of a seismic wave can be simply described by the factor $e^{-\varkappa D}$, where $\varkappa$ = absorption coefficient and D = the distance along the wave path. It has been found that $\varkappa$ is inversely proportional to the wave period T for a given material. This means that the absorption for a distance equal to the wavelength is constant for a given material. We can write this as follows for body waves:

$$e^{-\varkappa\lambda} = e^{-\pi/Q} \tag{8}$$

where $\lambda = Tv$ is the wave-length, T = period and v = wave velocity. The material constant Q is termed the *quality factor* and it is related to $\varkappa$ by the relation:

$$Q = \frac{\pi}{\varkappa Tv} \tag{9}$$

The lower the absorption $\varkappa$ is, the higher is the quality factor, hence its name.

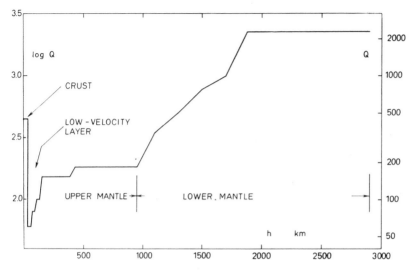

Fig. 81. The quality factor (*Q*) for *P*-waves in dependence on the depth (*h*) in the mantle.

The depth variation of *Q* in the earth has been displayed by several authors, all with some discrepancies from each other. However, the main features are common. Figure 81 shows the depth dependence of the *Q*-factor for *P*-waves in the mantle, as given by IBRAHIM (1971) in connection with his *PcP*-amplitude studies. Clearly, the poorest quality is to be found in the upper-mantle low-velocity layer. Comparable low *Q*-values are found for the uppermost granitic layer and in sediments. The lower mantle exhibits a steady increase of *Q* between about 1000 and 2000 km depth, below which a constant high value prevails. The quality factor *Q* no doubt depends on a number of material properties, such as pressure, temperature, chemical composition, structural properties, degree of homogeneity, etc. Further detailed studies of absorption will certainly prove very useful as additional constraints on our knowledge of the earth's interior.

Temperature and chemical composition. Seismology is not able to give any direct information about the temperature in the earth's interior, nor about the chemical composition. The temperature is a quantity about which our knowledge is particularly meagre. A number of temperature-depth curves has been proposed by different authors, some of them quite

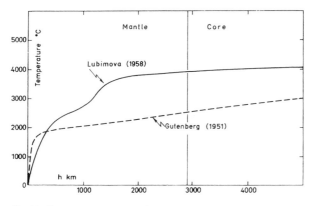

Fig. 82. The temperature as a function of depth in the earth after B. GUTENBERG (1951) and E. A. LUBIMOVA (1958).

divergent. The two curves shown in Figure 82 seem to represent more reliable information. The data are mostly based on measurements of temperature increase with depth in boreholes, on measurements of radioactivity, and for the rest of the earth mostly on hypotheses. The sharp bend of GUTENBERG's temperature curve at a depth of 70 km (Fig. 82) corresponds to an assumed transition from a crystalline to a plastic state of the material. The temperature at this level should be equal to the melting point of the material. Below this level, GUTENBERG's temperature curve exhibits a relatively small gradient. LUBIMOVA's curve increases steadily in the upper 1500 km of the earth, asymptotically approaching a temperature of about 4000 °C. According to the two curves shown in Figure 82, the temperature at the centre of the earth should be between about 3000° and 4000 °C. This is regarded as much more likely than the considerably higher values (up to 10 000 °C) which had been presented in some earlier studies.

The distribution of temperature in the earth's interior is still largely unexplored, but renewed interest in these problems derives partly from the new global tectonic theories (Chapter 6). This has been stated very clearly in a U.S. Program for the Geodynamics Project (May, 1971), from which I quote the following:

'Temperature is probably the most important parameter concerning the state of motion of the earth's interior. The earth's interior is, in effect, a heat engine, almost certainly involving heat transfer by large-scale transport of matter. Movement of matter from hot regions to cooler ones will tend to decrease the thermal imbalance unless it is regenerated by some process. The temperature distribution in the earth's interior is an indicator not only of the magnitude of the driving force for the heat engine, but also the mobility of matter (because viscosity is temperature dependent).'

'Seismic shear-wave velocities represent indirect indicators of temperature: these velocities are reduced in the vicinity of partially molten or near-molten conditions.'

We have seen that the earth's interior behaves like solid matter in relation to seismic waves, except for the outer core which behaves like a liquid. While this has been reliably confirmed, there is greater uncertainty about the chemical composition. The only guide that determinations of the physical parameters can yield, concerning the chemical composition, derives from comparisons with the behaviour of matter under high pressure and high temperature in the laboratory. However, until recently it has not been possible to reproduce conditions further down than about 300 km, and, therefore, this possibility for comparison was usually not available for greater depths. A recent improvement has occurred in this connection. Through explosion experiments in the laboratory it has been possible to reproduce pressures which are greater than at the earth's centre. Certain information on composition can be obtained from the relative abundance of different elements in meteorites. The earth's core, at least the outer one, is believed to consist of nickel and iron, the upper mantle of basic silicates and the lower mantle of metallic oxides and sulphides.

General comments. No doubt, the properties of the earth's interior are extraordinary, and it is not easy to get a good apprehension of them by comparison with well-known conditions on the earth's surface. Naturally, the most extreme conditions are to be found at the centre of the earth. Here, at practically zero gravity, we have a pressure of about 4 million atmospheres and a heat of a few thousand degrees in a material with a density comparable to that of mercury under normal conditions. And this condition prevails at a distance under our feet no longer than the distance from New York to Uppsala. By order of magnitude, the seismic wave velocities (P and S) are only about 10^{-5} of the velocity of light and about 10 times the sound velocity in air. It is in fact true that they fall within a range which is suitable for observations in station networks of continental size. As a point of interest it may be added that the P-wave velocity just under the Moho is comparable both with its velocity just inside the outer-core boundary (Fig. 75) as well as with the velocity of the satellites which encircle the earth.

Another interesting fact is that we can consider the earth's interior as quasi-stationary, i.e. practically independent of time. As a consequence, we can combine observations say from the year 1900 with those made in 1976 and still assume the conditions to be unchanged. However, if we were to combine observations from different geological epochs, the time factor has to be taken into account. In seismology it is essentially only in connection with earthquake prediction (Chapter 10) that the time factor is of significance.

Finally, let us apply a more general consideration of fundamental importance to the problem of determining the earth's internal structure. The following two factors dominate this work:

1. A number of given quantities, for instance, travel times, the earth's mean density and moment of inertia, the earth's free vibrations, etc.

2. A number of unknown quantities, which shall be determined, i.e. physical and chemical properties of the earth's interior.

In general, 1 does not permit a fully unambiguous solution of 2. Rather, 2 can at most be restricted within certain limits; in other words, there is a certain variation width of the calculated quantities. This variation can be diminished, i.e. the precision in our knowledge can be increased by an amplification of point 1, both qualitatively and quantitatively. The availability of large computers has made it possible to calculate a large number of earth models. This work has been considerably developed by F. PRESS (1970) in the USA.

7.2   The Earth's Crust

The earth's crust constitutes only a very small fraction of the whole earth (Table 19), but nevertheless it has been investigated in more detail than any other part. The reasons are obvious. The earth's crust is closest to us in space and hitherto the only part of the earth's interior which has permitted direct investigation, at least partly. The possibilities for profitable extraction of various deposits (salt, oil, minerals, etc.) have contributed efficiently to the exploration of the earth's crust. A number of different methods have been used in such investigations.

Deep drilling. This represents no doubt the most reliable method to determine the structure, not only the depth to discontinuity surfaces but also the chemical composition. However, such direct methods have a limited application, not only due to their high cost and the strictly local information they give but also due to their limitation in depth. The deepest boreholes of a few kilometres (about 0.1–0.2% of the earth's radius) penetrate only a fraction of the continental crust which has an average thickness of 30–35 km. Nevertheless, several relevant projects have been started, both in the USA and the USSR. In 1957, the Americans proposed the so-called *Mohole Project*, which aimed at a drilling down to the Moho (abbreviation of Mohorovičić discontinuity, i.e. the base of the crust). Since then, much effort has been put into realizing this project, especially in oceanic areas where the earth's crust is thinner (only about 5 km thick in typically oceanic areas). Even if it has not yet been possible to reach the Moho the drillings have been of very great value, both scientifically and technically. The Russians have started similar projects at a number of places in the USSR, all of them on continental structure and aiming at bore-hole depths of 15 km.

Even though bore holes do not reach very far when we are concerned with the earth's internal structure, they would still be sufficient to penetrate into the earthquake foci at several places. Many shocks occur at depths less than 5 km, and there a drilling could reach the focal region. In combination with stress measurements *in situ* such an undertaking would no doubt give very valuable information.

Apart from drilling experiments, all methods for investigation of the earth's crust – as well as the rest of the earth's interior – are indirect, i.e. they are observations on the earth's surface of various phenomena (seismic, magnetic, electric, gravimetric, etc.) which depend on the conditions in the interior. Obviously, indirect methods often involve great difficulties of interpretation, and discussions of such problems occupy the geophysical literature to a very large extent.

Near earthquakes ($\Delta < 10°$). Records of near earthquakes have been used extensively for the study of crustal structure, ever since the birth of

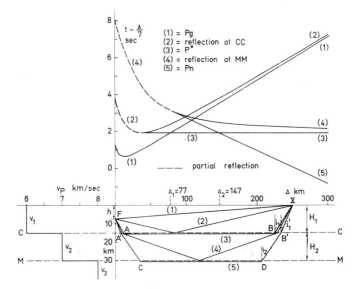

Fig. 83. Wave propagation paths and travel-time curves (quantitatively calculated) for an assumed velocity profile. CC = Conrad discontinuity, MM = Mohorovičić discontinuity.

instrumental seismology. Back in Chapter 3 we described the wave propagation through the earth's crust (Fig. 28). Figure 83 shows a case which has been calculated quantitatively. For distances $\Delta < \Delta_1$, Pg is the first arriving wave, while for $\Delta > \Delta_2$, Pn arrives first. Using the notation of Figure 83 we get immediately the following expressions for the travel times:

$$
\begin{aligned}
Pg: \quad & t = \frac{FX}{v_1} \\[2ex]
P^*: \quad & t = \frac{FA + BX}{v_1} + \frac{AB}{v_2} \\[2ex]
Pn: \quad & t = \frac{FA' + B'X}{v_1} + \frac{A'C + DB'}{v_2} + \frac{CD}{v_3}
\end{aligned}
\tag{10}
$$

In addition, we have the refraction law, equation (2) in Chapter 3:

$$\frac{\sin i_1}{v_1} = \frac{1}{v_2}$$

$$\frac{\sin i_1'}{v_1} = \frac{\sin i_2}{v_2}$$

$$\frac{\sin i_2}{v_2} = \frac{1}{v_3}$$

(11)

By a combination of (10) and (11) and by simple geometrical considerations of Figure 83, it is very easy to deduce relations for the travel times as functions of v_1, v_2, v_3, h, H_1, H_2, Δ. This is left to the reader as an exercise. The unknowns (partly source parameters, partly structural parameters) can be calculated from equations (10), when a sufficient number of stations is available. If in addition to the direct waves (Fig. 83), also those waves which have been reflected at the surfaces CC and MM have been observed, then additional relations are obtained. In Figure 83 we use a reduced time scale as ordinate, by which a better separation between the various travel-time curves is achieved.

In order to facilitate the understanding of the system of travel-time curves in Figure 83, some comments will be given. For the refracted waves there are certain minimum distances, inside which these waves cannot exist. This is a direct consequence of the refraction law. For the case shown in Figure 83 these minimum distances are 37 km for P^* and 80 km for Pn. The direct wave Pg as well as those reflected at CC and MM can exist down to $\Delta = 0$. For the reflected waves the dashed travel-time curves in Figure 83 (at shorter distance) correspond to partial reflection, while the full curves (at greater distance) correspond to total reflection. We can look upon the travel-time curves of the reflected waves as links between the different branches of the travel-time curves of the refracted waves:

1. The reflection at CC coincides with P^* at its minimum distance and it approaches Pg asymptotically at great distances.

2. The reflection at MM coincides with Pn at its minimum distance and it approaches P^* asymptotically at great distance.

These facts can also be envisaged in a purely intuitive way by considering the lower right-hand part of Figure 83.

A picture corresponding to Figure 83 holds also for the transverse waves *Sg*, *S**, *Sn*. Concerning *Pg* and *Sg*, different opinions have been expressed. According to GUTENBERG among others, *Pg* and *Sg* are not identical with the waves propagating upward in the upper layer, but instead they are so-called channel waves in the granitic layer. Related to this problem is the observation that there are two *Sg*-waves (*Sg1* and *Sg2*), also two *Pg*-waves (*Pg1* and *Pg2*), of which Figure 83 shows *Pg1* (cf. Section 3.3).

The structure dealt with so far consists of two homogeneous, plane and parallel layers over a homogeneous medium. This would correspond to average conditions on the continents. The upper layer is termed the *granitic layer*, the second layer the *basaltic layer*, although we have to take 'granite' and 'basalt' in a wide sense; in fact, these names have to be understood in such a way that the respective layers exhibit the same physical properties as granite and basalt would do. The surface CC is called the *Conrad discontinuity* and MM (the base of the crust) is called the *Mohorovičić discontinuity*, often abbreviated to *Moho* for simplicity. The average depths to CC and MM are 10–20 km and 30–35 km, respectively.

However, the near-earthquake method suffers from several limitations:

1. It is applicable only in areas where earthquakes exist. Non-seismic areas cannot be investigated in this way.

2. İt is limited to continental areas (at least as long as a network of ocean-bottom seismographs does not exist).

3. Inaccuracies in the location of the hypocentre F and the origin time of the shock, which both have to be determined from seismic records, are reflected as inaccuracies in the deduced crustal structure.

Explosions. Controlled explosions offer a considerably more efficient method to investigate the earth's crust and even the interior of the entire earth, provided the explosions are strong enough to be recorded at great distances (Chapter 11). None of the above-mentioned limitations applicable to near earthquakes are valid in case of the controlled explosion technique. There are mainly two procedures which have been developed for such investigations:

1. The refraction method. This can be illustrated by Figure 83, only by letting the focal depth *h* be zero. As a rule, observations are made

along a profile of seismographs (geophones) which extends from the shot point along a straight line to a distance approximately 10 times as large as the depths to be investigated. For exploration of a crust of 30 km thickness it is thus necessary to have a profile which is at least 300 km long.
2. The reflection method. In this method waves reflected at different boundaries are observed. This is a simpler method, because it is sufficient to record at only one or a few points. These may be located at various distances: within a few kilometres of the shot point or at a distance around 100 km – corresponding to critical reflection from the Mohorovičić discontinuity with a relatively high energy concentration in the reflected signal. On the other hand, the reflection method alone does not give as complete information as the refraction method. In general, it is advisable to combine both methods.

Compared to the near-earthquake method, the explosion methods clearly offer a superior flexibility. The explosions can be set off wherever they are needed and we are not restricted to earthquake sources. This makes a much more detailed structural study possible, with much higher precision also. Instead of the simple structure of Figure 83, we may assume sloping interfaces CC and MM. To determine such slopes, clearly, additional data are needed. By a relatively simple geometrical consideration of a case with a sloping CC and/or MM, we find that such a structure can be determined by means of *reverse shooting*. This means that the experiment is repeated with an explosion also at the opposite end of the profile. *Azimuthal shooting* provides for a more complete mapping of discontinuity surfaces.

As a rule, there are no problems in calculating a travel-time curve as soon as structure is given. But in practice we are faced with the inverse problem: to calculate the structure from a given travel-time curve. Here, as in many other similar cases in seismology, the question of the uniqueness of the solution has to be dealt with. To illustrate the difficulties which may arise, I have started in Figure 84 from an assumed structure. It is a slightly more complicated structure than in Figure 83, but still one which is very well possible. We have also shown some wave paths both for direct and reflected waves, together with the corresponding travel-time diagrams, all computed in a quantitatively exact way from the given velocity profile. We find that within certain distance ranges not less than

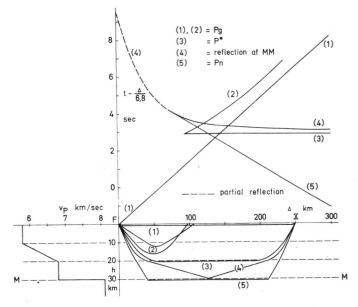

Fig. 84. Wave propagation paths and travel-time curves (quantitatively calculated) for an assumed velocity profile. Same notation used as in Figure 83.

Table 20. Examples of different interpretations of Figure 84.

Solution	Depth to CC km	Depth to MM km	v_P (from the top) km/sec
a) Only (1) and (5) measured	—	24.5	5.8, 8.1
b) (1), (3) and (5) measured	16.1	28.6	5.8, 6.8, 8.1

five waves arrive within short intervals of time. In general, it is a hard task to disentangle so many closely arriving phases in a record. The first phase is generally the easiest to measure, followed next by those later phases which rise markedly above the background of the already existing motion. In Table 20 I have given two possibilities in the practical evaluation as examples. We see that none of these efforts leads to correct result, in other words, none of these efforts has been able to reproduce the velocity profile we started from in Figure 84. These examples illustrate the fallacies which may be encountered in the interpretation of such records. It is important to be well aware of possible sources of error in the interpretation. It has often been discussed whether structural differences between continents (for example, between America and Asia) depend on real differences or if they just depend on differences of interpretation. Even though real structural differences seem to exist, we should not overlook the essential influence that the interpretation has on the results.

In spite of such difficulties, a large amount of important information about the earth's crust has been collected by the explosion methods. The most important result is the pronounced difference found between the typically oceanic and the typically continental structures (Fig. 28). Some representative results are summarized in Figure 85. The boundary between continental and oceanic structure is in general to be found around 2000 m

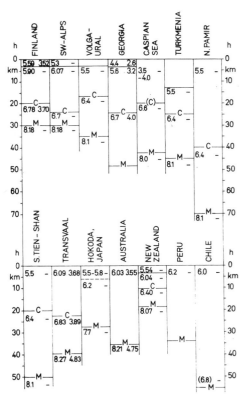

Fig. 85. Some typical results of seismic explosion investigations in Europe, Asia, Africa, Australia, New Zealand, and South America. C=Conrad, M=Moho. Each column gives to the left the longitudinal wave velocity, to the right the transverse wave velocity (in km/sec); h=depth below the earth's surface.

water depth, i.e. where the shelf zone ends and the slope towards the deep sea begins. Variations in the structure occur also within typical continents or oceans. Particularly noticeable in this respect is the thickening to about 50–60 km of the earth's crust (so-called *mountain roots*) which is generally observed under mountain ranges.

Surface waves. As we have seen already in Chapter 3, the velocity dispersion of seismic surface waves depends on the structure of the earth's crust and upper mantle. This method has been used extensively for structural studies. Also in this case we have to deal with a difficult interpretational problem: to translate an observed dispersion curve into some structure. In general, the solution is not unique. On the theoretical side, a great number of dispersion curves has been deduced for assumed structures, which can serve as a guide in interpretations of observed dispersion (one example is shown in Fig. 86).

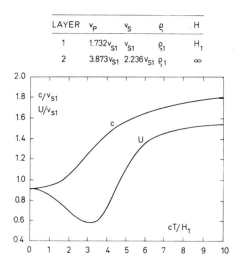

LAYER	v_P	v_S	ϱ	H
1	$1.732v_{S1}$	v_{S1}	ϱ_1	H_1
2	$3.873v_{S1}$	$2.236v_{S1}$	ϱ_1	∞

Fig. 86. Theoretical dispersion curves for phase velocity (c) and group velocity (U) for Rayleigh waves in a structure consisting of a top layer (H_1) over a semi-infinite substratum. The structural parameters are given in the figure. v_P and v_S = wave velocities for P- and S-waves, respectively, ϱ = density, H = layer thickness, T = wave period. After K. KANAI.

The two curves in Figure 86 are limited to the left by the abscissa $cT/H_1 = 0$, and to the right by $cT/H_1 = \infty$ (outside the figure). These limiting cases can arise under the following circumstances:

1. $cT/H_1 = 0$ for $H_1 = \infty$, if T is finite, or for $T = 0$, if H_1 is finite.
2. $cT/H_1 = \infty$ for $H_1 = 0$, if T is finite, or for $T = \infty$, if H_1 is finite.

In all the cases mentioned the structure is effectively reduced to a homogeneous medium (without a surface layer). Then, there is no dispersion and $c = U = 0.92 \times$ the S-wave velocity in the respective media. Between these extremes, there is dispersion, i.e. in this case $dc/dT > 0$ and $c > U$. The phase velocity (c) is determined by the physical properties of the medium, whereas the group velocity (U) corresponds to the net result from the addition of all elements with different periods and different velocities within the wave motion.

We distinguish two methods in the application of surface waves to structural studies:

1. The group-velocity method. The group velocity is obtained by dividing the epicentral distance by the travel time. It is calculated for each separate period that can be observed, and from this a dispersion curve can be constructed. A curve obtained in this way corresponds to an average structure along the path from epicentre to station. In order to obtain significant results, it is therefore important not to mix structures which are known to be different. For instance, if a path crosses an oceanic coast, we have to distinguish between the oceanic and continental parts of the path. Even much more detailed divisions of the wave path are possible. The Japanese seismologist SANTÔ was thus able to make a detailed mapping of dispersion properties, by using a large number of wave paths which crossed each other in as many different directions as possible (the 'crossing-path method'). See Figure 87. A typically oceanic dispersion is observed for the central parts of the Pacific Ocean and parts of the Atlantic and Indian Oceans. The most pronounced continental dispersion (with the lowest velocities) is observed for Tibet and also for a few other areas, partly due to thick sedimentary deposits.

The application of specially long-period seismographs (Chapter 2) has made it possible to extend observed dispersion curves to very long periods (Fig. 88). For the oceanic dispersion of Rayleigh waves the water layer (about 5 km thick) plays a dominant role. This is a low-velocity layer and leads to the steep curve in Figure 88, which implies a very pronounced dispersion. At a period of about 17 sec the velocities along continental and oceanic paths agree. For larger periods the oceanic basaltic layer with its higher velocity dominates over the granitic layer in the continent. The marked difference between continental and oceanic dispersion

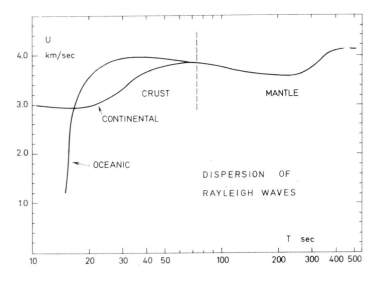

Fig. 88. Observed group velocity dispersion of Rayleigh waves for periods up to 500 sec. $U=$ group velocity, $T=$ wave period.

prevails up to periods around 75 sec. For longer periods the wave motion will extend to greater depths, and then the relatively superficial difference between oceanic and continental structure is of little or no importance.

2. The phase-velocity method. By observing the passage of a certain phase (e.g. a maximum) across a triangle of stations, it is possible to calculate both the velocity across the triangle and the direction of propagation for this maximum. The velocity thus obtained is the phase or wave velocity (Chapter 3). If the procedure is repeated for a number of maxima along a train of surface waves, we have the material for constructing a dispersion curve for the phase velocity. Since the 1950's, several such investigations have been made, as in North America, Japan, Scandinavia and elsewhere. In general, it has been found possible to interpret an observed dispersion curve in terms of a certain average thickness of the crust within the area occupied by the station triangle. The Scandinavian measurements thus gave a crustal thickness of about 35–40 km with only smaller variations between different parts of Scandinavia.

The phase-velocity method has the advantage of being as simple and inexpensive as the group-velocity method, and at the same time providing almost as much local and detailed information as expensive explosion methods. A necessary condition for an effective use of this method is that triangular station nets of proper size and with matched seismographs are available. According to other methods, phase velocities can also be determined from records at a single station by integration of group velocities or from phase spectra of surface waves.

Both surface-wave methods have also been applied to surface waves of higher modes. If for a given earthquake and a given station we have constructed dispersion curves both for Love and Rayleigh waves and both for fundamental and first higher modes, then we have four dispersion curves for the same structure. In favourable cases another two or three dispersion curves may be obtained for still higher modes. Then, there seems to be a good possibility to restrict the solution in terms of structure, as one and the same structure has to satisfy all observed dispersion curves. However, different modes of the surface waves depend to different degrees on different layers in the crust and the upper mantle. It is particularly surprising that higher modes apparently depend more on the upper mantle than on the crust. Therefore, different dispersion curves rather supplement each other than give information about exactly the same structure in all cases.

Recently (IBRAHIM, 1968), it has been found that dispersion observations of *PL*-waves are able to lend more restricting information as a supplement to other dispersion observations. By *PL* we mean a long-period wave which arrives between *P* and *S* and also after *S* and which is characterized as a leaking-mode wave. Like other surface waves, it exhibits both fundamental mode and higher modes. Its dispersion is governed primarily by the longitudinal wave velocity, whereas other dispersion curves (Rayleigh and Love) are determined in the first place by transverse wave velocities. Therefore, combined dispersion observations of *PL* and Rayleigh waves provide many more constraints on possible solutions in terms of structure than combinations of Love and Rayleigh waves.

7.3   *The Earth's Core*

We shall now pass on to a discussion in some detail of the nature of the boundary of the outer core at 2900 km depth. The discussions were particularly lively around 1950. The following account may have its interest mainly in demonstrating the difficulties which are inherent in the indirect methods of geophysics, but also in elucidating the sharpest discontinuity in the interior of the earth. It is of particular interest to notice the differences between studies of the earth's crust, where more direct methods are applicable to a high degree, and of the earth's core, where only indirect methods are available.

From seismic observations it is clear that the core boundary at 2900 km depth represents a very sharp discontinuity in physical properties. However, the question how this discontinuity should be explained cannot be answered by seismology alone. For a solution of this problem cooperation is needed not only between different branches of geophysics, but also with theoretical and experimental physics, chemistry, astrophysics, etc.

WIECHERT'S *iron-core hypothesis.* As early as 1897, the German geophysicist WIECHERT proposed the so-called iron-core hypothesis, sometimes also called the NiFe hypothesis after Ni = nickel and Fe = iron. According to this hypothesis, the earth's core consists of iron and nickel, while the mantle (outside the core) consists of silicates. Thus, the sharp discontinuity at the core boundary should be a discontinuity in chemical composition. As evidence in favour of the iron-core hypothesis, the following observations have been quoted:

1. Iron is the only one of the more abundant elements which would be able to explain the high density of the core (which is larger than compression alone would yield).

2. Besides stone meteorites (silicates) also iron meteorites have been found.

3. An iron core would be able to explain the earth's magnetism.

This idea about the earth's core was practically the only one in existence until the beginning of the 1940's, and until then the core was not considered a problem. However, since then several new and bold hypotheses have been proposed, which deviate both from the iron-core hypothesis and from each other. The conflicting ideas have had a stimulating effect on geophysical research. We have become accustomed to scrutinizing even so-called accepted truths, and to approach new theories with more scepticism than earlier. In spite of all attacks, it appears as if the iron-core hypothesis still has the most adherents, even though the original hypothesis has been modified in parts.

KUHN–RITTMANN's *hypothesis*. In 1941, two Swiss scientists, the physical chemist KUHN and the geologist RITTMANN, proposed an entirely new hypothesis on the earth's interior and the earth's origin from a homogeneous primitive state. Their objections to the iron-core hypothesis can be summarized in the following points:

1. The chemical composition of the earth according to the iron-core hypothesis deviates from the composition both of meteorites and of the sun.

2. The rapid decrease of the longitudinal wave velocity at the core boundary and the absence of transverse waves in the core do not necessitate a chemical discontinuity as an explanation, but can have a physical cause within a chemically homogeneous medium.

3. There is no plausible explanation of the processes which would have led to the chemical differentiation of different elements that the iron-core hypothesis assumes.

KUHN and RITTMANN developed a completely new hypothesis, which pays special attention to the origin of the earth from homogeneous solar matter. According to this idea, the primitive earth should have had the same composition as the sun, i.e. initially at least 30% hydrogen by weight. At the prevailing high temperature and the relatively low gravity, compared to the sun, large amounts of lighter elements, particularly hydrogen and helium, were lost to outer space from the primitive earth. The outer parts of the earth became heavier by cooling and loss of lighter gases and they began to sink inward. At the same time hot and hydrogen-rich gas masses ascended from the interior, i.e. convection currents and turbulence arose. The convection currents are assumed to have extended only over a limited depth range and not to have comprised the whole earth. During the continued cooling of the outer parts, the boiling point was first reached for the least volatile constituents, among which metallic iron played a great role. Drops were formed due to condensation, which fell down, but were soon again evaporated in inner zones with higher temperature. The gas mixture here became enriched in components with high evaporation temperatures. Upon continued cooling, crystallization occurred in the outer parts, and then also the crystals sank to some extent. As a consequence of the convection, gas bubbles, especially of light gases, were transported up to levels with lower hydrostatic pressure (so-called gas transport differentiation). Within the range of the convection currents, according to KUHN and RITT-MANN down to 2400–2500 km depth, we would thus get a differentiation with heavier elements being accumulated at lower levels. Below the range of convection currents, there should be unchanged solar matter.

The hypothesis of KUHN and RITTMANN created considerable discussion and, not the least, negative criticism. On account of the numerous serious objections which can be made against this hypothesis, its adherents have always been few in number. In spite of the many objections, KUHN and RITTMANN have no doubt pointed to certain problems which were earlier considered as self-evident. For easier review we shall summarize the objections in the following points.

1. The basic assumption – that the earth has originated from solar matter – is by no means reliable. And if this assumption is incorrect, the whole hypothesis becomes invalid. Several scientists maintain that observed physical and chemical properties of the planets can be satisfactorily explained by assuming an origin by condensation in a medium with everywhere about equal composition, but with varying temperatures and densities. It is found that the inner (terrestrial) planets have almost identical composition and the differences in densities depend primarily on different compression due to gravity. But in order to explain the densities of the major planets (which are close to 1 g/cm³), we have to assume an abundant occurrence of light elements. Thus, Saturn is calculated to consist of hydrogen and helium to 60% by weight and Jupiter to consist of the same elements to 87%. Further advances of our knowledge about the structure of the planets depend on further theoretical studies of the behaviour of *cold* matter at very high pressure. A number of scientists have proposed hypotheses according to which the earth and the other planets originated by accretion of cold matter in outer space.

2. A direct consequence of KUHN–RITTMANN's hypothesis is that no discontinuous change will occur in the physical properties at the core boundary. On the whole, the sharp discontinuity which the seismic data indicate

has been erased in this hypothesis. This is probably the strongest objection that can be made from the seismological point of view. In a paper dealing with seismic waves which have traversed the core GUTENBERG (1951) says among other things that 'no hypothesis can be considered a good approximation which disagrees with the result that the boundary of the outer core is very sharp'. 'Waves having lengths of 10 km or even less are reflected at both sides of the boundary.'

3. When KUHN and RITTMANN maintain that no separation of different elements was possible (as claimed by the iron-core hypothesis), then they imply that an equally high viscosity, as at present, existed at the beginning of the development when the earth was supposed to consist of gaseous solar matter. The separation according to the iron-core hypothesis has been compared to the process in a smelting-furnace, where iron and slag are separated.

4. EUCKEN (1944) objected that the high densities found for the core (10–13 g/cm³; see Fig. 77) cannot be explained if it were to consist of solar matter (mainly hydrogen), whereas they can be easily explained by an iron core. In fact, if the earth's core should consist mainly of hydrogen, its density would hardly be more than about 1 g/cm³. For comparison, the major planets (Jupiter, Saturn, Uranus, Neptune) have densities around 1 g/cm³ as mentioned above. It does not appear impossible that this hypothesis is better applicable to them than to the earth.

RAMSEY's *hypothesis.* We now pass over to the hypothesis for the earth's core that was proposed by the British scientist RAMSEY in 1948 and 1949. Like KUHN and RITTMANN, he started from a criticism of the iron-core hypothesis. The main points of his objections are summarized here.

1. The existence of the earth's magnetism has been taken as evidence for the iron-core hypothesis. The elements with the strongest magnetic properties, i.e. iron, nickel, cobalt, so-called ferromagnetic elements, lose their magnetism when their temperature exceeds a certain value (the Curie point): for iron 768 °C, for nickel 360 °C, for cobalt 1137 °C. In all probability, the core temperature is higher than the highest of these values (cf. Fig. 82), and, moreover, the pressure influence on the Curie points seems to be insignificant. As a consequence, it appears as if these elements cannot be magnetic at the conditions prevailing in the core. In other words, the existence of the earth's magnetism is no evidence in favour of the

iron-core hypothesis. However, the pressure influence on the Curie point does not seem to have been fully clarified. According to some results, this temperature increases so rapidly with increasing pressure for some iron alloys, that the ferromagnetism would remain even in the inner, solid core.

2. Measurements of the earth's magnetism in mines have shown that it decreases in intensity downward. This would indicate that the earth's magnetism derives to some extent from the outer parts of the earth and is not due exclusively to the core. According to more recent theories, the earth's magnetism is due to a great extent to convection currents in the earth's core. These currents are considered to arise as a consequence of the instability caused when the radial temperature gradient in the core exceeds a certain critical value.

3. The mean density of the earth (5.52 g/cm³) is the largest found in our planetary system. The four major planets have low densities, because their large masses have been able to retain light constituents, such as hydrogen. But it is a problem why among the terrestrial planets the earth has at the same time the largest mass and the largest mean density. This could be explained either by assuming these planets to be of different composition (which is improbable) or by assuming core and mantle to have the same composition, but with a gradual concentration of heavier elements inwards. According to RAMSEY, the iron-core hypothesis leads to planetary densities that are too high, if the planets are assumed to have iron-nickel cores in the same mass proportion as in the earth.

4. According to the iron-core hypothesis, the iron content of the earth should be about 40% by weight, while the average iron content in all meteorites is only about 25% by weight. An iron content according to the iron-core hypothesis appears to be too high, even if the earth is assumed to have originated from solar matter. Obviously, this is the same objection as KUHN–RITTMANN made (point 1 above).

5. According to RAMSEY, the densities in the earth's core cannot be explained, if this is assumed to consist of iron or an iron-nickel alloy. The density just inside the outer-core boundary is assumed to be 8.92 g/cm³ (this value does not agree exactly with BULLEN's curve, Fig. 77; instead this value, used by RAMSEY, has resulted from a recalculation of the densities under the assumption of a gravitational increase inward of the content of

heavier elements). Besides on the material composition, the density depends above all on the pressure, which amounts to about 1 million atmospheres at the core boundary (see Fig. 79); on the other hand, the temperature has only a minor influence on the density. If we now let the high pressure disappear and be replaced by the pressure of one atmosphere, the above-mentioned density will diminish to 4.8 g/cm^3. However, the densities at atmospheric pressure are 7.9 and 6.9 g/cm^3 for solid and liquid iron, respectively, and 8.9 g/cm^3 for solid nickel. The clear deviation from 4.8 g/cm^3 is not due to any uncertainty in the extrapolation to atmospheric pressure. Under the most extreme assumptions which can be made (i.e. no density discontinuity at the earth's core and no inward increase of the content of heavier elements) the density at atmospheric pressure will be 6.4 g/cm^3, i.e. still considerably lower than the densities of iron and nickel at atmospheric pressure. Therefore, according to RAMSEY, a core of iron or iron-nickel will give densities which are too high.

6. Just like KUHN–RITTMANN, also RAMSEY considers a separation according to the iron-core hypothesis as impossible.

However, RAMSEY does not adhere to KUHN–RITTMANN's hypothesis, mainly for the reason listed under 4 among the objections to KUHN–RITT-MANN. The high density of the earth's core cannot be explained, if the hydrogen content is assumed to be the same as in solar matter.

RAMSEY proposes an entirely new hypothesis to explain the earth's internal constitution, in particular the core. According to this hypothesis, the material of the earth's core is chemically identical with the material of the mantle, apart from a gradual increase inwards of the content of heavier elements, especially iron. The sharp discontinuity in the physical properties, required by seismology for the core boundary, is not due to any chemical discontinuity. Rather, it is due to a phase change (from the molecular to a metallic phase) in the same material at a certain high pressure and a certain high temperature. Similarly to KUHN–RITTMANN's hypothesis, there is thus no need to assume separation, as required by the iron-core hypothesis. However, compared to KUHN–RITTMANN's hypothesis, RAMSEY's model has the indisputable advantage of providing a good explanation for the seismic discontinuity at the core boundary.

In order to elucidate what we mean by metallic phase, let us take helium as an example. In the molecular state, the positively charged nucleus of a helium atom is surrounded by two negatively charged electrons. These form a so-called closed electron shell with the consequence, among other things, that the electric conductivity and the ability to react chemically are equal to zero. If one of the electrons is raised to an outer electron shell or is set completely free, it no more belongs to a closed shell. Then even helium becomes electrically conductive, thanks to the outer looser or completely free electrons and it behaves in many respects like metals. If all atoms are in this higher energy state, the element is said to be in its metallic phase. By means of thermodynamic relations, it can be shown that the molecular phase and the metallic phase are in equilibrium with each other (i.e. can exist in permanent contact with each other) only at a certain temperature and a certain pressure. RAMSEY assumes that such an equilibrium condition exists at the core boundary, naturally not in helium but in some silicate, probably olivine $(Mg, Fe)_2SiO_4$. At the phase change an increase of the density occurs also. The transition to the inner core would mean a further phase change according to RAMSEY.

This phase change has been calculated in detail for hydrogen. For this element the pressure at the phase change is 700 000 atmospheres and the density jumps from 0.4 to 0.8 g/cm^3, provided the temperature is at absolute zero. These calculations were modified by RAMSEY, who found that the pressure at this phase change is 800 000 atmospheres and that the densities of the two phases are 0.35 and 0.77 g/cm^3, respectively. Another well-known example of the phase transition is offered by tin. Grey tin is a non-metal and insulator, while white tin is a typical metal with about 25% higher density. They both exist at atmospheric pressure which demonstrates that the phase change does not require as high a pressure for all substances as for hydrogen. Similarly, arsenic has both metallic and non-metallic modifications. The density of metallic arsenic is 5.1 g/cm^3 and of non-metallic arsenic it is 2.0 g/cm^3. The cases mentioned are only to be looked upon as a few examples of transitions to a metallic phase.

Great difficulties are encountered in efforts to apply the hypothesis of phase changes to the earth's interior in a quantitative way. The pressure at the outer-core boundary is 1.4×10^6 atm. (see Fig. 79) and the density contrast probably about 50%, which could be taken as being in fair agreement with conditions at a phase change. RAMSEY assumes the earth to consist mainly of olivine with a gradual increase inward of the proportion iron:magnesium (gravitational separation). The latter assumption is based upon calculations of the densities of the planets. For instance, the density of Mars will be too small if a constant composition is assumed, but its density can be easily explained under the assumption of a central concentration of heavier elements. RAMSEY has also investigated the internal constitution of the major planets from the viewpoint of the phase-change theory. The result is that hydrogen constitutes 80% of the mass of Jupiter and 60% of the mass of Saturn, i.e. hydrogen contents comparable with that of the sun. The density distribution of the interior of Jupiter and Saturn has been calculated under the assumption of chemical homogeneity; however, there seems to be some increase of heavier elements towards the interior. According to RAMSEY, the planets can be divided into three distinct groups by their composition:

a) Terrestrial planets (Mercury, Venus, Earth, Mars) with about the same composition as meteorites.

b) Uranus and Neptune, consisting mainly of water, methane and ammonia.

c) Jupiter and Saturn, consisting mainly of hydrogen and helium.

Even in 1938 some scientists had arrived at the result that Jupiter and Saturn contain cores of metallic hydrogen.

Additional evidence in favour of his hypothesis was found by RAMSEY in the velocities of the elastic waves immediately outside the core and especially outside the inner core, as the curve was drawn by JEFFREYS (Fig. 75). These velocity features are explained by RAMSEY with the assumption that a certain percentage, about 1%, of the molecules are thermally excited just outside the cores.

RAMSEY's hypothesis is worth very careful consideration, as it points to phenomena which are probably of great significance in the earth's interior. It has also been emphasized from the physical-chemical side that such phase changes are to be expected under the pressures and temperatures which prevail in the earth's interior. Still, the hypothesis does not provide a fully satisfactory solution, above all for the following reasons:

1. A complete development of the hypothesis still remains. Theoretical calculations of the pressure and the temperature at which the two phases of olivine are in equilibrium with each other involve considerable difficulties. The same holds for calculations of the corresponding density change.

2. At the temperatures and pressures prevailing in and near to the

cores, chemical compounds could hardly exist. Possibly, a mixture of oxides of silicon, magnesium and iron might exist. These mix with each other and cannot be separated. As different substances in general will change their phase at different temperatures and pressures, we would expect a number of corresponding discontinuities following each other at different depths. On the other hand, it is conceivable that they are located within such a narrow depth range, that it has not been possible to separate them seismically. Some recent seismological results, described later in this section, may be relevant to this circumstance. In his calculations of the density distribution in the major planets RAMSEY assumes a mixture of hydrogen and helium and the same phase-change pressure for these two substances.

3. A third question concerns the fact that transverse waves do not penetrate the outer core and how this should be explained by this hypothesis. Is the change into a metallic phase accompanied by a change from solid to liquid state?

4. From a geochemical viewpoint, a density ratio of 1.6–1.7 at the outer-core boundary is considered as far too high to be explained by the phase-change hypothesis. However, other suggestions for this ratio have resulted from observations of *PcP*.

If is of interest to note that the same hypothesis as proposed by RAMSEY was also proposed by the German scientist HAALCK as early as 1931, but not at all in an elaborate way.

On the whole, phase changes have played a great role in the discussion of the earth's interior. There are essentially three cases where different kinds of phase changes have been introduced:

1. At the boundaries of the earth's cores.

2. In the upper mantle, especially in the range from 413 to 1000 km, where the velocities increase rapidly inwards. Here a more continuous phase change is envisaged, extended over the whole range.

3. At the Mohorovičić discontinuity.

In summary, the present situation seems to be that phase changes are believed to be of importance above all in case 2, followed by case 1 and lastly case 3. Observations on other celestial bodies, in the first place on the moon, will probably contribute effectively to the solution of these problems.

BULLEN's *compressibility-pressure hypothesis.* Next, we shall deal with the compressibility-pressure hypothesis proposed by BULLEN in 1949. This is based upon the values of compressibility ($=1/k$) and pressure in the earth's interior (Fig. 78—79), calculated from seismological and other observations. According to this hypothesis, the compressibility (as a function of pressure) and its first derivative (with regard to pressure) are both continuous across the boundary of the outer core. In contrast, density, etc., are discontinuous. This result is generalized by BULLEN, who maintains that the compressibility is independent of the chemical composition at the temperatures and pressures in the earth's interior. This generalization presupposes a change of the chemical composition at the earth's core boundary. On the other hand, we may combine the hypothesis concerning the continuous compressibility at the core boundary with RAMSEY's hypothesis. Then, our conclusion will be that the compressibility for a given substance is the same in its molecular phase as in its metallic phase.

After RAMSEY had proposed his hypothesis, BULLEN as well appears to be inclined to combine the two hypotheses in this way. However, RAMSEY is of the opinion that the compressibility-pressure hypothesis is only a consequence of BULLEN's assumption of constant composition between the discontinuities. This is not correct according to RAMSEY who maintains that there is a gradual, inward increase in the content of heavier elements. BULLEN's earlier density curve was modified both by BULLEN himself on the basis of the compressibility-pressure hypothesis as well as by RAMSEY on the basis of gravitational increase in heavier elements towards the interior. Among the results we may mention that, according to BULLEN, the inner core cannot be chemically homogeneous. Otherwise, there are only minor corrections to both these modifications. Information about the terrestrial planets has been of certain importance in this connection. It is common to BULLEN and RAMSEY, that these planets are all assumed to have the same chemical composition. The known mean densities for these planets are explained both by BULLEN by means of the compressibility-pressure hypothesis and by RAMSEY by means of the hypothesis of gravitational increase of heavier elements towards the interior. However, it has to be observed that the degree of this gravitational increase was determined by RAMSEY from observations of planets, while BULLEN's hypothesis in fact receives independent support from the planetary data. Another difference between the two hypotheses is that RAMSEY considers the surface of the inner core also to correspond to a phase change, whereas according to BULLEN this is a chemical discontinuity. One consequence of BULLEN's compressibility-pressure hypothesis is that the inner core is solid in all probability.

Some recent seismological results. As already emphasized, a clarification of the nature of the core-mantle boundary requires cooperation from many branches of science. Seismology established the existence of this sharp boundary, including information on changes of wave velocities, density and other properties at this discontinuity. But beyond this, seismology is not able to yield information on the exact nature of this transition.

However, recently seismology has again been put in a position to improve our knowledge about this boundary. And this is mainly due to detailed studies of waves, especially of the core-reflected P-wave (PcP), which have been made possible by use of well-controlled sources (nuclear explosions). In Figure 75 we find that even earlier a flattening of the mantle wave velocities had been found just outside the outer core, in some cases even a velocity decrease. This picture can now be established with more reliability and more detail. A. K. IBRAHIM, while working at the Seismological Institute at Uppsala, found on the basis of PcP-observations from nuclear explosion sources the following structural features of the core-mantle boundary:

1. There are two low-velocity layers at the bottom of the mantle. The thicknesses are 16.10 km (outer layer) and 19.96 km (inner layer), the P-wave velocities are 12.17 and 10.94 km/sec, respectively, and the S-wave velocities are 6.29 and 5.33 km/sec, respectively.

2. The core has in its outer part a finite rigidity with an S-wave velocity of 2.20 km/sec.

3. The density ratio core/mantle is 1.07.

Continued studies, especially of corresponding S-waves (ScS and other), will hopefully shed further light on this structure. A recent paper by MÜLLER et al. (1977) demonstrates that the nature of the core-mantle boundary is still not fully settled.

Concluding remarks. We have seen in this section that the nature of the boundary of the earth's core presents us with very complicated and very comprehensive problems. Let us here summarize some of the questions in a few points:

1. Has the earth originated from solar matter or from an accretion of cold matter in outer space?

2. Has a separation been possible into an iron core and a silicate mantle (the iron-core hypothesis) or not (KUHN–RITTMANN, RAMSEY)?

3. Can the density in the earth's core be explained if it is assumed to consist of iron-nickel (ELSASSER) or can it not be explained in this way (RAMSEY)?

4. What influence do high pressures have on the temperature at which the ferromagnetism disappears?

5. If the terrestrial planets are assumed to have the same composition, is it then possible to explain their densities with the assumption of an iron core (BROWN) or is this not possible (RAMSEY)? Does BULLEN's compressibility-pressure hypothesis or RAMSEY's assumption of gravitational increase inward of heavier elements provide the correct solution?

6. Does the boundary of the inner core correspond to a phase change (RAMSEY) or to a chemical discontinuity (BULLEN)?

In spite of all contradictory opinions and hitherto unsolved problems, we can state that, compared to the iron-core hypothesis, RAMSEY's hypothesis has the advantage of not requiring any separation of iron in an early stage of the earth's history. And compared to KUHN–RITTMANN's hypothesis, RAMSEY's hypothesis has the indisputable advantage of providing an explanation of the discontinuities at the core boundaries which is satisfactory even for seismologists. Recent seismological studies, such as those just described, deserve both to be followed up on the seismological side and to be fully considered in every endeavour to explain the nature of the core-mantle boundary. In spite of all efforts, it will certainly still take a very long time before such a picture of the internal constitution of the earth has been constructed that will remain unchanged in detail in the future–such a goal will perhaps never be reached. ...

Chapter 8

Methods to Improve Seismological Observations

8.1 Data Exchange and Data Improvement

It is no exaggeration to claim that seismology (as well as the whole of geophysics) depends more on a continuous international cooperation than most other branches of science. In seismology we are not only concerned with an exchange of scientific results but also to a very high extent with an exchange of primary data. Most seismograph stations report their readings regularly to several world centres. Frequently, reporting is made from a central station or institute responsible for a whole network. For instance, the Seismological Institute at Uppsala is the centre of a seismograph network in Sweden (Table 2), and both the reading of records and reporting are done for the whole network at the institute. Table 21 gives a review of the bulletin service and Figure 89 shows a sample page from the monthly bulletin. This may serve as an example of an activity which now goes on in most countries.

Among the world centres the following should be especially mentioned: NEIS = National Earthquake Information Service, U.S. Geological Survey, Denver, Colorado, USA, replacing the earlier USCGS = United States Coast and Geodetic Survey and USNOS = United States National Ocean Survey; ISC = International Seismological Centre, Newbury, UK, since 1964 replacing the earlier ISS = International Seismological Summary; CSEM = Centre Séismologique Européo-Méditerranéen, Strasbourg, France, since 1976 replacing the earlier BCIS = Bureau Central International de Séismologie. When such centres have accumulated enough data, the source parameters are calculated, since around 1960 almost exclusively by means of large computers. The results are then distributed to the seismological institutes the world over (cf. Section 4.2). This information then serves as a necessary basis for many special investigations in seismology.

We have seen in earlier chapters of this book that advances within seismological research to a very high degree depend on the availability of instrumental observations. Then, it is quite natural that efforts to improve

Table 21. Seismological reports issued by the Seismological Institute, Uppsala.

Type of report	Frequency of report	Stations reporting	Content of report	Start of service
Telex to NEIS[1]	Daily	Uppsala Kiruna	Arrival times of *P*, *PKP, pP*	Uppsala: May 20, 1952 Kiruna: October 13, 1952
Telex to CSEM[1] (shocks within 5000 km)	Daily	Uppsala Kiruna	Arrival times of *P*, occasionally also *PP, S*	Uppsala: January 7, 1953 Kiruna: January 1, 1977
Weekly air letter (at present 114 addressees)	Weekly	Uppsala	All short-period readings (arrival times)	June 15, 1961
Punched cards to ISC (air mail)	Monthly	Uppsala Kiruna Umeå	Arrival times of *P*, *PKP, S, SKS*	January 1, 1964
Monthly bulletin[2] (air mail; at present 149 addressees)	Monthly	All stations	Complete analyses of short- and long-period records (arrival times, amplitudes, periods, magnitudes, etc.)	All stations are reported together in the same bulletin since January, 1956

[1] Telex messages were started in January, 1977; before that we sent telegrams to NEIS and air-mail letters to CSEM.
[2] Up to 1959, inclusive, also an annual seismological bulletin was issued, which started back in 1904. Because of the very much increased demand for rapid reports, annual seismological bulletins have no longer the same importance as earlier. Their function is now fulfilled by the monthly bulletins.

this research have to be applied primarily to an amplification of the observations.

Up to the Second World War, nearly all seismographs in operation were of an old construction, many of them having been installed around the turn of the century. In those days there were no severe requests for high sensitivity. Even an old-time instrument, put up in any place, records in course of time a great number of larger earthquakes, and these can provide good material for the calculation of travel times and the earth's internal structure. In fact, the travel-time tables for seismic waves, which are still in use (JEFFREYS and BULLEN, 1940; GUTENBERG and RICHTER, 1936), were constructed almost exclusively by means of data obtained from old-time instruments. Considering this, their accuracy is remark-

Up = Uppsala, Ki = Kiruna, Sk = Skalstugan, Um = Umeå, Ud = Uddeholm,
De = Delary

1969						
Jan.	2	(cont.)				
		Sk	eP	15 23 40		
		Ud	iP	15 23 08.8		
		De	iP	15 22 38.9		
		Rhodes Island (h = N).				

" 2 Um iPKP 16 05 44.0 C
Santa Cruz Islands
(h = 640 km).

"	2	Up	iPKP	18 09 28.4	
		Ki	iPKP	18 09 43.9	C
			iSKP	18 12 53.8	
				micr	sec
			SKP	Z' 0.2	1.5
		Um	iPKP	18 09 36.6	
			iSKP	18 12 40.9	
			i	18 13 01.7	
		Ud	iPKP	18 09 26.7	

South Sandwich Islands
(h = 80 km).

"	2	Up	iP	18 18 05.4	
		Ki	iP	18 17 22.2	
		Um	iP	18 17 41.3	C
		Ud	iF	18 18 12.1	C
		De	eP	18 18 28	

Japan (h = 70 km).

" 3 Ud iPKP 00 37 47.6
Tonga-Kermadec Islands
(h = 70 km).

"	3	Up	iP	03 23 25.5	
			i	03 23 28.1	
			iPn	03 24 31.3	
				micr	sec
			P	Z' 0.2	1.3
			Mx	E 1.0	11
			Mx	N 2.2	14
			Mx	Z 1.7	11
		Ki	iP	03 23 51.3	
			i	03 23 53.9	
			iPn	03 25 04.5	
				micr	sec
			P	Z' 0.3	1.0
			Mx	E 4.0	12
			Mx	N 3.3	17
			Mx	Z 4.6	12
		Sk	iP	03 23 57.7	
			i	03 24 00.7	
			iPn	03 25 20.9	
		Um	iP	03 23 31.5	
			i	03 23 34.9	
		(cont.)			

1969						
Jan.	3	(cont.)				
		Um	iPP	03 24 55.6		
			iSS	03 31 45		
		Ud	iP	03 23 41.9	C	
			i	03 23 44.6		
			iPn	03 24 56.7		
		De	iP	03 23 33.6		
			iPn	03 24 45.7		

Iran-USSR (h = 10 km).
m = 5.9, M = 5.2 (Up,Ki).
Double P with P2 - P1 =
= 2.9 sec in average.
Clear Pn phases.

" 3 Ki iP 06 48 40.7 C

" 3 Ud iP 07 52 23.7
Aleutian Islands (h = N).

" 3 Up iP 10 10 19.4

"	3	Ud	iP	11 31 40.0	
			ipP	11 32 02.8	

Guatemala. h = 90 km (Ud).

"	3	Up	iP	11 43 03.0	
		Um	iP	11 42 40.1	
			i	11 42 55.5	
		Ud	e(P)	11 42 54	

" 3 Um iP 13 07 05.9

"	3	Up	iP	13 39 13.2	C
			i	13 39 34.3	
				micr	sec
			P	Z' 0.9	0.9
		Ki	iP	13 38 20.1	C
			iX	13 38 24.6	
				micr	sec
			P	Z' 0.1	0.8
		Sk	iP	13 38 52.0	C
		Um	iP	13 38 46.2	C
			iX	13 38 50.4	
		Ud	iP	13 39 13.7	C
			iX	13 39 18.4	
			i	13 39 29.5	
		De	iP	13 39 35.8	C
			iX	13 39 39.9	

Aleutian Islands
(h = 30 km).
X - P = 4.4 sec in average.
If X were pP, h would be
about 15 km only.
Remarkable PZ'-amplitude
(cont.)

Fig. 89. Sample page of the monthly bulletin from the Seismological Institute at Uppsala.

ably high, partly thanks to the application of an advanced statistical procedure.

However, the situation has changed considerably, especially in the last few years. There are now very severe demands both for increased accuracy in time and amplitude readings and particularly for increased sensitivity of the seismic recordings. These requests come partly from the seismologists themselves, who in recent time have become more and more interested in small earthquakes (so-called microearthquakes) for accurate information on seismic risk, and partly from the political side in connection with problems to detect nuclear explosions (Chapter 11). Especially for underground explosions, seismology seems to provide the only trustworthy method and, in course of time, the politicians have been requesting higher and higher sensitivity from the seismic records.

In the following I shall concentrate on some methods by which research people at the Seismological Institute, Uppsala, have tried to meet the modern requirements for increased accuracy and sensitivity. The methods used may be taken as representative of similar efforts at some other places.

Naturally, there are several obvious means to meet the demands. For example, increased time accuracy presupposes that an accurate clock (preferably a quartz crystal clock), accurate time signals and a constant drum rate on the recorders are used. Likewise, increased accuracy in calculation of the ground amplitudes presupposes accurately known response curves of the instruments. These obvious prerequisites for better recordings are assumed already fulfilled, and they will not be further discussed here. Our problem goes beyond this, i.e. to make further improvements in the recordings. This includes selection of good recording sites, both on the surface and underground, furthermore combination of seismometers in special geometrical patterns (arrays) and more sophisticated techniques in the evaluation of the records. On the latter point, the application of methods already developed in other related fields has proved to be very fruitful (for instance, from communication theory and radar techniques which are also concerned with the problem of discovering signals in the presence of background noise).

8.2   *Signal-Sensitive Localities*

The selection of a suitable site is probably the most important thing to consider when a new station is installed. A good site should be located on very stable ground (preferably unweathered granite) and as far away as possible from all kinds of noise (such as from traffic, industry, water falls, windy places, high trees, etc). It is no problem to increase the sensitivity of a modern seismograph almost without limit. But the main trouble is the existence of disturbances (seismic noise or microseisms), which also increase if the instrument magnification is increased. Therefore, quiet locations have to be searched for. And even then it is necessary to perform test recordings in order to get reliable information about a selected site.

The conditions available for seismic recordings vary considerably. In a country like Sweden, partly located in a shield area, there is generally no problem to find good rock. Therefore, our experience from recordings on loose ground is relatively little. But in 1957 we were operating a Grenet–Coulomb short-period vertical-component seismograph for a few months at the old seismograph station at Lund in south Sweden. There the ground consists of sand and clay, according to one source to a depth of about 60 m, below which there is limestone. The consequence was very disturbed recording, particularly from traffic.

Since the end of 1961, six to eight seismograph stations have been in continuous operation in Sweden. See Table 2. Although they are all located on very good, unweathered rock (granite, gneiss, porphyry), it was soon obvious that their seismic sensitivity was very different. This naturally means that in addition to the properties of the bedrock other characteristics are coming into play. Both in order to investigate what these factors may be, and in order to find out whether other localities exist in Sweden with still higher sensitivity than at the permanent stations, test recordings were made in 1964–6 at 11 places distributed over the country. These tests were made in cooperation between the Research Institute of National Defence, Stockholm, and the Seismological Institute, Uppsala.

By seismic sensitivity we mean the ability to record longitudinal (P) waves by means of high-magnification short-period vertical-component seismographs. This means that we are concerned with the period range of about 0.5–1.5 sec. The *P*-waves no doubt constitute the most important information from any seismic station. In addition, in this period range conditions vary considerably from place to place, much more so than for the longer-period transverse (*S*) waves or for the surface waves.

For future developments of station networks and especially for installations of more advanced seismic stations, it is of considerable importance to accumulate knowledge of factors significant for favourable seismic recording. This research, as carried out in Sweden in 1964–6, is both time-consuming and expensive, and therefore if some results of general applicability can be found, these could prove useful in seismograph station planning in other parts of the world, without the need to repeat in every area the same or similar investigation as in Sweden.

The records of the temporary stations are measured by visual inspection in the same way as the records from the permanent net. According to my experience, the number of events (usually the number of *P*-phases) recorded in any given time interval is a good measure of the signal sensitivity. Exceptions to this rule will occur when some special source dominates at one of the stations, but is not observed at other stations. Avoiding such instances, the numbers in Figure 90 represent the sensitivities at the various stations. They mean that while, for example, Uppsala has recorded 100 events, Kiruna recorded 86 on similar type instruments, etc. The figure well serves the purpose of illustrating the method and it gives at any rate preliminary results. Figure 90 also includes readings from stations in Finland and Norway, as obtained from their seismic bulletins.

In Figure 90 I have also tried to generalize the results, as shown by the sketched curves. The sensitivity numbers in Figure 90 can be considered to be influenced by a regional factor and a purely local factor. The generalization made has sense only if the regional factor dominates over the purely local influences. The latter could be traced only with a much denser network of stations.

We have to think of the curves in Figure 90 as being a result of several effects of which the most important are the following:

1. Geological structure: the eastern side of the country (mostly on granite) exhibits higher sensitivity than the western part (mostly on gneiss).

2. Distance from oceans: the greater the distance, the higher is the sensitivity.

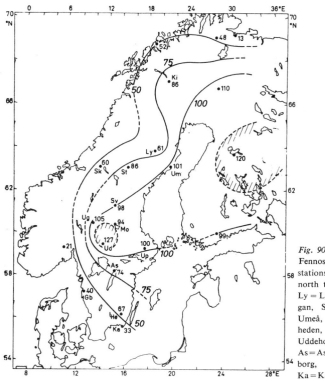

Fig. 90. Signal sensitivity in Fennoscandia. The Swedish stations are the following from north to south: Ki = Kiruna, Ly = Lycksele, Sk = Skalstugan, St = Strömsund, Um = Umeå, Sv = Sveg, Ug = Uggelheden, Mo = Mora, Ud = Uddeholm, Up = Uppsala, As = Askersund, Gb = Göteborg, He = Hermanstorp, Ka = Karlskrona.

It is then no great surprise that the central interior part of Sweden (on granite) has a high signal sensitivity. The hatched areas in Figure 90 indicate where the most sensitive seismic recordings can be made. One of them is located in the interior of central Sweden, the other in the interior of Finland. The good receiver conditions in these areas depend primarily on decreased noise. The geographical distribution of short-period microseisms has been studied by means of the same network as used for the sensitivity study. The results clearly demonstrate that these microseisms decrease inland from all the surrounding ocean coasts.

The importance of selecting a good site for seismic recordings is illustrated by the fact that while the elaborate Eskdalemuir array in Scotland records about five events per day, the ordinary Swedish network records on the average 10–15 events per day.

8.3 *Underground Recordings*

The source of seismic noise is to be found on the earth's surface, and this is true for the whole period range of microseisms, from those of periods above 20 sec down to local noise with periods around 0.1 sec or less. In addition to looking for quiet places on the surface, as already described, there is another way to avoid the noise, especially that of short period, namely by recording underground (in bore holes or in mines) instead of at the earth's surface.

In the summer of 1965 we recorded the seismic noise in the period range of 0.03–0.13 sec (or frequency range of 33–8 cycles/sec) in three mines (Haggruvan, Idkerberget, Stripa) in central Sweden at different levels down to about 750 m. The measurements were taken in competent hard rock (leptite), i.e. not in the ore veins but in the surrounding rock. There have been some related measurements made, especially in the USA, but these usually concern lower frequencies and sedimentary rock. These two facts have to be taken into account in any comparison of the results.

In every individual case we found a clear amplitude decrease with depth. As an average of 28 individual amplitude-depth curves, we found that the noise amplitude decreases to 25% of its surface value at a depth of 50 m, to 13% at 100 m, to 6% at 200 m depth, and that it is less than 1% of the surface value at depths exceeding 500 m. However, individual deviations from these averages may be large, depending on various effects. The amplitude-depth diagrams obtained (Fig. 91) suggest an interpretation in terms of body waves rather than surface waves, and the depth effect is mainly to be explained as due to increasing distance from the noise source.

Curve C in Figure 91 represents body waves and its equation is

$$a = 263 \frac{e^{-0.262\,b}}{b} \tag{1}$$

where a is the amplitude in 0.1 mμ (1 m$\mu = 10^{-3}\,\mu = 10^{-6}$ mm), b is the distance from the noise source in units of 100 m. This curve gives by far the best explanation of the observed values (crosses in Fig. 91), except for the uppermost 100 m. In this top layer the amplitude decrease with depth is much more rapid than can be explained by any simple theory for

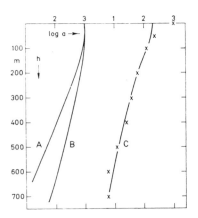

Fig. 91. Observed amplitude-depth variation (crosses) and theoretical curves: A for Rayleigh waves of 0.10-sec period in a homogeneous medium, B the same as A but for 0.15-sec period, C for body waves. The unit for the amplitude a is 0.1 mμ.

homogeneous media. This also demonstrates clearly the advantage of getting away from the free surface.

As a result, the short-period local noise of periods around 0.1 sec will be effectively removed at depths of 500 m or more in hard rock. However, in seismic recordings of distant events P-waves have periods around 1 sec, and then it is also of great importance to be able to remove noise in the same period range. Obviously, the noise reduction with depth will be slower the longer the wave-length is, and this holds whether the noise consists of body waves or of surface waves. The results above can be approximately extrapolated to other periods as follows. The depth effect depends on the ratio of depth to wave-length. This means that a given value of h/T (h=depth, T=period) corresponds to a certain percentage amplitude decrease within a limited range. For a period of T=1 sec we then find that the noise amplitude has decreased to about 25% of its surface value at a depth of 500 m, to 13% at 1000 m and is less than 1% of the surface amplitude at depths exceeding 5000 m. This is naturally a rough extrapolation from our results but could represent the conditions in hard rock.

The noise reduction with depth has already been utilized for more sensitive seismic recordings, e.g. in the USA. Specially constructed borehole seismometers are now available on the market, some made for operations at a few hundred metres depth, others for depths of several kilometres. In most cases only the vertical component of the ground motion is recorded, but recently three-component bore-hole seismometers are also available. The obvious thing to do is to combine such installations with a surface survey as outlined in the preceding section. That is, a deep bore hole should be made in a locality which on a surface survey has already been found to have a high signal sensitivity. Such a single bore-hole installation could under good conditions provide even higher sensitivity than an elaborate array station on the surface and also at a much lower cost.

8.4   Array Stations and Special Methods of Analysis

The discussion so far has concerned the selection of the best sites on the surface or underground. Another important method to improve sensitivity is to combine records from a number of seismometers spread out over an area, instead of using just a single channel. The idea is not new. Recordings by means of a profile of seismometers (geophones), usually ranging from a few hundred metres to a few kilometres in length, have been used for many years in field investigations of crustal structure and seismic prospecting. Sometimes, an areal distribution of the seismometers has also been applied, in addition to straight-line profiles. The new development—called array stations, i.e. an array of seismometers—dates from about 1958 and has been inspired by the interests in nuclear explosion detection by seismic methods. The profile or array technique is not new, but its application to continuous recordings of distant events is new.

The seismometers in an array form a regular geometric pattern, which facilitates the combination of the different channels. At present such arrays exist in the USA, Canada, Scotland, India, Australia, Norway, Sweden and in a few other countries. Figure 92 shows the geometrical pattern in some of the American arrays. The largest arrays so far built are LASA (=Large Aperture Seismic Array, Montana, USA) and NORSAR (=Norwegian Seismic Array, Norway), both consisting of over 500 seismometers at their maximum activity.

The advantage of an array compared to a single channel is that it makes use of the combination between the different channels. For example, by adding up the individual outputs in an array with variable time lags between the different seismometers, we get constructive interference for the P-wave at a certain lag and constructive interference for the S-wave

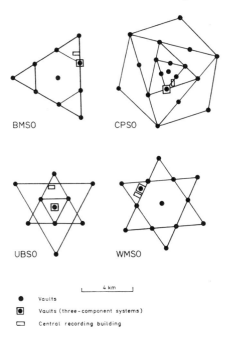

BMSO

CPSO

UBSO

WMSO

| | 4 km | |

● Vaults

◉ Vaults (three-component systems)

▢ Central recording building

Fig. 92. Geometrical patterns of some of the American arrays. BMSO = Blue Mountains Seismological Observatory, CPSO = Cumberland Plateau Seismological Observatory, UBSO = Uinta Basin Seismological Observatory, WMSO = Wichita Mountains Seismological Observatory. According to bulletin issued by Geotech, Garland, Texas.

at another lag, whereas the noise may be more or less uncorrelated. In other words, the signal-to-noise ratio is increased, by a factor of approximately $N^{\frac{1}{2}}$, if N is the number of outputs. And this naturally increases the sensitivity of such a station over a single station in the same place. In addition, the direction of approach of the waves can be easily determined, and this can be done with higher accuracy the greater the distances between the seismometers are. Obviously, the time differences between the different seismometers (channels) depend both on azimuth and on epicentral distance (due to the curvature of the travel-time graph). Therefore, it is possible to 'focus' or direct an array to special areas of the earth, for which a high detection capability is wanted. Tape recording is used and in combination with computers this also facilitates a number of operations which otherwise would take a much longer time.

The array stations have supplied seismology with an enormous amount of observational data. In fact, the material is often so large that nobody has the time or opportunity to fully utilize all of it. In addition, it is impossible to store all magnetic tape records in archives, and then this material is lost for any future research. LASA, as well as the other American array stations, and NORSAR used to publish their readings until some time ago; LASA and NORSAR also used to calculate source parameters by means of their own records. Similar determinations have been published also by some other array stations. However, these determinations often demonstrate the limitations of the method rather than being of any real value, even though improvements in course of time are noteworthy. On the other hand, a few array stations well distributed over the earth would be able to make reliable source determinations in a cooperative program. See Section 8.6.

The installation and operation of array stations are usually beyond the economic possibilities of university institutes. In addition, the limited size of many arrays was experienced as a serious handicap, which one tried to overcome by increasing their size more and more. Still, the ratio of economic input and scientific output could hardly reach agreeable values. Partly with this in mind, partly knowing how well the waves at the different stations in Sweden correlate with each other, we tried to explore the possibilities of combining the signals from the whole Swedish network of seismograph stations, i.e. applying array-data processing techniques. Of course, this network covers an area many times as large as the biggest array. LASA is roughly ten times as large as any of the other average-sized arrays in linear extent, but the Swedish network is roughly ten times LASA.

An important assumption which has to be made in applications of array-data techniques is that the pulse shape is at least approximately identical at all seismometer outputs in the array, apart from time shifts. This assumption is not necessarily valid since the character of a recorded signal is determined by the source, propagation path, receiving station and the seismograph. The effect of the seismographs is straightforward and assumed known. Although the influence of the path is considered linear, i.e. only phase distortion and amplitude variation of the pulse are possible, the crust and upper mantle can strongly change the character of a seismic wave. For example, an incident P-wave can be converted both to body and surface waves at discontinuities in the earth. It is significant that the stations to be compared are located on geologically homogeneous structure. For example, earthquakes recorded by California stations show only little

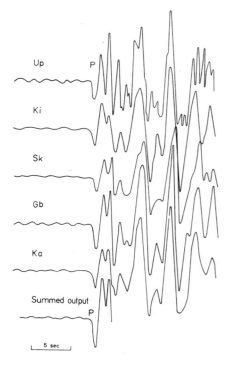

Fig. 93. Short-period *P*-wave signals from the Panama earthquake of July 26, 1962, magnitude *M* = 7.4, demonstrating the possibility of treating our network as a 'super-large' array. Up = Uppsala, Ki = Kiruna, Sk = Skalstugan, Gb = Göteborg, Ka = Karlskrona (see Table 2). After E. S. HUSEBYE and B. JANSSON.

similarity from one station to another, even if the mutual station distances are much shorter than in Sweden. This is explained by the complicated structure of California with a much dissected crust, also by sediment layers of varying thickness and structure. The Swedish stations, on the other hand, are located on more homogeneous ground, and application of array-processing techniques to this network has been successful.

Figure 93 shows a good case with excellent correlation between the short-period *P*-wave signals all over the Swedish network. Records at the Swedish stations generally exhibit the signal resemblance required by the array methods. By combination of the records (usually a summation after appropriate time shifts) the signal-to-noise ratio can be increased by a factor of about two and the accuracy of the *P*-readings is improved. Also the direction of the first motion of *P* (compression or dilatation; see Chapter

6), which is important for source-mechanism studies, can be determined more reliably by the combination of records.

Later, these investigations have been continued to include stations in the whole Fennoscandian region. The results are equally good despite some differences in crustal structure at the stations and in their instrumentation. Such methods will greatly improve the seismological information which can be extracted from ordinary station records. Another recent extension of this project includes the study of S-waves from a Fennoscandian array of five long-period stations. This study also suggests that considerable improvement in the readings of S can be achieved by a summation of records, with appropriate time shifts. In this case, SH and SV (see Chapter 3) are summed separately, and then comparisons between these two wave types are facilitated.

Computer calculations of hypocentres from arrival times of P and pP at the Swedish and Swedish-Finnish station networks (SHAPIRA and BÅTH, 1976, cf. Chapter 4) compare very favourably in accuracy with those from the largest arrays. The smaller number of stations in our networks is compensated by the larger array size and the homogeneous bedrock. So far, we have applied the array techniques to special cases (earthquakes and explosions) mainly in order to test the applicability of the methods, whereas routine application has not yet been introduced.

As a conclusion from our research in Fennoscandia, we believe that many other station nets around the world could apply the same techniques with advantage. In my opinion a group of stations form an array as soon as their records can be combined and studied ensemble in the way that array techniques require. A regular geometrical pattern of the stations is certainly helpful, but not of such a discriminating nature as correlations between different channels.

Application of techniques already developed in other fields, such as communication theory and statistical analyses of time series, have considerably improved the analyses of seismic records. Especially, I would like to mention the method of *filtering*. Filtering means separation, i.e. we separate out the part of the record we want to study and eliminate other simultaneous but unwanted parts. In order that filtering should be successful, it is necessary that the wanted and the unwanted wave motions differ in some property. Then, this property can be made the basis of the filtering technique.

This is the reason why we distinguish between different filtering techniques, such as velocity filtering and frequency filtering. Velocity filtering depends on different velocities of the wave trains which should be separated, for instance a seismic body wave (P) and microseisms. Frequency filtering may be more common and depends on different frequencies of the wave trains to be separated.

Figure 94 shows a striking example of frequency filtering. Higher-mode surface waves are usually seen as short-period waves riding on the longer-period surface waves of fundamental mode (upper part of Fig. 94). By applying a band-pass filter which permits waves of periods 6–15 sec to pass but which excludes other periods, it is possible to isolate the higher modes from the given record (lower part of Fig. 94). Obviously, the accuracy in measuring higher modes from the filtered record is far better

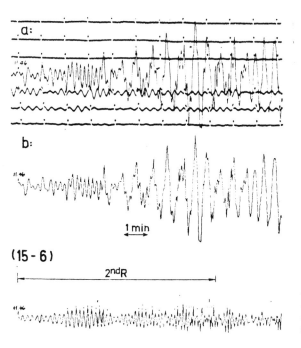

Fig. 94. Digital filtering of higher-mode surface waves. a) Long-period record at Uppsala of an earthquake in the Ryukyu Islands on August 17, 1963, magnitude $M=6.6$, b) the same record played back, before and after passage through a band-pass filter (15—6 sec). $2^{nd}R=$second-mode Rayleigh waves. After S. CRAMPIN and M. BÅTH.

than from the original records directly. By application of a series of different filters, it is even possible to separate different orders of the higher-mode waves from each other. As another example, the technique can be used with advantage to separate different kinds of microseisms from each other.

Spectral analysis of seismic waves is another promising method, which has come into much use during the last decades, as a consequence of the general availability of large computers. The usual seismic records display the amplitude as a function of time (in the 'time domain'), whereas a spectrum, just as in optics, shows the amplitude as a function of frequency or period (in the 'frequency domain'). Also the phase of a seismic wave can be displayed as a spectrum and, therefore, in seismology we usually talk about amplitude spectra and phase spectra. The application of spectral analysis, both to body waves and surface waves as well as to the earth's free vibrations, means a breakthrough in seismological research. There is no doubt that the evaluation of spectra yields very important information both about the source and the medium traversed by the waves. The problems may, for instance, concern the dependence of a spectrum on epicentral distance, focal depth, magnitude, etc. Spectral comparisons between earthquakes at normal and greater depth as well as with nuclear explosions give significant information for source identification. At the same time as the spectral analysis provides a powerful improvement to the ordinary seismogram analyses (in the time domain), there are also more difficulties involved in the interpretation of spectra. The reason is that an observed spectrum as a rule depends on a number of factors, such as the source mechanism and the wave propagation (e.g. selective absorption, reverberations, etc.). Ingenious combinations of observations may be necessary in order to separate different effects from each other. Figure 95 gives an example of *P*-wave spectra of an underground nuclear explosion. The regular occurrence of minima in these spectra is explained by destructive interference between *P* and *pP* in the source region.

In addition, there are a number of other methods to improve seismic data which can be applied to single channels or to combinations of several channels. It has been known for many years that local corrections to observed arrival times of seismic waves exist, due to variations of the crust and upper mantle under the stations. Several such determinations of station corrections can be found in the literature. However, it has been found

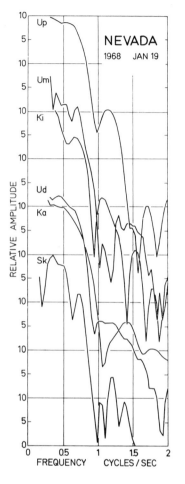

Fig. 95. Amplitude spectra of short-period vertical-component *P*-waves recorded at Swedish stations (Table 2) from an underground nuclear explosion in Nevada on January 19, 1968 (Table 30). After O. KULHÁNEK (1971).

that more progress in this field can be made if the correction is separated into a station correction and an epicentre correction (just as for magnitudes; cf. Section 4.3). So far the same travel-time tables are applied for the whole earth (except for regional studies). Information, especially from records of some nuclear explosions, has shown that regional corrections are needed to the presently used travel-time tables, of the order of −2 sec but different in different regions. Therefore, the results suggest that regional travel-time tables, different for different parts of the earth, will be the next development in this field.

Other geophysical observations, for example of gravity, are always reduced in some way before being studied. This is not the case with arrival times of seismic waves which are used directly, apart from exceptional cases when station corrections are applied. One could think that it would be better to use arrival times reduced to the Mohorovičić discontinuity at the base of the crust. The great difficulty is that before such a reduction can be made with confidence we need to have a good knowledge of the crustal structure under each station.

8.5   The World Network of Seismograph Stations

Nearly a century has elapsed since the first seismograph record of a distant earthquake was written. This happened on April 18, 1889, in Potsdam, where an earthquake in Japan was recorded (see Chapter 1). This marks in a way the birth of instrumental seismology in its world-wide sense. Since then, stations have been set up in many parts of the world, usually on initiatives from single countries, universities or individuals, with little or no coordination of the efforts. The development has been very different in different countries, reflecting in a striking way the variation in possibilities, especially financial, and in individual enthusiasm for the subject. Many countries had seismograph stations already around the turn of the century. On the other hand, several seismic countries developed slowly in this respect. For instance, Iran got its first seismograph station not until 1958, in connection with the increased efforts during the International Geophysical Year 1957–8, and Greece had until recently only one seismograph station, in Athens. Lately, both Iran and Greece as well as Turkey, Morocco and a number of other seismic countries have got well equipped station networks.

With all appreciation of the important contributions of seismologists in the establishment of stations, it must nevertheless be mentioned that as a consequence of the usually non-coordinated efforts we have to face a very inhomogeneous network, both regarding instrumentation and spacing between stations. Table 22 gives a review of present stations based on the list published by the U.S. Geological Survey in March 1974. The list also

Table 22. Presently operating seismograph stations according to the list published in March 1974 by the U.S. Geological Survey.

No.[1]	Region	Number of stations n_0	Surface area A 10^6 km^2	Station density n_0/A
1	Japan	155	0.37	419
2	New Zealand	31	0.27	148
3	Italy	22	0.30	73
4	Central Europe, south of Denmark, west of USSR, north of Spain, Italy, Greece	134	2.24	60
5	USA, including Aleutians, Alaska, Hawaii	480	9.36	51
6	United Kingdom–Ireland	13	0.33	39
7	Greece–Turkey	34	0.90	38
8	Spain–Portugal	15	0.60	25
9	Formosa–Indonesia–New Guinea	64	2.58	25
10	Scandinavia–Finland	25	1.15	22
11	Central America–West Indies–Mexico	57	2.74	21
12	India	25	3.15	7.9
13	Australia	42	7.70	5.5
14	Eastern Mediterranean–Pakistan	15	3.33	4.5
15	South America	69	17.80	3.9
16	Canada	36	9.97	3.6
17	Soviet Union	79	22.40	3.5
18	China–Mongolia–Korea–Tibet	38	15.06	2.5
19	Africa	59	30.29	1.9
20	Greenland	3	2.18	1.4
21	Antarctica	11	14.1	0.8
22	Pacific Ocean	39	180	0.2
23	Atlantic Ocean, including Iceland	12	106	0.1
24	Indian Ocean	3	75	<0.1

[1] Arranged in order of decreasing station density.

gives station density, expressed as number of stations per million km^2. The area A is only the land area, except in the case of the three large oceans. In some cases especially for island areas, such as Formosa–Indonesia–New Guinea and Central America–West Indies–Mexico, this method naturally leads to high values of station density. Quite generally, we can state that station density n_0/A can be made large by making n_0 large or by making A small. It is easier to get a high density over a smaller area. Therefore, a more strict approach would be to divide the earth's surface into equal areas

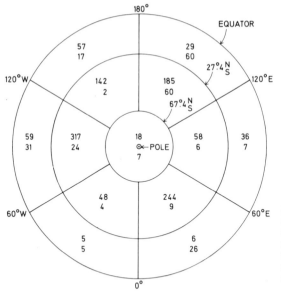

Fig. 96. Number of operating seismograph stations per 19.62 million km², obtained by dividing the earth's surface into 26 equal areas. Upper numbers refer to the northern hemisphere, lower numbers to the southern hemisphere.

A and to count the number of stations n_0 in each of them. This is illustrated in Figure 96.

It is obvious from Table 22 that the highest station density is to be found in some of the most seismic countries of the world, especially Japan, but there are notable exceptions to this rule. The large, practically empty areas are naturally the three large oceans with only 1–2 stations per 10 million km². The table contains a total of 1461 stations, which divided by the land area of the earth, i.e. 145 million km², gives 10 stations per million km².

In recent years, important steps have been taken towards the creation of a more homogeneous network. This has at least two sides:

1. Standardization of seismographs. The U.S. Coast and Geodetic Survey installation of 120 stations with standardized instruments marks the most important contribution in this field. This program started in 1960. Comparisons between these instruments and types developed and used in the Soviet Union will extend the standardization further still.

2. Creation of networks of equal spacing between stations. Although the importance of such a project has been recognized, only very little

progress has so far been possible in a world-wide sense. We may distinguish between three kinds of networks:

a) World-wide net of about 20 array stations.

b) World-wide net of fully equipped teleseismic stations, i.e. with three component long- and short-period seismographs.

c) Local or regional networks in seismic areas, usually only with short-period vertical-component seismographs. Seismic areas require a higher concentration of stations, especially for studies of earthquake mechanism and tectonic relations, as well as to be able to reach a much lower magnitude limit than is considered necessary on the world-wide scale.

The present section will mainly be restricted to a discussion of points 2a and 2b, but the ideas may be applied to 2c as well. The requirement of equal spacing between stations may be fulfilled by choosing some appropriate geometrical model. Different models differ only in less significant details. Among a number of different patterns I selected a triangular one, which means that the earth's surface is divided into spherical equilateral triangles of equal size and with one station at the centre of each triangle. Figure 97 shows part of such a grid and gives also the pertinent formulas with notation explained in the figure. The three curves give the number of stations (n) which are situated within any distance (D) from any point on the earth's surface, for three different spacings (x) between stations. n_0/A is the number of stations per million km^2 and connects therefore this figure with Table 22. A station spacing of $x =$ about $10°$ has been recommended internationally for fully equipped teleseismic stations (of type 2b). This corresponds only to $n_0/A = 0.6$, which is surpassed already by most regions of the world. However, it must be noticed that this requirement is connected with certain requirements concerning instrumentation, which are not fulfilled by most stations included in Table 22. Moreover, even within the different areas listed in Table 22, the spacing may be very different in different parts.

The plan as demonstrated in Figure 97 includes the installation of unmanned stations in remote parts of the continents as well as of ocean-bottom seismographs. At present very promising results on an experimental basis have been achieved in both these respects. Several organizations in the USA have thus been operating ocean-bottom seismographs. For instance, seismographs have been constructed for 30 days' automatic opera-

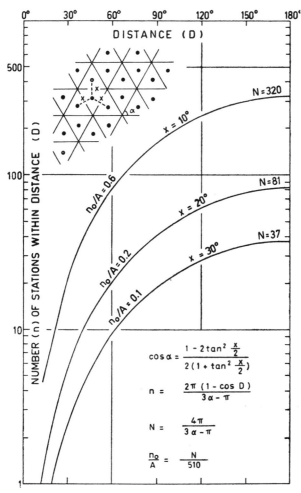

Fig. 97. Some numerical information on world-wide station networks with equal spacing. The assumed grid is shown in inset in the upper left-hand corner, where there is one station (marked with a dot) within each equilateral triangle. $x =$ station spacing, $n =$ the number of stations within distance D from any point on the earth, $N =$ the total number of stations and $n_0/A =$ the number of stations per million km^2.

tion at water depths of 7700 m. Several problems still remain to be solved, such as to develop a more reliable operation, to improve determinations of the exact position of the seismometer on the sea bottom and to provide for a sufficiently good coupling to the bottom. The Russians also operate a number of deep-ocean seismographs (down to 4000 m depth) in the North Atlantic–Arctic Ocean area, mainly for microseismic research. In Japan, several similar experimental set-ups are operated at ocean depths down to 5600 m, mainly as a contribution to the earthquake prediction problem (Chapter 10).

The application of diagrams like those in Figure 97 is best illustrated by a specific case. Figure 98 shows a relation between magnitude (*m*) and the maximum recording distance of longitudinal waves on the most sensitive short-period seismographs, excluding array stations. The diagram in Figure 98 is valid only for surface sources, i.e. mainly explosions. For earthquakes at some depth within the earth the maximum range is expected to be greater. As an example, say that we are interested in recording explosions of magnitude 5.0 and over, irrespective of the location of the explosion on the earth. Figure 98 tells us that the maximum range for magnitude 5.0 is about 35°. Comparing then with Figure 97, we see that if we want recordings of this event from 30 stations we need a station spacing (*x*) of 10°. If we are satisfied with recordings at 8 stations, a station spacing of 20° will be sufficient. Finally, with a spacing of 30° the event will be recorded at only 3 stations. This is valid under the assumption that all stations are very sensitive; otherwise, the number of stations with positive indication will be correspondingly decreased.

This example demonstrates a matter of principle in planning a station net. We have to decide the magnitude limit, above which we want to have complete information, but below which the material will be incomplete. The chosen magnitude limit then defines the station spacing which is necessary to achieve this objective, subject to modifications of local origin. In some international agreements certain station spacings have been recommended, but corresponding magnitude limits have usually not been mentioned. The same principles as laid down here also hold for local or regional networks, only with the difference that we then want to proceed to much lower magnitudes.

Some seismologists have expressed the opinion that stations should not be located within the seismic zones, whereas others prefer them concentrated to just these zones rather than to have them at equal spacing. As already outlined, I am convinced that we need both, i.e. both a worldwide net with equal spacing for the earth as a whole and denser nets within the seismic regions. The equal spacing, which has been emphasized here, has a number of advantages, which may be summarized as follows:

1. Seismicity studies (Chapter 5). Accurate location and timing of

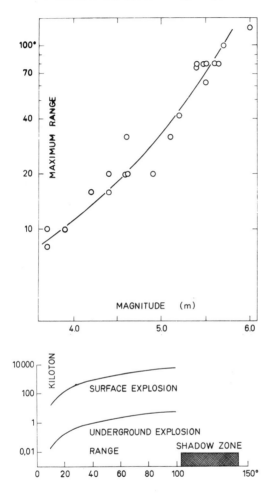

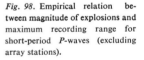

Fig. 98. Empirical relation between magnitude of explosions and maximum recording range for short-period *P*-waves (excluding array stations).

earthquakes require good azimuthal and distance distribution of stations. It is especially difficult to separate corrections of focal depth from those of epicentre location in case all stations are concentrated within a small range of azimuth. In principle, this is obvious from Figure 42. There are still a number of seismic regions which are poorly covered with observations. This is true especially for parts of the southern hemisphere, for which, however, the Antarctic stations installed in 1957 and later, meant an essential improvement. The oceanic areas, which are seismic in many parts, are generally lacking observations from near stations.

2. Magnitude determinations (Chapter 4). The equal spacing permits consistent recording of all events, whether seismic or not, above a certain magnitude limit, for all parts of the earth. In addition, a good coverage of azimuths is of significance in magnitude determinations, because of unequal radiation of seismic waves in different directions.

3. Detection problems, whether of small earthquakes or of explosions (Chapter 11). In this connection great use can be made of the caustic effect of the earth's core which produces very large sensitivity to *PKP1* at a distance of 144° and a little beyond (Chapter 3), in addition to the observations at near stations. At the caustic it is possible with the instrumentation of the Swedish network to detect events with magnitudes (m) down to 4.5 in spite of the large distance. With equal spacing on a world-wide scale, there would always be some stations close to the caustic at 144° from any point on the earth, provided the spacing is dense enough. It is now considered that with a world network of 20–30 array stations, located in areas with good recording possibilities and each one equipped with about 100 seismometers, it would be possible to detect and to locate all events down to the magnitude $m = 3^1/_2$.

4. Mechanism studies (Chapter 6). Studies of mechanism of events is of great significance, both for the identification of the source (whether earthquake or explosion) and for tectonophysical studies (in the case of earthquakes). Such studies are often hampered by insufficient azimuthal coverage, which could only be remedied by a homogeneous net with equal spacing.

5. Studies of path properties from dispersion of surface waves (Chapter 7). In such studies we clearly want the best possible coverage of large areas, preferably in many crossing directions for more detailed use of the group-velocity method. In phase-velocity determinations, networks are essential with station spacings small enough to permit phase identification from station to station.

All examples demonstrate the usefulness of homogeneous networks, although the required spacing may vary according to the special problems to be solved. For magnitude determinations, for instance, a denser net is needed for body waves than for surface waves, as the latter carry a larger part of the energy, at least for shallow-focus earthquakes, and thus travel farther. Ten to twenty stations, well selected considering their spacing and their quality, would be a very satisfactory net for magnitude determinations.

Even though a scheme as outlined here with its numerical evaluation has to be kept in mind in any planning for network developments, it does not represent more than a first approximation. Going a step further, it is very important in choosing locations of new stations to take due regard to the noise background. New sites should preferably be tested before expensive installations are made. This means in practice that it will not be possible to use a scheme like the one outlined here for more than a first orientation of station locations. Tests on each spot will indicate their suitability, and as a consequence we have to consider shifts of the positions up to the same order as the station spacings. In addition, only experience based on recordings will define the lower magnitude limit which can be reached. Also, in practical applications, economic considerations often dictate possible network developments.

Although it will certainly take many years before we have a homogeneous world-wide station network, considerable progress is already under way in this direction. On the technical side we note very promising tests as mentioned above. On the organizational side, we note that among others UNESCO has taken an interesting action in such matters. Since a very important part of the project concerns international territory (the large oceans), the active cooperation of an international body like UNESCO would be very significant.

8.6   A World Network of Array Stations

The accuracy of source parameter determination has increased considerably in the last decades. This is witnessed among others by world seismicity maps of the USCGS covering the interval 1961–9. They demonstrate very clearly the mid-Atlantic rift zone and its branches into adjacent oceans, and particularly that these structures are very limited in width. Similarly, the arrangement of hypocentres in the circum-Pacific zones is well demonstrated. A more accurate knowledge of the earth's seismicity is of great significance for many studies and applications, for example, the modern hypotheses on global tectonics as well as in engineering.

The increased precision is based upon a favourable cooperation between a number of factors, of which the following are most important:

1. A better distribution of stations over the earth.

2. Installation of special array stations.

3. Increased time accuracy, by general use of well-controlled quartz crystal clocks.

4. More accurate travel times, especially of *P*-waves, derived from records of nuclear explosions.

5. Application of large computers in the evaluation of the observations.

In the present section, I shall limit the discussion to points 1 and 2 above. Even though advances have been made in these two respects, much still remains to be achieved. In the preceding section, I advocated the idea of homogeneous networks over the globe. Here, I will apply the ideas to the development of a world network of array stations. I will base my proposal upon already existing array stations rather than suggesting a homogeneous network (which may be unrealistic).

A world network, preferably as homogeneous as possible, is required for numerous seismological studies. I shall here limit the discussion to the determination of source parameters, especially hypocentre coordinates and origin time. With regard to this problem, we can summarize commonly used methods of determination as follows, considering only teleseismic events:

1. Using only one station, by combination of times and of three-component amplitudes of different waves.

2. Using a limited network, say of 5–10 stations over an area of roughly $10° \times 10°$ in extent, primarily by combination of *P*-wave readings.

3. Using one array station, say of $1°–2°$ in extent and with numerous sensors, essentially by combination of *P*-wave readings.

4. Using a world network of ordinary stations, by combination of *P*-wave arrival times. This refers to the determinations made at some world centres, as for example NEIS, ISC, CSEM.

Let us consider the accuracies achieved by the different approaches mentioned. Method 1 can only provide a preliminary location, sometimes with wide margins, and its purpose is only to deliver rapid messages to certain institutions and to news agencies. Method 2 is usually not applied to teleseismic events, but only to regional ones. Method 3 has been practised by several array stations, and it has become customary that each array makes its own determinations. They have often large error margins. Method 4 is still superior to any other method and yields data of great value to seismology.

The question now appears if there is any possibility to improve source data information and then, in what ways. Method 1 has always a justification for rapid information on epicentral area (and magnitude, etc.) but has to be left out from the discussion of improvements. Comparisons of methods 2 and 3 would be very informative. Some comparison of this kind has been made; see Section 8.4. For increased accuracy these methods alone are insufficient. In order to achieve higher precision, method 3 could be amplified by the use of a world-wide network of array stations instead of just one station at a time. Method 4 could be improved by extended and preferably homogeneous networks as discussed above.

Let us consider a world-wide network of array stations. With greatest emphasis on an equal geographical distribution of such stations we could list the following locations:

Europe (several arrays exist)

Asia (one array exists in India)

Australia (one array exists)

North America (arrays exist both in Canada and the USA)

South America (one array exists in Brazil)

Africa (this represents a gap, which could preferably be filled by an array in southern Africa)

Antarctica (one array would be necessary)

This system would include about 10 array stations with quite a good coverage of the world. In applying this system to hypocentral location, all information available from array stations should be used, i.e. they should be treated as arrays and not as single stations. In other words, all data which single array stations put into their source location programmes should be combined for this world system. Then the accuracy of the results should be compared with results already obtained from method 4. I have not seen any such detailed and systematic study, and therefore I would like to make a strong suggestion for this project to anybody with access to primary data from all existing array stations (including calibration data) and to a large computer.

Comparative studies of the achievements of the world net of array stations and the usual method 4 are very significant for further development of the whole project, for instance, if the additional array stations (in southern Africa and Antarctica) would fill a need or not. A combination could also be envisaged of the world net of arrays and method 4, i.e. the world net of ordinary stations, and the accuracy achieved by this combination should be compared with either of the methods applied separately.

We have discussed accuracy of source determination by different methods. Closely related to this problem is the sensitivity of any seismograph system. The networks suggested should be compared also from this point of view and particularly the lower magnitude limit should be specified, above which a homogeneous world-wide material can be guaranteed.

8.7 Concluding Remarks

In conclusion, we can summarize the different aspects of improvements of seismic recordings in the following points:

1. Use of better seismographs. This is only a small part of the recent improvements, as sufficiently effective seismographs had already been developed about 30 years ago. On the other hand, the combination of these seismographs into arrays with magnetic-tape recording has meant a considerable improvement.

2. Location of seismographs in specially selected sites.

3. Special techniques to evaluate records.
4. Development of homogeneous world-wide seismograph networks.

The efficient combination of different seismometer channels, as made in arrays, whether these consist of ordinary station networks or not, means a great step forward. Earlier combinations of different outputs were generally limited to the arrival times of P for epicentre computations. However, many problems remain before the new techniques can be fully evaluated. For instance, the interpretation of a P-wave spectrum in terms of source properties and propagation effects is a problem of great significance today.

On the other hand, we could ask what the aim is of the increased accuracy and sensitivity of seismic recordings. Obviously, an increased accuracy of the arrival time of P- and other seismic waves will result in increased precision both in epicentre location and ultimately in our knowledge about the earth's internal structure. But this is true only if the increased accuracy is carried through at many stations, because increased accuracy at only few stations would not help. Therefore, world-wide collaboration is necessary. Still, it is very likely that the increased precision will not lead to any revolutions in our knowledge about the earth's structure, but rather to refinements.

Let us consider the accuracy of hypocentre location also from another viewpoint. Due to its inhomogeneous structure, the earth itself puts a certain limit to the obtainable accuracy in location of teleseismic events. And this limit can be approached but hardly surpassed by improved stations and networks, excepting the cases where an earthquake is well surrounded by near stations. Therefore, in the teleseismic case, it appears reasonable to try to work up to this limit. The problem is which method will lead to this goal in the easiest and most practicable way: a world net of array stations, a (homogeneous) world net of ordinary stations or a combination of the two.

Another justified question is what use is it to record smaller and smaller events, in other words, what is the use of a very much increased sensitivity. From the nuclear explosion side the answer is obvious, and in fact this side has prompted much of the development in seismology since late 1950's. But seismologists have also shown increased interest in small events (microearthquakes). However, in this case recordings are best performed within and near the seismic area studied and not by distant stations, which may be the only method in nuclear detection work. Observations of small earthquakes have revealed interesting facts about earthquake mechanisms and also given clues to the secular stress variations. For other problems in seismology not much use can be made of very small events. For ordinary seismological research, events of intermediate magnitude (around 6–7.5) have always proved to be most useful.

For an efficient improvement of the seismological observations (both increased sensitivity and increased accuracy) a combination of the different procedures described will be necessary. In addition, world-wide cooperation is of profound significance to seismology, not only in the daily exchange of observations but also in the planning of future development of station networks.

Chapter 9

Model Seismology

9.1 Fundamental Problems

As emphasized earlier (Chapter 1), seismology works on a broad front: it draws upon observations in nature, laboratory investigations and theoretical research. In general, the seismological problems are so difficult that close cooperation between all these disciplines is necessary in order to arrive at reliable results.

Laboratory seismology comprises a number of different branches. Such laboratory subjects as solid mechanics and solid state physics, which are usually represented at universities and technical colleges, are closely related to laboratory seismology. Concerning strength of materials, in seismology the interest is concentrated on the behaviour of various rocks under different stress conditions, not only at room temperature but also at high temperatures, prevailing simultaneously with high pressure. Hitherto, there are relatively few investigations of the behaviour of rock specimens, which have been subjected to high pressure and high temperature at the same time. But this method offers the only procedure by which it would be possible to simulate conditions in the deep interior of the earth. The highest pressures until recently being used in the laboratory corresponded only to about 300 km depth in the earth. But by the application of shock waves it has now been possible in the laboratory to reproduce pressures even exceeding those at the centre of the earth.

It is of special interest to seismology to investigate the stress conditions around cracks in various materials, corresponding to the stress conditions around a fault in the solid earth. The German seismologist DUDA performed at Uppsala a series of such stress measurements by means of photoelastic methods. These experiments concerned the static stress field around a fissure in a plate of araldite. There appeared to be a high concentration of stress around the two ends of the fissure. Such investigations and, even more so, the corresponding dynamic experiments can give valuable clues to our understanding of the stress conditions around a fault as well as their changes when an earthquake occurs. Model tectonics, which is conducted at several institutes, e.g. at the Geological Institute at Uppsala, provides another method for studying stress conditions and slow motions in the solid earth.

In model seismology some efforts are concentrated on studies of the wave propagation in various model structures, corresponding to real or assumed structures in the earth. On the whole, it can be said that wave propagation studies by means of models have come into use from two lines of approach:

1. As a supplement to mathematical studies. Theoretical investigations of more complicated structures may be very difficult and then a laboratory model may be able to demonstrate the properties of the wave propagation.

2. As a supplement to observations in nature. On seismograms we find from time to time waves and wave forms which deviate from the usual and well-known ones. Then, we may try to construct a model which in the laboratory may reproduce the empirically found wave or wave forms.

Let us here summarize in a few points some problems in laboratory seismology which are of current interest:

1. Wave propagation in inhomogeneous media. In particular, thorough investigations of various diffraction effects are needed, which have not yet been sufficiently analysed theoretically. Wave guides (low-velocity layers) play a significant role in model seismology. Also wave propagation in layered media, where the layers are not horizontal but sloping, is an important branch of research.

2. Wave propagation in media with properties which deviate considerably from the exact elastic case. Measurements of attenuation of P- and S-waves at high pressures and high temperatures are of great significance.

3. Stress release in sources of different types: explosion sources, dislocation sources (=earthquake sources), etc. The creep effects of the material are of interest. Laboratory experiments on phase transformations are of special value in connection with the new global tectonics (Chapter 6). These investigations are closely related also to the problem of predicting earthquakes (Chapter 10).

This list is far from complete. In fact, there is an equally great variety of problems within model seismology as within the rest of seismology. Japanese laboratory studies of Rayleigh waves as early as 1927 are probably among the earliest model experiments made. The most important development has taken place since around 1950, and nowadays model seismology is conducted at several institutes around the world. Both surface waves and various kinds of body waves are now studied. Earlier models were usually three-dimensional. Such models can entail certain complications from unwanted wave reflections at the different sides of the model. Therefore, it was an important improvement when Russian and American scientists in the beginning of the 1950's introduced two-dimensional models. These consist of plates of different materials, frequently plexiglass. If the plate thickness is one-tenth or less of the wave-length, the plate can be considered to be two-dimensional. Important technical improvements were made possible by the application of crystals of barium titanate. These have the piezoelectric property of being able to oscillate in phase with an electromotive force applied between two electrodes in the model; in other words, the crystal acts as an electroacoustic transducer. This substance is more efficient for this purpose than Rochelle salt, which was used earlier in such experiments. The crystals are used both as transmitters and as receivers of the waves. Recording is generally made by means of oscilloscopes.

9.2   *Seismic Wave Propagation*

In model seismology it is important to use models which represent true structures and true wave propagation, scaled by certain well-known factors. By appropriate transformations, wave equations and boundary conditions can be expressed in dimensionless quantities. Such equations and conditions can then be used for transformation from one model to another or to the real earth. With regard to simulation of true conditions, we may distinguish between two kinds of models:

1. The model parameters (wave velocities and density) are equal to the corresponding parameters in nature.

2. The model parameters bear a certain proportion to the parameters in nature. This is the more usual case.

By means of 2, a possibility is offered to reproduce the wave propagation, when wave velocities would be too high for direct reproduction in the laboratory. However, sometimes it may be difficult to find materials which satisfy any simulation requirements. Then one has been compelled to make experiments with model material with properties lacking obvious relations to the corresponding properties in the earth.

Two-dimensional models are designed in different ways in order to simulate different structures in the earth:

1. Bimorphic and polymorphic models. Two or more plates are glued together to correspond to a layered structure.

2. Perforated models. Perforated plates correspond to heterogeneous media. Both absorption and wave velocities for both body and surface waves have been investigated with perforated plates. The wave velocity can be diminished by 50% depending on the hole density.

3. Thermal models. Certain parts of a plate are heated. By using plates consisting of a mixture of paraffin and polyethylene and heating some part of the plate, it has been possible to simulate low-velocity layers. Then, comparisons are made between the wave propagation in such a plate and in the corresponding homogeneous plate.

4. Models of variable thickness. The variation of the plate thickness corresponds to density variations. In combination with the preceding method, density and wave velocity can be varied independently, which clearly multiplies the possible experimental conditions. See Figure 99, which illustrates the refraction of a wave front caused by a change in plate thickness.

A continuously varying velocity in a three-dimensional model can in principle be achieved by composing the model of two or more substances with different wave velocities. A continuous velocity variation can be produced if the substance with the lowest velocity is mixed with varying concentrations of other substances with higher velocity. The latter then determine the maximum velocity in the model. A usual method of constructing such a model is to combine a large number of thin plates, of which adjacent plates exhibit only insignificant velocity differences. The thickness of each plate should be less than the wave-lengths used. Among substances which have proved to be suitable for such models, we can mention a gel consisting of water-glycerol-gelatine. It has proved to be particularly useful for investigations of P-waves. A disadvantage with this gel is that

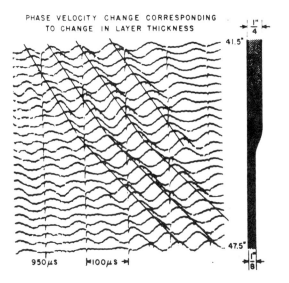

PHASE VELOCITY CHANGE CORRESPONDING
TO CHANGE IN LAYER THICKNESS

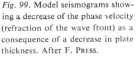

41.5°

47.5°

950 μs   ←100μs →

Fig. 99. Model seismograms showing a decrease of the phase velocity
(refraction of the wave front) as a
consequence of a decrease in plate
thickness. After F. PRESS.

the velocity variation amounts to 26% at most, which is insufficient to
simulate the earth's interior. Therefore, a large number of other gels of
different composition have been tested, by which it has been possible to
raise the velocity variation to 54%. Such investigations have been made
by scientists in Czechoslovakia. Among other things the velocity variation for *P*-waves in the upper mantle down to about 550 km depth has
been modelled. Good agreement has been achieved between model and real-
earth velocities.

Cement has also been used for construction of three-dimensional
models with several layers. It is especially useful when greater contrasts
between the different layers are desired. As the wave velocities depend on
the age of the cement, stabilized values cannot be expected earlier than a
month after the solidification of the cement.

Just as in ordinary seismology, it is necessary to record in three directions perpendicular to each other for a more complete study of the wave
motion. In model seismology this is achieved by placing the piezoelectric
receivers in three mutually perpendicular directions.

Special technical problems are involved in the construction not only
of the models but also to an equally great extent of the transducers. These
are used both as transmitters and receivers of seismic waves. The linear

extent of a transducer is in many cases of the order of one-tenth mm to satisfy the scaling laws. However, such small dimensions entail difficulties not only in the manufacturing but also because of their small power. The properties of the piezoelectric material play a great role, e.g. for the free period of the transducer. The damping and the directional properties of the transducer are other characteristics to take account of. For the receiver we have in addition to require a number of properties, among others:

1. Distortion-free recording of signals within the entire frequency range in which we are interested, e.g. from 100 kc/sec to 2 Mc/sec (1 kc = = 1 kilocycle = 10^3 oscillations; 1 Mc = 1 megacycle = 10^6 oscillations).

2. Amplitude measurements possible with an accuracy of 3% and stability of the sensitivity.

3. Time measurements in the interval from 0 to 1000 μs possible with an accuracy better than 0.1 μs (1 μs = 1 microsecond = 10^{-6} sec).

Among a number of wave propagation problems, which have been studied by model seismology, we like to mention particularly those connected with low-velocity layers in the upper mantle. By methods outlined above models have been constructed with different velocity profiles, including a layer with lower velocity. The wave propagation in the presence

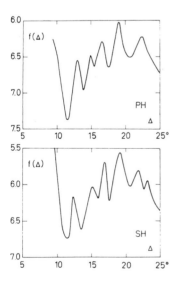

Fig. 100. Amplitude-distance curves for P- and S-waves (horizontal components) for southeastern Europe. The ordinate is $f(\Delta) = m - \log(a/T)$, equation (1), Chapter 4, i. e. for a given m the ordinate gives the variation of $\log(a/T)$ with distance. After J. Vaněk.

of such a layer is often difficult to deal with theoretically. This is particularly true for the diffraction effects inside the shadow zone. In such cases models are able to furnish valuable information. Moreover, seismographic observations in nature have revealed variations of amplitudes of body waves between about 5° and 25° epicentral distance, which are usually ascribed to special velocity anomalies in the upper mantle of the earth (Fig. 100). Also for the investigation of such amplitude variations, model seismology seems to offer the most practicable method. Such experiments have suggested the following results:

1. The amplitude increase near 20° (Fig. 100) may depend on a low-velocity layer (for P) at about 100 km depth.

2. The amplitude oscillations near 20° indicate the existence of a low-velocity layer near 100 km depth and a discontinuity at about 400 km depth; a relatively small negative velocity gradient in the low-velocity layer and a second-order discontinuity, respectively, may be sufficient.

3. The amplitude oscillations at distances less than 12° can be interpreted as due to interference between waves reflected at different discontinuity surfaces of the first or second order and located between the Moho and 100 km depth.

These results have been proposed by the Czech seismologist VANĚK. The method certainly seems to be effective in exploring the details of the velocity profile in the upper mantle.

Seismic soundings of the earth's crust (Chapter 7) have for a very long time been subjected to various interpretational difficulties. Also in this area, model seismology offers a method which is able to give further information and limit the uncertainties. See Figure 101.

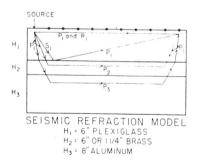

SEISMIC REFRACTION MODEL
H_1 = 6" PLEXIGLASS
H_2 = 6" OR 1 1/4" BRASS
H_3 = 8" ALUMINUM

Fig. 101. Waves observed in a three-layer model: H_1 plexiglass, H_2 brass, H_3 aluminum. After F. Press, J. Oliver and M. Ewing.

9.3   *Earthquake Mechanism*

Studies of source mechanism is another aspect of model seismology. Experimentally, one has tried to simulate various source mechanisms and to study the wave propagation in the model. I should like to mention the experiments which the American seismologist PRESS conducted in Pasadena (California) some years ago. Two-dimensional models of plexiglass in the shape of circular plates were used. The source (simulating an earthquake) was located at the centre of the plate and the receivers were placed along its circumference. They recorded amplitudes, wave shapes and directions of motion both for *P*- and *S*-waves (i.e. just like ordinary seismograms). Crystals of barium titanate were used as transmitters and receivers.

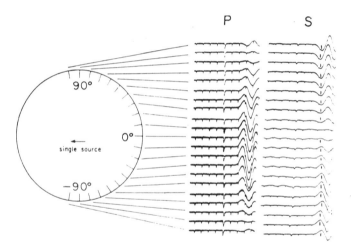

Fig. 102. P- and S-waves from a single source in a circular plate. After F. PRESS.

As wave generators PRESS used both a single source (Fig. 102) and a dipole. A dipole source is made of two barium titanate crystals located close to each other but oriented so as to produce opposite phases. The amplitude diagrams and the directions of motion show good agreement with theory both for *P*- and *S*-waves. A third experiment was also performed, including a dipole with a fissure between the two forces, in order to simulate nature still better. For this case the *P*-waves again behaved in a normal way, whereas for the *S*-waves a motion was observed even in those directions where a node was expected. The cause was considered to be a transition of Rayleigh waves, propagating along the fissure, into S-waves at the end of the fissure—an observation which may be of importance to consider also in focal mechanism studies from seismograms of natural events. Similar results have been found in Japan. On the other hand, caution is required in translating model results into conditions prevailing in nature. What has been found may be representative of the model only and may not represent nature.

In this connection, it is of interest to note that the American seismologist HOWELL did not find the S-wave anomaly in his experiments which were conducted later. Some of his results, which may be of significance in the determination of earthquake focal mechanisms, are summarized here:

1. The first motion, observed for homogeneous models, confirms the theory for a single force, a couple and for two couples (perpendicular to each other).

2. P-waves exhibit compression everywhere and with nearly constant amplitude in all directions for a radial force (explosion), as expected, while weak S-waves also appear.

3. In a homogeneous model, an unsymmetric couple may give rise to P-waves which are similar to the distribution from a radial source, but S-waves are able to discriminate. Such findings are of importance in connection with detection of explosions (Chapter 11).

4. A discontinuity may entail reduced amplitudes for P and S within the layer which has the higher velocity. As a consequence, the first displacement may become too small and may be overlooked in this medium. In turn this may cause erroneous readings of the first motion and incorrect conclusions about the focal mechanism. These results indicate the significance of a velocity discontinuity at a distance of the order of one wave-length from the source.

In other experiments, the wave propagation has been studied from thermally induced cracks in a glass plate. Comparison with theory has shown serious discrepancies, especially for the S-wave. These deviations are ascribed to non-elastic effects, particularly at the end of the crack. On application to real conditions in nature we have to face the question: which approach is most suitable, theory or model experiment.

Earthquake mechanism can be studied by means of the emitted seismic waves, as in the methods mentioned so far. But in addition, direct stress measurements in a simulated earthquake source are of great significance in laboratory studies. Such research is currently conducted at several institutes, among others in the USA. BRACE (1969) has summarized the objectives of such laboratory studies in three points, which I reproduce here:

1. To discover and investigate all possible mechanical instabilities likely to cause earthquakes in the crust.

2. To study changes in rock properties as stress is increased to the point of instability.

3. To explore the possibility of conversion of damaging forms of sliding motion to undamaging forms.

The importance of such studies for earthquake prediction (Chapter 10) is quite obvious, especially as it has been shown that many properties of rock under stress change drastically just before fracture.

9.4 Seismic Wave Velocities

In seismology we are frequently faced with the problem of comparing seismic velocities measured in the field with those determined in the laboratory on samples of rocks and minerals. The purpose of such comparisons is, if possible, to determine the chemical composition of the earth's interior. In this way, we understand that laboratory measurements of wave velocities have a great importance.

For laboratory determination of seismic wave velocities there are essentially three methods which have been applied:

1. Resonance methods, in which free vibrations are generated in the specimen and the corresponding frequencies measured.

2. Pulse-transmission methods, in which the travel time is measured of a high-frequency pulse propagating through the specimen.

3. Ultrasonic-interferometry methods, in which internally reflected waves are made to interfere.

We shall discuss these three methods each in its turn, a little more closely.

Resonance methods. For many years elastic properties of materials have been determined by various static experiments: bending, torsion, compression, etc. While such methods are suitable for metals and glass, they proved to be unreliable for rocks because of their inhomogeneous structure. About 30 years ago the resonance method was introduced, in which cylindrical specimens were used. The frequencies were observed for different free vibrations of the specimen: longitudinal (along the cylinder axis), transverse (perpendicular to the cylinder axis) and torsional vibrations. In each case the phase velocity in the specimen can be calculated from the length of the cylinder and the observed frequency. The free vibrations are induced by means of a transducer of some kind (electromagnetic, magnetostrictive, electrostatic or piezoelectric). The receiver can be placed at one end of the cylinder or can be moved along the cylinder to locate nodes and antinodes. These measurements can be performed also under high pressure and high temperature. The method is relatively reliable even for porous samples. The errors in the velocities amount to about ±5%. The method has proved particularly suitable to the determination of transverse wave velocities at high temperatures and high pressures.

Besides cylindrical specimens also prismatic specimens have been used, but these require more comprehensive corrections depending upon their shape. In a newer development of the resonance method, one observes free vibrations of small spherical specimens—in a way corresponding to the free vibrations of the earth observed after larger earthquakes.

In a variation of the resonance method, single crystals or crystal aggregates are used as specimens instead of the cylindrical or prismatic samples. When a single crystal is used as the specimen, it may be brought into immediate contact with a piezoelectric crystal in such a way that desired vibrations are induced in the specimen. An alternating potential applied to the electrodes in the piezoelectric crystal will induce vibrations in the combined oscillator. The frequencies generally used are in the range from 90 to 250 kc/sec, depending on the frequencies of the specimen itself. This method can be used with small crystals of only a few mm in length. It is also easy to apply at high temperatures, but it has not been used at high pressures.

Pulse-transmission methods. It was not until sufficiently accurate electronic equipment had been developed during and after World War II that this method could offer a possibility for laboratory seismology. The pulses (50 kc/sec to 10 Mc/sec) are generated by means of a crystal transducer which is attached to the specimen. The travel time through the specimen is measured either absolutely or by comparison with another specimen (Fig. 103). A circular-cylindric shape of the specimen has proved to be

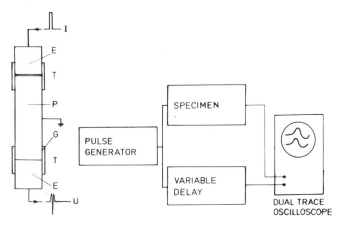

Fig. 103. Principle of the pulse-transmission method. I = input signal, E = electrode, T = transducer, P = sample, G = rubber tubing, U = output signal. After F. BIRCH (1960) and G. SIMMONS (1964). See ANDERSON and LIEBERMANN (1966).

most suitable. Its geometrical shape and dimensions have to fulfill certain conditions to avoid unwanted effects, such as disturbing reflections, scattering of waves, etc., and thus to make it possible to reach reliable results.

Most applications of this method have concerned the direct waves, P and S. In some experiments also transformed waves, such as PSP, have been used. By PSP we mean a P-wave which has been partially transformed into SV by reflection at the cylinder side and then by another reflection has been transformed into P. The travel-time difference PSP–P depends on both the P-wave and the S-wave velocities in the sample. This method has proved useful for metals, limestone, marble. But for more coarse-grained rocks, serious errors arise with the transformed pulse method because of interference and reflections at the grains in the sample.

Pulse-transmission methods have been in frequent use since the beginning of the 1950's. They do not require homogeneous specimens, but can be used also for porous specimens, and moreover they can be easily applied at high temperatures and high pressures. The accuracy of the method is about 1–3%.

For very fine-grained specimens (with grain sizes less than 0.5 mm), a pulse-echo method has also proved to be useful. In this method a pulse passes through the sample and is reflected back to the receiver and the travel time is measured.

Total reflection of a P-wave at the boundary between a liquid medium and a solid medium (the specimen) has also been applied for the determination of wave velocities (Fig. 104). When the angle of incidence i_1 is

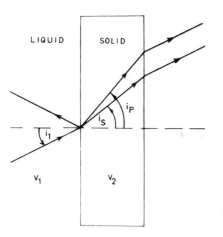

Fig. 104. Reflection and refraction at the surface of a specimen. i = angles of incidence, v = velocities. After G. SIMMONS (1965). See ANDERSON and LIEBERMANN (1966).

varied continuously, total reflection will occur at a certain value (i_{crit}). By application of the refraction law, equation (2) in Chapter 3, we find the following relation for calculation of the wave velocity in the sample:

$$v_2 = \frac{v_1}{\sin i_{crit}} \tag{1}$$

It is considered possible to observe the critical angle of incidence with an accuracy of 0.1°, which corresponds to an accuracy of the P-wave velocity of 0.5–1.0%. It has been observed that the velocity decreases with increasing grain size of the specimen, and the results become unreliable when the grain size approaches the wave-length of the signal used. For the frequencies usually used (1–5 Mc/sec) the applicability of the method is thus limited to samples with an average grain size of less than 1 mm.

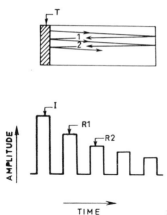

Fig. 105. Principle of ultrasonic-interferometry methods. T = transducer, I = direct wave, R1 = once reflected wave, R2 = twice reflected wave. After O. L. ANDERSON and N. SOGA (1965).

Ultrasonic-interferometry methods. The principle of these methods is obvious from Figure 105. A transducer, acting as a transmitter, is cemented to one side of a specimen. The same transducer is used to receive waves which have travelled through the specimen with one or several reflections from the opposite side. The echo waves will arrive at equal time intervals but will exhibit gradually decreasing amplitudes. The time differences yield the velocity immediately. Various methods have been introduced to solve the problem of accurately measuring the time interval between reflections. Interferometric methods are used either between successively reflected pulses or between reflected pulses and regularly superposed pulses. Both P-wave and S-wave velocities may be determined in this way, by application of transducers generating P-waves and S-waves, respectively, through the sample.

This method is especially suitable because it renders very exact measurements possible even on small samples (with an extent of a few millimetres). In addition, it can be used under high pressure, but hardly for temperatures exceeding 300 °C. Severe requirements have to be placed on the condition of the specimen and its geometrical shape.

The three methods, which we have described briefly, are those in most general use. In addition, there are several other methods which are still for the most part in a stage of development, but which may be more generally used in the future.

Chapter 10

Prediction and Artificial Release of Earthquakes

10.1 *Background and Purpose*

The problem of earthquake prediction is of primary concern to seismological research, but it is only in the last few years that more optimistic hopes have been expressed about the possibility of solving this problem. For many years it has been considered impossible to solve it in a way which has practical significance, i.e. to predict the location and time (at least to the nearest day) and magnitude of a coming shock.

There is still a long way to go before this goal is reached, but recently active interest in the subject has been displayed both from governments and international bodies like UNESCO. This interest has been activated by a number of disastrous earthquakes in the last few years, as for instance Morocco (Agadir), Iran (Lar) and Chile in 1960, Iran in 1962, Yugoslavia (Skopje) in 1963, Alaska and Japan (Niigata) in 1964, Turkey (Varto) in 1966, Iran in 1968, Peru in 1970, Nicaragua in 1972, Guatemala and China in 1976 (Table 11).

In Japan, where this problem is always of acute significance, a detailed plan was made in 1962 by TSUBOI, WADATI, and HAGIWARA. Although there was no doubt that very much would be learned if their proposal of about 10 years' intensified observational research was carried through, it was still an open question as to what extent this would solve the earthquake prediction problem. But naturally the beginning has to be made at some end of this enormously large complex of problems, and then we must agree that the observational side is the one which should receive the greatest attention. This plan was limited to a 5-year plan, and in the spring of 1969 it was communicated from Japan that the Japanese government had approved another 5-year plan with a beginning in 1969. The total budget for the 9-year period 1965–73 amounted to 18 million dollars. Early in 1977 it is reported from Japan that they are presently busy with their third 5-year plan.

In the USA there was a special committee for the development of plans for a solution of the prediction problem. Also here, a 10-year plan had been proposed for intensified observations in three seismic areas: California, Nevada, and Alaska. Cost estimates for the original project appeared very high (about 140 million dollars), and economic considerations were a major obstacle to the realization of the original plan. On the other hand, it has been emphasized quite correctly that this cost would correspond only to a very small fraction of the losses in a destructive earthquake in California. And this holds still true, even though later requests amount to 3 to 4 times as much. Thus, for a 10-year plan proposed in 1976 the sum has risen to 490 million dollars, to be shared about equally between the U.S. Geological Survey and the National Science Foundation in a broad attack on the whole problem. In the USSR, investigations aiming at earthquake prediction have been started in several seismic areas in Siberia, as in Garm and Tien-Shan and on Kamchatka.

In planning for a solution of the prediction problem, we have at present a better background than ever before, partly because of the interest from national and international bodies, partly because of the very much increased effectiveness in modern seismological observations and data handling techniques. It is also true to say that the time has passed when earthquake prediction was left only to amateurs, who underrated the inherent difficulties, and unfortunately made the field one of bad scientific reputation. The consequence was that for many years anyone who pretended to be able to predict earthquakes was placed in the same category as those who tried to construct a perpetual motion machine. Such prediction efforts have correctly been dismissed from the scientific side. Many so-called predictions were also lacking the necessary precision. For instance, if I say that an earthquake will occur tomorrow, I am one hundred percent right. This is obvious from the earthquake statistics (Chapter 5). But such a 'prediction' is completely useless. Some other viewpoints from a responsible seismological side may also have acted as an obstruction to the development of scientifically based earthquake prediction. In 1941, the famous geophysicist GUTENBERG wrote among other things the following: 'Even if earthquake predictions could be as accurate as weather forecasts, they probably would do more harm than good since they would create a wrong feeling of security and neglect of precautions against earthquake effects where no prediction has been made and, on the other hand, they might cause unnecessary worries where an imminent earthquake has been predicted.' In spite of such obstructive factors, the prediction problem has since around 1960 been seriously adopted both among seismologists and by responsible authorities. It is now recognized everywhere that only a strictly scientific approach to this problem will have any chance of success. This must be based on very thorough testing of every idea, especially as we meet a field where our knowledge is practically non-existing.

The prediction problem has a practical importance (to save at least a large fraction of the 10 to 20 thousand lives lost annually in earthquakes and their aftereffects; Table 11), but the problem has also a scientific importance (to increase our knowledge about the earth's interior and the acting forces). It is true that most large earthquakes are located in coastal regions (especially around the Pacific Ocean) and many of these regions are among the most populated in the world, e.g. Japan. As emphasized in Chapter 5, economic interests play a far greater importance in the distribution of population than earthquake danger, at least on the whole. It is considered more practical to take measures of precaution against earthquakes and their effects. Such precautions can be developed along two main lines:

1. Application of building codes: in this way, buildings and other constructions, like dams, power plants, etc., can be made earthquake-proof. Such codes are enacted by law in several countries, as, for instance, in parts of the USA, USSR, Italy, etc. However, there are still many seismic areas lacking such codes. For the preparation of building codes reliable seismic maps are necessary. These are based in the first place on recorded and observed earthquakes within the area in question. But such information may often be insufficient, because the statistics may cover too short a time. Then one may have to resort to historical information on earthquake effects and, moreover, try to see relations between the tectonic structure of an area, especially tectonic motions, and earthquakes. It is a common experience that earthquake risk is always greater in the borderline zone between different tectonic units than within each unit. Such studies may result in *seismic zoning maps* (Chapter 5), which show the maximum intensity which is likely to occur within each area. See also Chapter 4. As additional precautionary measures against earthquake effects, it is of great importance to prepare protection against fires, considering that fires, following major

earthquakes, often have caused greater damage than the earthquakes themselves (for example, San Francisco 1906, Tokyo 1923). 2. Earthquake prediction: through this, people could be protected from catastrophes by evacuation and by other measures of precaution. But evacuation with ensuing interruption of many activities is terribly expensive, and no scientific seismologist would be willing to take that responsibility before he had 100% certainty in his prediction. The most suitable procedure would probably be that seismologists for a long, long test period make their own 'secret' predictions, and that official predictions not be made until the technique has been developed to a sufficiently high precision. Otherwise, there may again be a great risk of this whole problem being discredited among the general public.

To be of value, a prediction should tell the time of a coming shock (preferably to the nearest day), its location (e.g. towns, villages etc., likely to be damaged) and its intensity. From the practical point of view the problem can be limited to damaging earthquakes, i.e. those above a maximum intensity of 7–8 in case of badly designed structures and of 9 for specially designed structures, and the problem is naturally limited to populated areas. From the scientific point of view prediction has significance also for unpopulated areas and also for deep shocks. In order to understand the problem, it will be necessary to include also these shocks in planned projects. We have also to consider that many seismic areas, which are uninhabited today, will be built-up areas in the future. During such expansions of the population into new areas, it is of the greatest advantage if the earthquake engineer is able to give advice in the planning of the building projects.

Items 1 and 2 represent two different aspects of the prediction problem; while 1 is concerned with *statistical prediction*, item 2 concerns *deterministic prediction*. Statistical prediction is based on past seismic history, and can be made with fair reliability as soon as this is available. On the other hand, deterministic prediction, i.e. prediction of isolated events, is a much more difficult problem and for its solution a large variety of phenomena has been investigated, as we shall learn in following sections.

There is hardly any problem that has so far-reaching consequences as earthquake prediction. This is not only a scientific problem, but one with the greatest concern for the general public and all its occupations. The public experience of earthquake disasters is unique for at least two reasons, compared to other natural disasters: firstly, earthquakes are sudden events which usually occur without previous warning (at least nothing observable to people); secondly, in any given place, the intervals between earthquake disasters are frequently much longer than human lifetimes, whence most people have no previous experience from such catastrophes. The repercussions of earthquake predictions on numerous activities are enormous, especially those of industrial and commercial as well as social nature, not to mention earthquake insurance, etc., in short, the whole operation of a modern community. In the following two sections, we shall see how efforts are made to approach the scientific side of the prediction problem.

10.2   *The Physical and Observational Side of the Problem*

Let us first approach the problem from the physical side and consider the simplest possible model: a vertical string (this is essentially a one-dimensional body) is loaded at its lower end, and the load is increased continuously until the string breaks (for example, the load may be a vessel, into which we are pouring mercury). If the initial load (at time $t=0$) is zero, the string will break at a time $t_0 = F_0 : (dF/dt)$, where F_0 is the load at the breaking-point and dF/dt is the rate of increase of the load, this rate assumed constant. Therefore, in order to predict the time t_0 when the string breaks we have to know both the rate with which the load increases with time (dF/dt) and the properties of the string (which define F_0).

The string corresponds to the earth and the load to the stress- and strain-generating forces in the earth. Professor H. BENIOFF in California proposed some years ago a possible scheme for earthquake prediction which corresponds essentially to this model. By observations of stress and strain at an active fault and of the points in time when it breaks (i.e. when earthquakes occur) it would be possible to predict coming events. But according to BENIOFF, observations would be needed for every single fault and extending over at least two centuries, before reliable prediction could be made. Experiences from one fault cannot simply be applied to other faults. However, not only is the long time-scale of this project discouraging, also the possibilities are uncertain. The model is extremely oversimplified, since for the real earth we have practically no reliable knowledge of the origin and nature of the acting forces, nor can we assume that stress and strain vary linearly with time (rather the variation is apparently irregular, frequently speeded up just before larger earthquakes, sometimes even reversed); furthermore, there are possible influences from external factors of various kinds (such as activities on adjacent faults, triggering forces, etc.). The earth is of a complicated and inhomogeneous structure, especially in earthquake regions with numerous fractures, and we have practically no information on breaking strength.

A direct physical approach to the prediction problem is therefore practically impossible at present. As far as prediction is concerned, seismology is certainly in a worse situation than meteorology, where it is possible to make predictions of weather on a strict physical basis (using large computers). In the distant future, similar predictions could be envisaged in seismology. But there is a very long way to go before we know all parameters of importance and their numerical values for each region. Therefore, in seismology we are forced to try indirect methods, i.e. to observe various effects produced by increasing stress and strain, of all possible kinds, both mechanical (tilts, slow motion, etc.) and electromagnetic. Actually, the present efforts are mainly based on 'trial and error', and in a first attack one tries to include as many different effects as possible. To conclude from observations of such effects something about the underlying physical causes, is a very difficult interpretational problem, and has resulted in many controversial statements, especially with regard to the use of the various observations for prediction purposes. This difficulty is further augmented by the fact that the observations are restricted to the earth's surface or very near to it, i.e. we have observations only from one boundary of the medium, from which we would need observations throughout.

Above we have specified the earthquakes for which prediction is wanted. In view of the facts just mentioned, we could complete that statement by defining the geographical scale on which prediction is possible or wanted, and let us distinguish between small scale, intermediate scale, and large scale. In a small-scale project we are only concerned with a single fault. In this case the situation is closest to the physically simple model, and this would correspond to BENIOFF's scheme. On the other hand, in a large-scale (or global-scale) project we are concerned with interrelationships between earthquakes, first in the same earthquake belt, later maybe over the entire globe. The projects which have been proposed by some nations belong rather to the intermediate scale and concern regional properties. Then, we have to bear in mind that features of single faults will be obscured as we generally have to deal with an ensemble of a large number of faults with different characteristics. The intermediate-scale projects are certainly very useful, but in my opinion they could be even more valuable if extended both to small-scale investigations (of a few well-defined faults) and to the large-scale project.

Also the time scale of any observable changes is of crucial importance for the success of any prediction project. Changes which would be observable only over geological epochs or centuries are equally useless as very rapid changes during a few seconds or so. Intermediate-scale changes, say of days to years, are most likely to lead to success. In general, the larger the earthquake magnitude is, the longer is the premonitory period.

10.3   *Existing Projects*

I shall summarize the essential points in the projects outlined by Japan, China, the USA and the USSR and give some comments. The more important items are as follows:

1. Investigations of crustal deformation (slow motion), partly by geodetic measurements (triangulations) repeated at intervals of only a few years, partly by continuous recording with tide gauges, tiltmeters and extensometers (Fig. 106). In several cases, especially in Japan, it has been observed that greater earthquakes are preceded by relatively rapid deformation of the ground (sinking, upheaval, tilts) several days in advance. Even though the connection between such deformations and the following earthquake is beyond doubt in the reported cases, it is questionable to what extent this method can be generally used for prediction. There may be large deformations without a following earthquake and there are earthquakes without observable deformations beforehand. From Italy there are observations from time to time of slow vertical ground movements (Italian "bradisismo"), which are taken as forerunners of coming earthquakes or volcanic eruptions. Thus, the north Italian earthquake on May 6, 1976 (Table 11), is said to have been preceded by such movements.

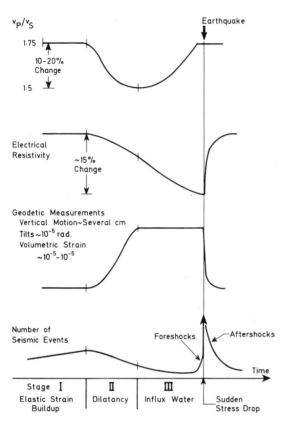

Fig. 106. Various premonitory phenomena ahead of the release of an earthquake. After MJACHKIN et al. (1975), modified.

At the well-known San Andreas fault in California it has recently been observed by means of accurate field measurements that stresses are released not only by sudden rupture (as earthquakes) but also by slow movements (which do not produce any recorded earthquakes, so-called *silent earthquakes*). Such motion, on the average amounting to 11 mm/year, does not occur continuously but only at certain times, but each time it can last for several hours and it is not sudden as in ordinary earthquakes. If such a slow stress release is of global importance, it would mean that energy calculations (based on seismic records) would only yield a fraction of the strain energy. Laser techniques offer the hitherto most accurate method to measure such motion over a longer distance (see Chapter 2).

Modern techniques permit accurate measurements of relative motion of points at large, even global, mutual distance, thus providing a link between earthquake prediction and plate tectonics (Section 6.3). There are essentially two techniques which are in use:

a) Laser ranging, using laser rays reflected from reflectors on artificial satellites or on the moon. Measurements of travel times to different points on the earth's surface yield information on their relative motion.

b) Very long baseline interferometry, using radio antenna stations to track quasars, i.e. very distant radio stars. Measuring the difference of arrival of signals to different stations yields information on their relative motion.

Other methods consist of direct measurements *in situ* of the absolute stresses in the solid earth. Such measurements carried out at many places on the earth by Professor NILS HAST (Sweden) exhibit a remarkable common property, namely very large horizontal compressional stresses, many times larger than the weight of the overburden. To map stresses and their time variations around active faults would certainly be very instructive. A disadvantage inherent in these as well as all other direct measurements is their limitation to the level nearest the earth's surface (for the stress measurements mentioned, to about 1 km depth), while earthquakes in general occur at considerably greater depth.

2. Intensified instrumental observations of earthquakes, especially of microearthquakes (magnitude less than 3; Fig. 106). Volcanic eruptions are often preceded by a series of small shocks, and the prediction of a volcanic eruption seems to be more reliable than the prediction of a large earthquake from series of small earthquakes. The difficulty is that so far we have no reliable criterion by which to tell if a small earthquake is a foreshock or a more or less isolated event. If such a criterion could be found, it would certainly be very helpful. The number of shocks in relation to magnitude could possibly give a clue, as such a relation seems to be dependent on the current stress conditions in the earth. In particular, such relations appear to be different for foreshocks (when the stress in the medium is high) and for aftershocks (when the stress is low). However, in order to establish such relations, a *series* of observable foreshocks is required which would permit a determination of the magnitude relation. Moreover, it is required that the conditions are known from earlier investigations for the region concerned. Another way of discriminating

between fore- and aftershocks may be offered by the observation, particularly by Russian seismologists, that the wave frequency, at least of *S*-waves, is noticeably higher for foreshocks than for aftershocks. This is presumably due to the higher stress in the rock during the foreshocks.

Within seismic areas it is possible to record hundreds of micro-earthquakes per day (as a rule 200/day in Japan, over 100 in Central Asia, etc.). There is therefore no need to wait very long until statistically useful data have been collected, to establish, for instance, a relation between number of shocks and their magnitude [see equation (1) in Chapter 5]. Then, the idea is to extrapolate such relations to considerably larger magnitudes and in this way to produce a statistics as would otherwise take many years to accumulate. However, caution is required in such extrapolation to large magnitudes. While this would appear to be permissible within certain areas, it is clearly unreliable in others. The reason for this is that the relation between number and magnitude of shocks may be different for different magnitude intervals. In such cases the kind of extrapolation mentioned leads as a rule to an overestimate of the number of large shocks.

It ought to be emphasized that thorough studies of series of aftershocks may give useful information about the rheological properties of the solid earth. Analysis of such time series may be made with the techniques developed in Chapter 6. There is no doubt that series may behave differently in different regions. Sometimes, such differences have been interpreted as indicating corresponding differences of the material properties or the stress variations in different regions. For a better approach to a solution of this problem we also need comparative studies of different aftershock series which have occurred in the same earthquake region. If predictions of the development of the aftershock activity were possible, these would also be of great practical significance. It has often been observed that buildings which have had their joints loosened by the main shock have subsequently collapsed completely as a consequence of a relatively insignificant aftershock. This has often given the impression that such an aftershock was considerably stronger than its magnitude indicates, sometimes even stronger than the main earthquake.

We usually do not discourage people from taking tourist trips or other journeys to seismic countries. We may only warn against visits to places where a disastrous earthquake has occurred, shortly after such an event. Such visits are only for rescue parties and for seismologists, engineers and others who investigate the effects. There is naturally always a certain risk in visiting seismic areas–even though this risk is smaller than in traffic–and the risk is in fact that much greater the longer such an area has remained quiet with no larger earthquakes.

3. Intensified research, mapping, etc., of active faults, using all available geological and tectonophysical methods.

4. Observations of radon emission (radon is an inert radioactive gas). For example, radon gas increased anomalously in well water before the earthquake swarm at Tashkent in 1966.

5. Observations of electromagnetic phenomena, especially variations in the geomagnetic field, in the electrical conductivity of rocks and in earth currents in seismically active areas (Fig. 106). These are based on the experience that stress changes in the earth are combined with electromagnetic effects. However, such variations due to stress changes are generally smaller than variations caused by other factors. In order to separate the seismo-magnetic variations, it is therefore most suitable to combine two or more stations and to investigate differences between them. In some exceptional cases motions have been observed in the ionosphere, which are considered as due to such disturbances in the earth's magnetic field. Russian investigations in Kamchatka, the Caucasus and Central Asia have revealed a relatively shallow origin (10–20 km depth) of some geomagnetic variations. Clear relations to some volcanic eruptions have been noticed and great hope is expressed that earthquakes also could be predicted by this means.

As in most other aspects of the prediction problem, there are controversial opinions about the usefulness of these observations. While some consider geomagnetic variations to be a reliable indication of a coming earthquake, others consider their value questionable.

6. Laboratory investigations of rock behaviour under extreme pressures and temperatures. It has been found that most rocks are relatively resistant to compression, but they soon break under a shear stress. One problem of great significance concerns the origin and propagation of cracks. Another important problem concerns the behaviour of various materials showing creep effects on the application or removal of a load (Fig. 69).

7. Determination of seismic wave velocity, especially of the *P*-wave, by field measurements or by regular recording of near and distant events. This idea is based on the fact that stress changes in the earth will give rise to velocity variations prior to an earthquake. Even better success has been experienced in investigations of the velocity ratio as well as the amplitude ratio of *P/S* and their variations with time for a given area.

This discovery is ascribed to Russian seismological experience, especially from investigations in the Garm area. Since around 1971, these ideas have been adopted and further developed by American seismologists, with field data primarily from California. It is a common feature of the Russian and American measurements that the local *P*-wave velocity (v_p) and the local ratio of *P*- and *S*-wave velocities (v_p/v_s) exhibit a characteristic variation for some time prior to the release of an earthquake (Fig. 106). The time scale varies from case to case and can be anything between days and years. Similar velocity variations are reported also from Japan and China. Somewhat different explanations have been produced to lighten this behaviour. The American explanation, usually called the *dilatancy-diffusion* model, assumes one stage (dilatancy) in which cracks open up, with seismic velocities decreasing, followed by another stage (diffusion) in which cracks are filled with water whence the velocities increase again. At the end of the diffusion stage, when velocities have returned to their original values, the main earthquake occurs. The Russian explanation is similar, including crack formation in intact rock. However, it does not assume any diffusion stage, but works exclusively with experience from fracture mechanics, explaining the velocity increase by crack closure. It is still unsolved which one of the two explanations is the best approximation, but they certainly both point to essential developments in the rock due to stress accumulation prior to its release.

There is no doubt about the observed velocity variation, but the suggested explanations may as always stimulate to some further questions. Considering the dilatancy-diffusion model, we may make the following comments:

a) The water is probably in a liquid state due to pressure increase with depth. The critical point of water corresponds to a pressure of 218 atmospheres (equivalent to the weight of the overburden at a depth of around 0.7 km) and a temperature of 374 °C (approximately corresponding to a depth of 11 km in the earth). In the depth range down to 11 km, the water should then be in its liquid state.

b) Considerably deeper penetration of water appears difficult to visualize. Particularly for deep shocks, it seems probable that the mechanism is different. So far, the method has apparently been tested only in areas of relatively shallow shocks.

c) The clear separation into two distinct stages — dilatancy and diffusion (Fig. 106) — is an empirical fact. But in the explanation it appears a little hard to realize such a separation, as one would rather be inclined to expect a more continuous influx of water simultaneously with the dilatancy process. In other words, if the dilatancy-diffusion model is accepted, there must be some additional mechanism whereby the clear two-stage separation comes about.

d) With the clear two-stage dilatancy-diffusion development, one would be tempted to expect a corresponding two-stage development of the electrical resistivity. But as Figure 106 shows, there is no such two-stage development, a fact which also raises certain questions concerning the exact mechanism.

Nevertheless, among the many different experiments undertaken to find reliable methods for earthquake prediction, it is now clear that the method of velocity variations is the best and most reliable one. Therefore, in many earthquake prediction projects, the main emphasis is nowadays laid on further investigations of such velocity variations.

8. In addition to the methods outlined above, which are generally of an instrumental nature, we also note non-instrumental observations of various earthquake forerunners. These consist for example of the behaviour of animals, which become uneasy, or of changes of water level and turbidity in wells, etc. Such observations, based on a long experience from history, play a significant role also today, especially in China, where a large number of amateur seismologists is engaged in such observations and their reporting to central institutions. From China there are several reports of successful predictions, based on a wide range of both instrumental and non-instrumental observations. A noteworthy case is the Liaoning earthquake on February 4, 1975 (Table 11), when a great number of people could be saved. This is considered as the first major earthquake that has been predicted successfully (ADAMS, 1976).

Finally, concerning all methods for earthquake prediction, it is very important that observations are made in a variety of seismic areas and that results are compared. This is important for areas with very similar seismic structure, as for example the San Andreas fault in California and the North Anatolian fault in Turkey. But it is also important to compare regions with quite different seismic regime.

In spite of all remaining problems, there is no doubt that the prospects for successful earthquake predictions are brighter today than ever before. The optimists in the USA, working in this field, estimate that within about 10 years from now (1977) it will be possible to issue public warnings for earthquakes on a rather regular basis, at least for the best investigated areas, like California. In recent (1977) statements from Japan, hopes are expressed that before long it will be possible to issue warnings for earthquakes with a magnitude $M > 7$ with an accuracy of about 10 km in location and 2 to 3 years in time. The low timing accuracy is, of course, still discouraging, and could hardly be of real value for evacuation purposes, but could be enough for taking other measures of precaution.

10.4   Oscillations and Migrations

In this section I shall outline some evidence for apparent relationships between earthquakes on a large scale, a field of study which has not received as much attention as it would deserve in the current projects. When an earthquake occurs at some point of the earth, the stress conditions are changed at this point. But as the whole earth constitutes one great stress system, a change at one point would be expected to have repercussions at all other points–even if small. This would be still more expected within one and the same earthquake belt, which more obviously constitutes one stress system. However, modern plate tectonics is able to offer a more plausible explanation of this behaviour.

It has frequently been observed that seismic activity apparently wanders from place to place along an earthquake belt. These wanderings are of two kinds: oscillation and migration. See Table 23. Oscillation patterns are found in aftershock areas and imply a more or less periodic shift of the highest seismic activity between the two extremes of the area. It was exceptionally well developed in the Aleutian aftershock sequence in 1957,

Table 23. Some examples of oscillation and migration.

Direction	Oscillation	Migration
Horizontal	Aleutian Islands (DUDA, 1962)[1]	Chile (DUDA, 1963)
	California (DUDA and BÅTH, 1963)	California—Chile (TAMRAZYAN, 1962)
	Chile (DUDA, 1963)	Carpathians—Kamchatka (TAMRAZYAN, 1962)
		Atlantic Ocean—Burma (BÅTH, 1966)
		Italy—Irian (first presented here)
Vertical	Kamchatka (TARAKANOV, 1961)	Southwest Pacific Ocean (BLOT, 1963)

[1] The information in the brackets refers to the discovery.

the oscillation period increasing with time from a few hours just after the main earthquake and approaching about 300 days three years later. Oscillations have been found not only in horizontal directions along a belt, but also in the vertical direction as, for instance, along the sloping planes shown in Figures 54 and 55. Migration patterns, on the other hand, are unidirectional and only found in ordinary series of earthquakes, excluding aftershock sequences. A repeated northward migration in Chile between 35 °S and 10 °S was found for the years 1957–60, the speed of migration increasing from 7.6 km/day in the first cycle to 41.4 km/day in the fourth. These cycles all began at the southern end of the belt and moved northwards. This scheme was interrupted by the large Chilean earthquakes in May 1960. Similarly, a southward migration from California to Chile has been found for 1951–8.

The east-west belt extending from the middle of the Atlantic over the Mediterranean area and southern Asia has a tendency to show an eastward migration of seismic activity. Especially clear sequences occurred in 1962 and 1976 (see Fig. 107). It should be noted that all these shocks are significant in the sense that their magnitudes are well above the average, except for the earthquake in Spain (August 14, 1962) which had a much smaller magnitude. Note that in this sequence the magnitude increases gradually to a maximum of 7.1 (Iran) and then decreases again. The Mediterranean

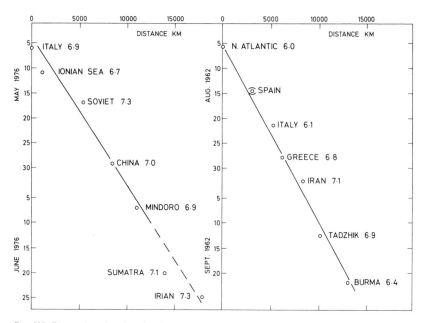

Fig. 107. Eastward earthquake migrations along the Mediterranean to southern Asia belt in 1976 and in 1962. Distances are measured along the earthquake belts starting from the epicentre of the initial earthquake. After each geographical name, the magnitude *M* is given.

and the mid-Atlantic belts join each other in the middle of the Atlantic. Such junctions (and also bends) are places of high stress concentration and could be assumed to be sources of migration patterns. As a matter of fact, an oscillation pattern can be looked upon from the same viewpoint. In this case–in an aftershock region–there are in general two 'poles' of increased stress accumulation, at either end of the belt. It is from these poles that the activity starts, and it becomes oscillatory with two poles, but migratory with only one pole. A migration very similar to the one from 1962 was observed in 1976 (Fig. 107) extending from Italy to Irian in western New Guinea.

As seen in Figure 107, the migration velocity is nearly constant and is equal to 12 to 14 km/h. With this velocity known, we could predict the time of occurrence of the shocks with an average error of ± 2 days, starting from the time and location of the initial shock and assuming the locations of all subsequent shocks known. However, the constancy of this velocity from case to case cannot be relied upon, and, what is still worse, we have no knowledge of where along the belt the releases will occur. Such global systems have to be supplemented by detailed investigations at the various earthquake sites.

The shocks in Agadir, Morocco (February 29, 1960) and Lar, Iran (April 24, 1960) may be looked upon in a similar way, but with more doubt: partly because of the small magnitudes of the shocks (even though devastating), partly because of the lack of shocks in between. The migration velocity would then be about 5 km/h.

On the basis of such indications there is reason to suspect similar relationships between earthquakes all along the circum-Pacific belt. Combined studies of time and space distribution of seismicity will certainly prove useful. As an addition to the current projects, it is therefore of importance to coordinate observations all along that belt as densely as possible, at least for a limited time. A systematic and comparative study of deformation and energy release in aftershocks (in relation to main shocks and foreshocks) is also of great significance.

Another related problem concerns the 'sympathetic' behaviour of earthquakes over the globe, implying that relatively quiet intervals alternate with intervals of higher seismic activity. It has been found that earthquakes, even in widely different parts of the earth, frequently occur in groups, separated by quieter periods. Statistical tests have suggested that the phenomenon is not real, but still it cannot be excluded that there is some real phenomenon behind this. On the other hand, in the oscillation and migration patterns there is implicitly contained an opposite tendency in neighbouring areas: when activity in one of them increases, it decreases in the next one. Also, when aftershocks following large earthquakes take place, there is apparently a decrease in the seismic activity over the rest of the earth. It would seemingly be easy to dismiss this as just due to a concentration of the attention on the aftershocks (disregarding other shocks) or that the records are so overcrowded by aftershocks that no others are discovered. Neither is a true statement (the latter only for 2–3 hours after the main shock) and therefore this may be attributed to some real phenomenon–but what? It should be added that there are notable exceptions to this as to most other observations in the prediction field.

Even though observations like these have been made for many years, it was not until the advent of the new global tectonics or plate tectonics theories (Section 6.3) that plausible explanations could be produced. That one earthquake can release another earthquake at a greater distance, even within the same belt, is an idea that has to be dismissed. But larger plates which move in relation to each other, provide an acceptable explanation. For example, the eastward migration in the Mediterranean and southern Asia could be explained by slow motion of the African–Indian plates towards the Euroasiatic plate, combined with a small rotation. Speeds of only 5 cm/year of the plate motion would be enough to explain the propagation of the contact point eastwards by 12 to 14 km/h. Likewise, the world-wide sympathetic behaviour of earthquakes can get a reasonable explanation from plate motions. A related problem is the alternation of quiet and active periods, which would require irregular motion of the plates. This cannot be excluded, but to explain the reason, we are rather referred to hypotheses. One is that the driving force, probably due to convection currents in the earth's mantle, would vary in intensity. Another possibility is that external forces vary, e.g. gravity action from the moon, sun and planets, according to their relative positions. Anyway, these ideas suggest that a fruitful combination could probably be made between the new global tectonics and large-scale (global-scale) earthquake prediction.

If the reader has got the impression that prediction is a field about which the specialist's knowledge is very limited–then this is a correct impression. But nevertheless it is gratifying to know that there is active interest in the problems and that current projects, preferably supplemented as outlined here, will make the prospects much brighter. In a decade or two, it will probably be possible to issue predictions regularly and generally.

Our discussion has so far been limited to purely natural processes, both for strain accumulation and release. With external influences, especially those of an artificial nature, the prospects of prediction are much more hopeful as we shall see in the following sections.

10.5 Trigger Effects

Another controversial point concerns the importance of triggering forces. There are numerous forces which possibly could act as triggers. Meteorological factors (especially variations in air pressure) have been correlated with seismic activity, and in some particular areas apparently good correlations have been obtained. The forces produced by body tides are about 10 times as great and have therefore attracted more attention. Whereas several scientists deny any correlation, usually on statistical grounds, others believe in some tidal triggering of some large earthquakes. Figure 108

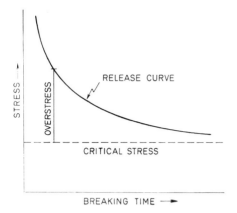

Fig. 108. Curve illustrating how the time for a break to occur (breaking time) increases with diminishing overstress.

illustrates a difficulty by virtue of which many efforts to demonstrate correlation between possible triggers and earthquakes fail: if the excess over the critical stress (breaking stress) is small, it will take a very long time before a release occurs; in other words, the greater the overstress, the sooner will the release take place. This is true whatever the trigger is. Quite generally, it is always difficult to prove the existence of a triggering force: there must be very careful and balanced cooperation between the trigger and the stress system in the earth. If the stress is far from release, no additional small force will have any effect whatsoever. And if the stress is near release, this will come anyway, irrespective of possible triggers. This circumstance is one example of the great difficulty which is often encountered in geophysical research: standard statistical tests will fail to give any positive evidence even if there might be an effect sometimes. Also several simultaneous effects generally tend to mask each other. In such cases it is a more reliable procedure to approach the problems from the physical side and not only from the statistical side.

A unique opportunity of tidal trigger action has been expected for 1982 when several of the planets line up and reinforce their gravitational action on the earth. Besides regular tidal triggering, there does not seem to be any reason for unusually many or large earthquakes, as testified by a number of scientists. The tidal action of planets is anyway quite small; also on an earlier similar planetary constellation, no exceptional earthquake activity took place.

Besides the triggers mentioned, which are of natural origin, there are artificial triggers, created by human activities, which have become of great significance in recent years. Since around 1940 it has been observed at a number of places around the world that the local seismic activity has increased some time after the construction of a dam and the impounding of a reservoir. This phenomenon has usually been ascribed to the increased load on the earth's surface due to the dam and especially the artificial lake. Later observations indicate, however, that other effects also may be involved. The observation that large underground explosions release strain and are thus followed by a local sequence of 'aftershocks' is much more recent. The reason is that it is only in the last few years, since 1965 and with greater intensity from 1968 onwards, that large underground explosions have been made (Chapter 11). Mining operations are still another trigger, releasing small local shocks, called rockbursts (see further Section 10.8).

10.6 Dams and Earthquakes

Let us consider the earthquake mechanism by applying what school physics tells us about friction. Figure 109a shows the well-known picture, where the block on the rough surface begins to move when the force F is equal to the product fN of the frictional coefficient f and the normal stress (pressure) N (M = the mass of the block and g = the acceleration due to gravity). The common contact surface corresponds to the fault surface in an earthquake, and the force F corresponds to the stress in the solid earth, due to the action of tectonic forces between blocks and structures. We assume the force F to increase linearly with time and we consider only purely elastic effects, i.e. we disregard elastic aftereffects and other more complicated cases. Then, we can illustrate an earthquake sequence as in Figure 109b. Every time $F(t)$ attains the value fN or slightly above, a sudden 'pull' occurs (i.e. an earthquake occurs). The stress is then released practically instantly, i.e. $F(t)$ falls almost immediately to a value considerably below

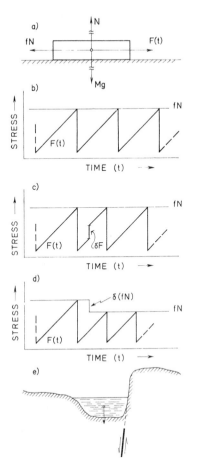

Fig. 109. Schematic picture of an earthquake process and its changes due to external influences of various kinds.

fN (in Fig. 109, fN is identified with the release level). After this, the phenomenon is repeated in a similar manner and this continues over geological epochs.

Obviously, this process and especially the condition $F(t) \geqq fN$ can be influenced, if any of the factors entering this relation can be changed due to some external effects. We consider two cases.

1. In Figure 109c we assume that at a certain moment F receives a constant increment δF, while f and N are kept constant. As we easily see from the figure, the effect is that the value fN is reached earlier. This means

that the next earthquake will be released earlier than in the undisturbed case. On the other hand, its magnitude (corresponding to the frictional force fN) will be unaffected as well as the frequency of shocks thereafter.

2. If instead we change the frictional force fN, keeping $F(t)$ unaffected, we get the situation shown in Figure 109d. If fN is decreased by $\delta(fN)$, then the next shock will again occur earlier, but its magnitude will be lower as fN is lower. Figure 109d also shows that the earthquake frequency will be increased. The total energy released during a longer time span will nevertheless be unchanged as long as the generating factor F is unaffected.

By combination of Figures 109c and 109d it is easy to deduce the development also under other conditions. This model is naturally a strong simplification of reality, but still it can serve well as a first approximation. In nature, $F(t)$ does often not vary linearly with time, but shows a much more complicated time variation with both accelerations and retardations depending on various influences. In nature, we have also to assume that, for some interval before the release, $F(t) > fN$ by a finite amount, corresponding to the unstable area above the fN-level in Figure 109, and that the release level is somewhat above fN (cf. Fig. 108). A more exact treatment must also take into account that f is lower when in motion than when at rest. Finally, in nature we have not to deal with just one fracture but generally with a system of fractures, which can be more or less parallel. The different fractures behave differently, especially with regard to their values of f and N, and they can influence each other. All this leads to a complicated time function.

Figure 109e gives a somewhat more realistic picture with an almost vertical fracture below a dam. In this case the load of the dam means an increase of the force F (as in Fig. 109c). The slope of the fault surface is of great significance, and we easily see that if this surface were horizontal the effect of the water load would only be an increase of the normal pressure N. Then, the water load would impede the release. For a sloping fault surface the water load affects both F and N, and these effects can be of opposite sign. If instead we had unloaded the surface, we could have diminished F (i.e. a negative δF in Fig. 109c), which would also have had an impeding effect on the seismic activity.

As is clear from Figure 109d, the process can be influenced not only by load changes but also by changing f or N. The penetration of water into existing cracks in the earth's crust can lead to a decrease both of the friction between the rock walls (lubrication) and of the normal pressure by the separation effect of a water system which is continuous from the surface downwards.

However, in every case we have to remember that the effects dealt with here are small in comparison with the underlying tectonic phenomena and that the additional effects are to be considered as triggers. Moreover, contrary to the tectonic forces, the effects of a dam reservoir do not continue. Once their effect (due to load or injection of water) has released stresses, there is no more stress generation or release due to the dam. In other words, the only forces then active are the natural ones, i.e. those which existed already before the dam was built. Therefore, in agreement with observations, the increased seismic activity due to a dam construction is only of a temporary nature.

The first observations of increased seismic activity caused by a reservoir were made by MEAD and CARDER (1941) at Boulder dam (now called Hoover dam) in the Colorado river in the USA. No earthquakes had been recorded from this area for 15 years, but after the construction of the dam in 1935 activity started: the first shock in September 1936, followed by about 100 shocks in 1937, and culminating in May 1939, with a maximum magnitude M of 5.0, after which conditions returned to normal, i.e. as they were before the dam existed. This seems to be an observation of general validity, i.e. that no lasting change occurs. This is also in accord with Figure 109c.

After this, CALOI (1966, with additional references) is among those who have worked most thoroughly on this problem since around 1950, particularly in relation to the dams Pieve di Cadore, Tolmezzo, Vajont, and others in Italy. CALOI has emphasized the importance of more extensive observations, in addition to the seismic ones, near dam sites. Thus, he refers to continuous observations of changes in the elastic properties of the bedrock due to a dam; furthermore vibrograph recordings to establish free vibrational periods of various earth layers (to avoid resonance effects in the dam) and inclinometer observations, by which slow changes of slope in the vicinity of a dam can be discovered at an early stage.

A number of Greek seismologists (COMNINAKIS et al., 1968) studied the conditions at the Cremasta dam in Greece. The impounding of water began on July 21, 1965. A few shocks were noted near the lake already in August 1965, and in December the shock frequency started an exponential increase up to February 5, 1966, when a bigger, damaging shock (magnitude $M=6.2$) occurred. This was followed by an aftershock sequence during the rest of 1966. It is their opinion that the foreshocks and the main earthquake were released by the water load in the Cremasta lake. But it is also emphasized that it was a coincidence that the impounding was made at a time when the tectonic energy was near release anyway.

On December 10, 1967, a destructive earthquake of magnitude $M=6.5$ occurred at Koyna, near Bombay in India. This is a relatively strong earthquake to be located so far south of the seismic belt through southern Asia, and a connection with the Koyna dam was suspected. Following the impounding of the reservoir in 1963, an increased frequency of small shocks occurred in the area, culminating in the large shock of December 1967. And this occurred in an area considered to be very stable, apart from, to some extent, the Indian west coast. The December shock of 1967 is rated as the largest which has occurred in historical time in the peninsular shield of India. This event attracted enormous interest and was the impulse to many geophysical and geological investigations. These investigations seem to suggest the existence of a relatively deep, north-south fault through the area. This dormant fault was probably reactivated by the reservoir load. However, several Indian seismologists are of the opinion that the dam did not release the earthquake, but that this was a purely natural phenomenon. Some scientists suggest that the large number of small shocks since 1963 were directly caused by the reservoir load, while these in turn released (by trigger action) the larger shock in 1967.

On seismicity maps of later years the area around the Kariba dam on the border between Zambia and Rhodesia exhibits a strong concentration of mainly smaller shocks. This activity is considered to be due to the artificial Kariba lake. The water began to be impounded in 1958. The strongest seismic activity came in September 1963, with among others a magnitude $M=6.1$ earthquake on September 23. After this, the activity has again decreased. A Bulawayo report of 1969 gives the following excellent summary of the development: 'In general terms, and up to the present time, the pattern at Lake Kariba appears to have followed the pattern of behaviour at other large dams at which seismic activity has occurred. The pattern is as follows: (1) The region contained pre-existing weaknesses in otherwise competent rock. (2) Prior to the building of the dam the region was apparently seismically inert. (3) Weak but persistent seismic activity commenced very soon after water started being impounded. (4) Some time later there occurred an outburst of intense activity which accounted for almost all the seismic energy released. (5) After the outburst the activity continued at a gradually decreasing level for a considerable time.'

To this scheme we may comment that item (3) is explained by the schematic picture in Figure 109, and these small shocks are probably a direct result of the dam reservoir. Item (4), however, corresponds to a larger shock, often of magnitude $M = 6-6^1/_2$ and occurring 3–4 years after the impounding of water. This is not directly contained in our Figure 109, and its relation to the dam and to the 'foreshocks' (item 3) is still an open problem. It is conceivable, though not at all proved, that the foreshock activity is due to the fact that dam reservoirs are filled gradually, over an interval of several years, and that the main shock only occurs when the water level has reached a certain critical value (about 100 metres).

The seismic activity developed after dam constructions exhibits a pattern quite similar to what nature itself produces: a minor foreshock sequence, one or a few major earthquakes, a long sequence of aftershocks. The question why, for instance, foreshocks occur is not unique to the dam problem, but concerns nature itself. A major earthquake requires the practically simultaneous release of strain over a larger volume of rock. Foreshocks are much smaller and are localized to zones of weaknesses and naturally the first to be released after the application of a load (Fig. 109c). Their effect is to create a more uniform stress field over a larger volume, a necessary prerequisite for a major earthquake to occur.

In the cases mentioned the increased load is usually believed to be the reason for the increased seismic activity. Quite generally, it appears that the water depth is more significant than the water volume, and that water depths of 100 metres and more are needed to cause appreciable seismic activity. On the other hand, observations of variations of fN (Fig. 109d), and thus released shocks, are of later date and are so far almost exclusively limited to Denver in Colorado. In 1962, injections of waste water into a well, 3670 metres deep, started. This led to a marked increase in shock activity in an area, where no shocks had been noted since 1882. For instance, from April 1962 to November 1965, more than 700 earthquakes were recorded in the area, the greatest magnitude M being 4.3. As a consequence it was decided at the beginning of 1966 to give up the water injection. The area appears to be underlain by steeply dipping faults. Later, the same results have been obtained through special water-injection experiments at Rangely in Colorado. Similar effects are probably of significance also in dam reservoirs, caused by water penetrating through cracks and fractures in the crust.

The cases reported exclude any 'doubt about the seismic effects of dam reservoirs. Cases with increased activity have naturally attracted most attention, whereas the opposite case (decreased activity) can also be expected due to a dam. One difficulty inherent in most observations of this kind is that seismic recording at dam sites has usually been started only at or some time after the construction of the dam. In order to make reliable and quantitative conclusions on the role of the dam as a seismic stimulator, we would need a long series of observations at the dam site before the dam existed. Teleseismic observation is too insensitive to small local shocks to act as a useful substitute. Special sensitive seismograph stations in the near vicinity of the dam site are necessary. This is certainly something to bear in mind in future dam construction. In the cases reported above the information from the time prior to the dam is for the most part incomplete, but in all cases the general opinion is that the increased seismic activity occurred well correlated to the dam construction.

In deciding upon the causative relations between reservoir loading and shock activity, the correlations between time series of these two parameters are most informative. This is especially true when the reservoir load has been changed several times and a significant correlation with the local seismic activity is obtained. Such diagrams, usually convincing, have been produced in most cases reported above.

Dams are usually built in narrow passages in order to create a lake from a river, flowing through the passage. But such gorges are only found in complicated topography, which – at least in seismic areas – may have originated through some earlier seismic activity. This gives us special reason to be careful in selecting a site for a new dam, and every dam construction, especially in seismic areas, must be preceded by careful geophysical and geological investigations. It is probably rather exceptional that the reservoir load gives rise to new cracks in the crust and, more usual, that already existing cracks are activated. This explains why seismic activity can reappear where shocks have occurred only far back in time. If in Figure 109d we assume the tectonic forces F to have been weakened so much that the original level fN is no longer reached (i.e. no shocks occur), then a lowering of this level can lead to a new series of shocks.

From time to time it has been said that the water load is the shock-generating factor and that tectonic stresses are only of secondary importance. This suggestion appears very unlikely for several reasons. A water column exerts a pressure which is only about one-third of that of the surrounding rock medium. Furthermore, even in stable areas horizontal pressures have been measured, which are about 10 times the weight of the overburden (HAST, 1958, 1967; cf. Chapter 2) and thus about 30 times the water pressure. Finally, in tectonically active areas stresses are probably orders of magnitude higher still than in stable areas. The water pressure is no doubt significant, but its role is only one of releasing or triggering.

In this problem we are certainly facing a version of the law of action and reaction: the dam influences seismic activity; by the same token, this activity influences the dam. The last factor is of course not the least important, and in potential seismic areas, the effects of seismic forces on the dam must be included in the calculations of the dam construction.

We can state that an accurate knowledge of fluid pressure and its variations on and inside the earth is of great significance in connection with the problem of earthquake prediction. But we should remember that the most essential factor is, after all, conditions in the earth's crust, i.e. whether or not stresses exist which can be released as shocks. Considering this, there is generally no reason to fear any seismic effects of dams in stable areas with no faults, neither active nor dormant.

10.7 *Explosions and Earthquakes*

Nuclear explosions will be dealt with in the next chapter. Table 30 presents the largest detonated underground explosions.

The Amchitka tests of October 2, 1969, and November 6, 1971, which were announced beforehand, caused some strong protests from Japan and Canada, referring to the possibility of earthquake release by such large underground explosions. Considering this problem, it is appropriate to distinguish between (1) release of smaller 'aftershocks' and (2) release of major earthquakes. As far as (1) is concerned we have already gained quite some experience from the large underground Nevada tests, especially since the beginning of 1968. This area has a natural seismicity, and these tests have led to the release of a large number of aftershocks, in many respects similar to the aftershock sequences following greater natural earthquakes. Local observations and recordings in Nevada have shown that several thousand shocks have been released in the largest tests. However, special investigations show that all these aftershocks are very small and that the largest has a magnitude m about one unit lower than that of the explosion itself (Table 24).

The Nevada cases give closely agreeing values of the magnitude difference Δm (about 1.2). However, caution is required because we are comparing magnitudes for explosions and earthquakes. Due to various effects

Table 24. Examples of underground nuclear explosions which have released aftershocks.

Explosion site and date	Observation period	Number of aftershocks	Magnitude difference Δm	Remark	Data source
Nevada:					
1966 Dec 20	Dec 20—23	12			U.S. Coast and Geodetic Survey
1968 Apr 26	Apr 26—				U.S. Coast and
	May 4	13	1.2		Geodetic Survey
Dec 19	Dec 19—31	33	1.3	30 aftershocks Dec 19—23; in all, about 10 000	U.S. Coast and Geodetic Survey
1969 Sep 16	Sep 16—20	6	1.2	115 aftershocks within the first 3 hours, with $\Delta m = 1.3$	U.S. Coast and Geodetic Survey
Novaya Zemlya:					
1973 Oct 27	Oct 27	5	2.6		Uppsala Seismological Bulletin
1974 Nov 2	Nov 2	2	(2.8)		Uppsala Seismological Bulletin

it is more correct to reduce the value of Δm by about 0.4 (Chapter 11), i.e. $\Delta m = 0.8$ gives a better estimate. Recalculated into surface-wave magnitudes, this gives $\Delta M = 1.4$, which is in good agreement with natural earthquake sequences. Expressed in energy, this means that even the largest aftershock of an explosion has an energy amounting only to about 1% of the wave energy released in the explosion itself. Such small shocks are almost always harmless.

An aftershock sequence occurred also after the Amchitka test in October 1969, as the Aleutians are also seismic. The sequence was smaller than anticipated on the basis of Nevada experience. None of these shocks, nor the explosion itself, led to tsunamis of any significance, as was feared in some of the protests. With this experience, it came as a rather great surprise that underground nuclear explosions in a new test area in southern Novaya Zemlya also produced aftershocks, especially as this area is generally considered aseismic. It is probably symptomatic that the explosions were unusually strong, also that the magnitude difference was more than twice as large as found for Nevada (Table 24). Lacking further information on these aftershocks, the following information in this section is based upon Nevada experience exclusively.

As a general rule it can be stated that significant increase in seismic activity occurs after underground explosions only with a body-wave magnitude m exceeding $5\frac{1}{2}$. This in fact is a threshold value for explosions corresponding to the threshold of 100 metre water depth in case of dam reservoirs, mentioned above. In addition, the activity is strictly local, usually not extending further than 20 km from the explosion site and only exceptionally twice this distance. A detailed account of effects of underground nuclear explosions in Nevada, especially strain release and various geological effects, is given in the Bulletin of the Seismological Society of America for December 1969.

Concerning point (2), i.e. release of a major earthquake by an explosion, some fear may be justified. However, an earthquake requires a source of energy, and this consists of the strain energy corresponding to the stresses in the solid earth. The earthquake occurs when the stresses have grown so much that the medium breaks (usually at a fault surface). The effect of an explosion in such a medium could be to release already existing strain energy, but not to create any strain energy. We also understand that release of an earthquake by an explosion could only happen if the earthquake was relatively near release anyway, i.e. the explosion would act as a trigger. This would require quite a delicate balance between the occurrence of the explosion and the state of stress in the earth. The condition is so special that, at least so far, no such thing has happened neither in Nevada nor in any other test area.

Moreover, we can state that the effect of the explosion in case (2) would only be to accelerate a development, which would inevitably take place anyway.

Finally, we have to observe that a major earthquake involves the practically simultaneous release of strain over a large volume–larger, the larger the magnitude is. Table 25 shows that a shock of $M=8.5$ would

Table 25. Earthquake volumes (assumed hemispherical of radius r) corresponding to given magnitudes [calculated from equation (1) in Chapter 6 and equation (3) in Chapter 4].

M	m	r km
8.5	7.7	180
8.0	7.4	100
7.0	6.8	33
6.6	6.6	20

require the almost simultaneous release over a volume with a radius of no less than 180 km, assumed hemispherical with the explosion at the centre. This must be considered as a very unlikely event, especially as experience shows that the aftershock area seldom extends beyond 20 km from the shotpoint. Moreover, in the aftershock sequence there is no simultaneous release, but instead this is spread out over an interval of days or weeks.

Obviously, the usual question whether large underground explosions can release earthquakes must be answered in the affirmative. However, as a rule there will be no catastrophic shocks. Such events could be released in a seismic area only due to a rare coincidence of various conditions. To be perfectly safe, it would be preferable to make large explosions in non-seismic areas and to avoid the seismic ones. There is no support for the belief, sometimes expressed, that nuclear explosions would influence the seismic activity at remote locations or world-wide.

The mechanism of shock release due to explosions is essentially different from release due to dams. In the latter case it is a static pressure which is the crucial factor, whereas explosions are transient phenomena of short duration. The exact mechanism[1] for release by explosions is less clear. For one thing there can be permanent rock displacements (which in their turn may have a releasing influence on the stress field, as for example by the opening of cracks), for another there may be the transient pressure pulse, which at places would cause excessive pressures with ensuing release. As an example, the Nevada explosion of April 26, 1968, caused a fracture in hard rock with a two-foot vertical displacement and a one-foot horizontal displacement almost three miles long.

Figure 110 shows schematically that the time pattern of deformation

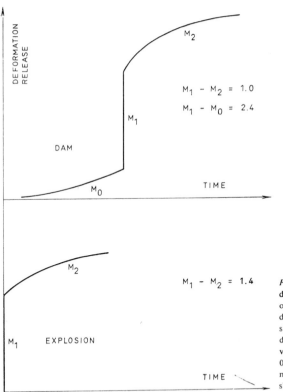

$M_1 - M_2 = 1.0$

$M_1 - M_0 = 2.4$

$M_1 - M_2 = 1.4$

Fig. 110. Schematic picture to demonstrate the different patterns of deformation release caused by dams and by underground explosions. The surface-wave magnitudes (M) refer to the maximum values in each sequence (subscript 0 for foreshocks, subscript 1 for main events, subscript 2 for aftershocks).

release due to dam reservoirs is essentially different from that of explosions. For dams the foreshock activity is relatively small compared to the major activity, which sets in with the main shock (M_1). The magnitude differences in Figure 110, all expressed in surface-wave magnitudes, are only approximate. For dams the values given are averages of the closely agreeing results for the Kariba and Cremasta dams.

The different shapes of the released activity (Fig. 110) result from the different time functions of the applied stress, transient in explosions versus steady, static pressure in the case of dams. Simultaneous activation of large volumes of rock is possible for dams, given due time for development of a uniform stress field, but generally not in the case of explosions.

It is also of interest to compare the 'aftershock' sequences of large explosions with those following natural earthquakes. At least superficially, the two kinds of sequences exhibit similar features; for instance, the largest aftershock is about 1.2–1.4 magnitude units lower than the main shock/explosion. Also the activated volumes are of similar size (Table 25). But the mechanisms are probably different. According to generally accepted views, proposed mainly by BENIOFF, the natural strain in the earth is only partly released in the main earthquake; the rest is released in aftershocks, corresponding to elastic creep recovery. For aftershocks of explosions the mechanism must be different from earthquake aftershocks and is most nearly tantamount to readjustments of the rock. It is not unlikely that further studies of explosion aftershock sequences will shed light upon studies of natural aftershock release. A few such comparisons have been made, which demonstrate different frequency-magnitude relations and different focal depths, the explosion aftershocks being shallower, at less than about 5.5 km depth.

Now and then it has been proposed that explosions could be used in seismic areas to release strains before these had grown to a size corresponding to a major earthquake. The idea is that this would be a means to avoid major earthquake disasters. First of all, we can state that such experiments would have some chance of succeeding, only if we had an accurate and detailed knowledge of the stress conditions in the earth, their natural variation with time, the material strength, etc. Without this knowledge, as yet almost totally lacking, tests at random would probably fail. For energy reasons release in terms of a large number of small shocks is too inefficient. A simple calculation shows that an earthquake of magnitude $M = 8.5$ corresponds to the total wave energy of about 100 000 shocks of magnitude $M = 5$. Thus, if we want to eliminate the large 8.5-earthquake by releasing magnitude 5 shocks, we must release no less than 100 000 of these. If by a large underground explosion we succeed in releasing 20 000 shocks of magnitude 5 (which far exceeds achievements thus far), we would still be left with 80 000 shocks to be released. If they are not released, we have still to expect a major earthquake, certainly smaller than magnitude 8.5, but not lower than 8.4, which without doubt is a practically insignificant lessening of the catastrophe.

If we instead imagine the release of a major earthquake all at once, i.e. the explosion used as a trigger, then the danger has not been eliminated. The only difference is that it happens at an earlier time than in the natural development, also at an approximately prescribed time. The latter would permit some measures of precaution beforehand, but still a major earthquake is a very uncontrollable event.

In an effort to use explosions to avoid earthquake catastrophes one has probably to aim at something intermediate between the extremes described here. Such tests are very difficult for many reasons. They are intimately connected with the problem of earthquake prediction. The same is true for applications of the water-injection experiences from Denver, mentioned above, to controlled release of earthquakes.

10.8. Mining Operations and Rockbursts

Quite generally, any change in the natural stress conditions in the earth's interior could lead to release. Examples are excavations within the crust, as in mines, which may cause rockbursts. Another example is the extraction of oil, especially in such seismically active areas as Iran with prestressed crustal layers. Whereas relatively little is known so far about the latter effect, rockbursts in mining areas have been known for many years. The existing literature on rockbursts from several different countries is concentrated on various trigger effects and on methods for statistical prediction of rockbursts.

Rockbursts are a daily experience in many mining areas, especially near coal mines in central Europe or gold mines in South Africa, and they are also reported from North America. As distinct from this, the Swedish iron ore mines have apparently experienced notable rockbursts only in the last few years. This is the case with Kiruna and Malmberget in north Sweden and with Dannemora, Grängesberg and other mines in central Sweden.

It is known from direct measurements that the horizontal stresses in rock generally increase rapidly with depth already in the uppermost kilometer of the crust (Section 10.6). The horizontal stresses reach amounts many times the weight of the overburden. Excavations in mining operations lead to disturbances in the natural stress field in surrounding rock masses, and release may take place at certain points, where the stresses exceed the strength of the rock material. This is demonstrated schematically in Figure 111. In modern mining operations, excavations are done at a generally much higher speed than earlier, with the consequence that the time variation of the stress field will be more rapid. The natural ability of the crust to adjust itself slowly to changing stress may then no longer be enough to avoid sudden breakage. In addition, mining has gradually reached greater depths with larger pressures. These are factors of significance in explaining why

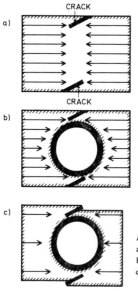

Fig. 111. Schematic illustration of rock failure under stress:
a) A rock specimen with cracks is subjected to compression.
b) A circular excavation in the specimen reduces the supporting element.
c) This element becomes too weak and a sudden rupture occurs along the cracks.

rockbursts seem to be more general in recent years than earlier in a country like Sweden.

On the basis of recordings at the Swedish seismograph network (Table 2 and Fig. 18), a rockburst sequence at the Grängesberg iron ore mines in central Sweden (60.1°N, 15.0°E) has been investigated for the interval from August 1974 to July 1976 with 340 events in all (BÅTH and WAHLSTRÖM, 1976). This is the first time that Sweden has produced a sequence of shocks, analogous to an aftershock sequence, that has permitted a detailed study. The magnitudes range from $M_L = 0.5$ to 3.2, with a complete coverage from $M_L \cong 1.0$ upwards. Tentatively applying equation (7) in Section 4.3 to the largest event (August 30, 1974) with $M_L = 3.2$, we find the seismic moment $M_0 = 3.5 \cdot 10^{20}$ cgs. This would correspond to an average dislocation of no less than 10 cm over an area of 100×100 m. However, application of equation (7) may not be very trustworthy in this case, as the equation was derived for a totally different region. An average dislocation of 1 cm over the same area corresponds to $M_L = 2.6$, which appears more reasonable. The focal depths of the events, derived from amplitude ratios of Rg to Sgl waves, exhibit a certain variation, with the first and largest event (August 30, 1974) at greatest depth (estimated as about 1 to 2 km), rising to much shallower depth during the summer of 1975 (probably around 200 m), after which the depths tend to increase again. The last phase a coupled with a decreasing rate of seismic energy release, resulting from a decrease of the active volume of rock.

Our curves for energy and strain accumulation and release have shapes similar to those of regular aftershock sequences, starting with a relatively mild phase, followed by a much more active period, finally again decreasing in activity (cf. Section 6.4). The accumulation and release are studied in relation to various triggering factors. Mining operations may influence the large-scale development, whereas rainfall seems to correlate well even in details. Semidiurnal lunar tidal action is found, and certain correlation appears between the fortnightly lunar tide and the occurrence of larger events. From the practical point of view, rockbursts are a significant factor in the planning and performance of future mining excavations. For this purpose, several mines (e.g. Grängesberg and Malmberget in Sweden) operate their own local geophone networks around the mines, to have the development under control.

The moral to be drawn from this and the preceding sections is that human impacts on the solid earth have now reached such dimensions as to fully require a much more detailed scrutiny than hitherto of the geophysical and geological properties of areas for planned works. This is especially true for areas which exhibit seismic activity, now or in the past. On the other hand, much exaggerated fear of the effects has been expressed, usually emanating from lack of knowledge of the conditions.

Chapter 11

Nuclear Tests and Other Explosions

11.1 Seismological Aspects

Since late 1950's, many seismological institutes around the world have been in the centre of public interest as never before. This is connected with the seismological possibilities of detecting nuclear explosions and the considerable interest in such events from the political and military points of view. Even though it has been of satisfaction to seismologists to be able to give significant information on such explosions on the basis of our records, still it has to be emphasized that this constitutes only a minor part of our work. Compared to a daily average of 10–15 recorded events, for instance, over the Swedish network, it is obvious that the number of nuclear explosion records is only a small fraction.

However, in parallel with the reporting activity, seismologists have also an interest of their own in such recordings. Thorough studies of nuclear explosion records are able to furnish very important contributions to basic seismological research, mainly from the following two viewpoints:

1. Controlled explosions with accurately known location and explosion time eliminate the difficulties inherent in every earthquake study. This considerably increased precision together with explosions so strong that waves right through the earth can be recorded are necessary to increase our knowledge about the earth's interior.

2. Explosions provide us with a seismic wave source, which is quite different from the one encountered in earthquakes. The different mechanisms entail different appearances of the records, and a thorough study of explosion records can contribute to increased knowledge about earthquakes and their mechanism.

We have learnt earlier (Chapter 7) how controlled explosions can be used to get considerably more reliable information about the structure of the earth's crust than is possible from earthquake records. The explosions generally used for crustal research are chemical and with few exceptions no larger than to produce records out to distances of at most a few hundred kilometres. With ever increasing chemical explosions and nuclear explosions, the possibility of obtaining records to greater distances is offered and with sufficiently large yields to any point on the earth. This entails a considerably increased precision in investigations of the deeper structure of the earth's interior, including the cores. Among the first records of explosion-generated *PKP*-waves are those which derived from American tests in the Marshall Islands in the Pacific Ocean. The Australian seismologists BULLEN and BURKE–GAFFNEY studied *PKP*-records from these tests. Figure 112 demonstrates in a striking way the increase of precision that explosion records permit compared to earthquake records.

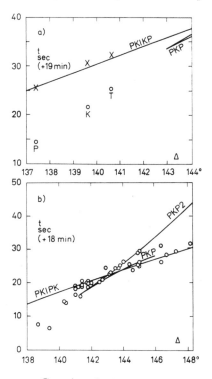

Fig. 112. Readings of *PKP*-waves: a) From nuclear explosions in the Marshall Islands (P=Pretoria, K=Kimberley, T=Tamanrasset), b) from a deep-focus earthquake in South America. t=travel time, $\varDelta$=epicentral distance. After K. E. BULLEN and T. N. BURKE-GAFFNEY (1958).

Convinced about the increased accuracy of our knowledge of the earth's interior that seismic records of explosions could lead to, Professor BULLEN proposed in his introductory speech at the assembly of the International Seismological Association in Toronto in 1957, that a number of atomic bomb explosions should be performed for scientific purposes. BUL-

LEN proposed four such explosions: one in the USA, one in the USSR, one near Australia and one in the Pacific Ocean area. The plan included also the requirement that satisfactory precaution should be taken against radioactive fallout. The proposal stimulated a very lively discussion with arguments both for and against the idea. The plan was never carried out and presently the prospects of its realization are probably more distant than ever before.

Although explosion records would be able to give seismology large possibilities, it is not for that reason that public interest has been focussed on nuclear explosions. Instead this is related to the political and military aspects of the problem. As early as 1958, a conference of experts was held in Geneva to explore the possibilities to detect and to identify nuclear tests. One of the proposals from this conference concerned installation of so-called array stations (Chapter 8), also called Geneva-type stations.

Even at an early stage, seismology proved to offer one of the most efficient methods for discovery of large explosions. From the political and military sides many questions were addressed to the seismologists, and in many cases sufficiently accurate answers were not immediately available. As a consequence, responsible authorities in several countries, especially in the USA, found it necessary to give strong economic support to seismological research on a broad basis. Since around 1960, the Advanced Research Projects Agency in Washington, D.C., has been in charge of the American project in this connection, called VELA, as a common name for all detection research. Within the frame of the VELA project, many institutions have been conducting basic seismological research, both within the USA and abroad. Even though the ultimate goal of this project is to improve detection possibilities, it was obvious that only considerably enhanced basic research in seismology could lead to this goal.

Besides their seismological and their military-political interests, nuclear explosions have also another, peaceful interest, i.e. to perform large works, as in connection with river regulation, building of canals, etc. Technically such a use of large explosions is very well possible and has also been tested, both in the USSR and in the USA. However, the permission to conduct such tests is also connected with the disarmament discussions in Geneva. Construction of nuclear power plants is another peaceful application of nuclear energy.

11.2 *Seismological Results*

Numerous seismological results have emanated from the study of explosion records, some of which we shall describe here.

An exact knowledge about explosion time and explosion location permits the most accurate determinations of seismic travel times. It has been shown that the most commonly used travel-time tables, for instance, those of JEFFREYS–BULLEN, are in need of a reduction by a few seconds for surface foci. Thus, according to records at eight American stations in the distance range of 69.0°–78.6°, the first underwater nuclear explosion (on July 24, 1946, in the Bikini atoll) gave a difference of -1.8 ± 0.8 sec for the *P*-wave, compared to the JEFFREYS–BULLEN tables for a surface focus. Similar differences have been obtained in several following investigations with much larger material. The Russian seismologist KOGAN studied records of the American tests in the Marshall Islands and found an average difference between observations and JEFFREYS–BULLEN tables of -1.8 ± 0.6 sec for the *P*-wave and between -4 and $+5$ sec for the *S*-wave. The negative differences for the *P*-wave were usually explained by the structure under the Pacific explosion area which deviates from the continental structure. The latter structure corresponds better to the tables. If the velocities beneath the explosion area are greater than in the continental crust, this will naturally entail earlier arrivals or shorter travel times.

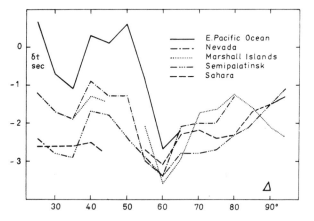

Fig. 113. Travel-time differences (δt) between explosion data and JEFFREYS-BULLEN tables for *P*-waves. After ESSA Symposium on Earthquake Prediction, 1966.

Later, these investigations were continued and Figure 113 gives a summary. From this an interesting general result is obtained, namely that negative differences prevail almost consistently and not only for oceanic but also for typically continental explosions (such as Semipalatinsk in Kazakh, Siberia, and Algeria). Therefore, it would hardly be justified to explain the differences as solely due to the higher velocities in the oceanic structure. Even the continental structure has to exhibit higher velocities than assumed in the construction of the travel-time tables. In the JEFFREYS–BULLEN tables, the P-wave velocity is assumed to be 7.8 km/sec just beneath the Mohorovičić discontinuity, whereas later investigations have shown that this velocity is usually higher in typically continental areas, i.e. between 8.0 and 8.5 km/sec.

The significance of these corrections and of more exact travel-time tables is quite obvious. The more accurately the travel times are known, the more accurate calculations of the sources are possible, whether the source is an earthquake or an explosion. Records of explosions with well-known source data have thus been used for recent revisions of the travel-time tables. Up until now, the same tables have been used for the whole earth. In the future, it appears to be necessary to replace these by a number of regional tables, depending on the location of the source and the station.

But the improvement of travel-time data has also another aspect. We generally assume the travel time to be the same in all directions from the source. However, most earthquake epicentres are located in coastal areas, especially all around the Pacific Ocean (Chapter 5). The structure down to great depths, at least 300 km, is probably different in different directions from such an earthquake source: on one side we have the oceanic structure and on the other the continental structure. In order to attain higher precision, it would then be necessary to use azimuth-dependent travel times. Natural events (earthquakes) can hardly give any reliable information on this point, whereas suitably located explosions of sufficient strength would be able to do so. Such a seismic experiment was carried out by the Americans on October 29, 1965, on Amchitka Island in the Aleutians. A nuclear explosion of 80 kilotons (1 kiloton = 1000 tons TNT) was detonated at a depth of 660 m. Records were obtained all over the earth. For instance, Uppsala recorded besides P also clear PcP- and $P'P'$-waves. On October 2, 1969, another Amchitka test was performed (Fig. 114), this time with a

a)

b)

c)

d)

megaton yield (1 megaton = 10^6 tons TNT), followed by a five-megaton explosion on November 6, 1971. The importance of the studies of these records is quite obvious. They can lead to much more exact information about the location of earthquakes within island arcs than hitherto available (Fig. 54–55). For a more complete knowledge of azimuthal variation of travel times, it would be necessary to record controlled explosions performed all around the Pacific Ocean as well as in other seismic areas, where the structure is different in different directions.

Other seismological results also concern the structure. As an example, we may mention a series of atomic bomb explosions carried out at Maralinga in central Australia in 1956. The seismic records permitted a determination of the crustal structure within this area, which had not been possible by means of earthquakes. Within the USA, the explosions have given very interesting information on the crustal structure. Most of these explosions have been performed in Nevada (at Nevada Test Site = = NTS). This site is located in the western USA within an area whose structure can hardly be considered as typically continental. From a structural point of view, this area is a landward continuation of the Easter Island ridge structure, i.e. with relatively low wave velocities in the upper mantle. An extremely elucidative test was made on December 10, 1961, further to the east (in New Mexico; yield = 3.5 kilotons; depth = 800–900 m; this test has the code name Gnome). This experiment was made in an area where the structure is considered to be more typically continental. The wave propagation from this test exhibited very interesting properties: the amplitudes of the P-waves diminished strongly during propagation westwards, where the waves had to pass under the Rocky Mountains. In California, only weak and unclear records were obtained. But towards the east and northeast, the amplitudes were exceptionally large, and clear records were obtained even at Scandinavian stations. In addition, the travel times were different in different directions from the test site. Towards the west, the relatively low velocities for the upper mantle were confirmed (about 7.8 km/sec for the P-wave), whereas towards the eastern and north-

Fig. 114. Records of some underground explosions:
a) Aleutians, October 2, 1969. 1.2 megaton, 1220 m depth, $\Delta = 68°$, $m = 7.0$ (Uppsala short-period Z).
b) Nevada, January 19, 1968, 1 megaton, 980 m depth, $\Delta = 74°$, $m = 7.0$ (Uppsala short-period Z).
c) North of the Caspian Sea, July 1, 1968, $\Delta = 23°$, $m = 6.6$ (Uddeholm short-period Z).
d) Novaya Zemlya, October 27, 1966, $\Delta = 15°$, $m = 6.6$ (Umeå long-period Z).

eastern USA, the corresponding velocities were found to be as high as 8.4 km/sec. The latter value is of some interest because it agrees with some determinations of upper-mantle velocities under Scandinavia. It has to be taken as a more typical value for the upper mantle in continental structures.

Nuclear yields of about 20 kilotons or more are able to provide observational material for more detailed studies of the upper mantle. Again the most important information derives from the travel times of *P*-waves, but also *S*-waves and surface waves are of significance in this study. Surface waves in the period range of 5–50 sec have been recorded from nuclear explosions and then regional structural variations can be mapped with higher accuracy.

By means of still larger yields, it is possible to extend the investigations to depths greater than 1000 km into the earth. Major explosions–for instance, one of 10 megatons in a deep-sea trench as was proposed by some scientists–would be able to give *P*-waves comparable to those of the largest earthquakes. In addition, such explosions would presumably be able to generate free vibrations of the earth–similar to those after the earthquakes in Kamchatka in 1952, in Chile in 1960 and others. Through experiments of this kind we would be independent of large earthquakes for such studies; we would have a source mechanism which is well known and we would be able to vary the source conditions within certain limits. Possibly it could be envisaged that different modes of the free vibrations could be generated by variation of the source mechanism. Combinations of several explosions – with appropriate time and space shifts – could probably contribute to diversify the experimental conditions.

An additional advantage with controlled explosions as wave sources for the exploration of the earth is that they can be performed anywhere, at least theoretically. In any case, we would not be restricted to the localities where earthquakes occur.

In the present situation, methods for seismological research with high precision are available, but other considerations–especially military and political – prevent their application. As a consequence, seismologists have had to study records of tests which have been performed for other purposes. In those cases where source data have been communicated, such studies have yielded useful results. Even those cases, where direct information about the source has been lacking, have permitted interesting studies, but the increased precision may not be attained. If exact source data could be published for all test explosions, this would undoubtedly stimulate seismologists to further studies of their records. And these in turn would increase our knowledge of the earth. Primarily, data on the exact position of the source and the explosion time are needed and for atmospheric tests also their height above the earth's surface. Even if the yield is not published (this is apparently the most secret data), the information mentioned would be of the greatest scientific value. Seismological studies, thus made possible, would certainly also contribute to providing a better foundation for the disarmament discussions in Geneva. Even though explosions provide seismic sources of high accuracy, it should be emphasized that seismologists would not suggest their application in cases which would be hazardous to mankind, from any point of view whatsoever.

11.3 Detection: General Problems

We shall start by giving a review of the detection problem as a whole. We are most concerned with seismological detection methods, but as we shall see, difficulties are so great that all possible methods have to be included. In particular, smaller tests are difficult to discover, especially if only records from greater distances are available. For successful detection it is frequently necessary to combine indications of various kinds.

As a general rule, we can state that we should preferably 'listen' in the same medium as that where the explosion is set off. In this way, the greatest amount of radiated energy is likely to be received, while in listening in another medium we have to expect that losses of energy occur in the passage of waves from one medium to another. In this discussion, the word medium refers to the solid earth, the sea, the atmosphere and the ionosphere. Thus, we arrive at the following review:

1. The solid earth, i.e. underground explosions. The seismological methods are practically the only ones that can be used. Examples of records are shown in Figures 114 and 115. Sometimes electromagnetic effects have been mentioned as useful indications, but this is probably true only for relatively short distances from the source.

2. The sea, i.e. underwater explosions. Hydrophones are the best reception equipment in this case. This is particularly true if both the ex-

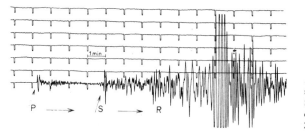

Fig. 115. Underground explosion in Novaya Zemlya on September 27, 1971, as recorded by Uppsala long-period Z, $\varDelta = 19°$, $m = 6.7$.

plosion and the hydrophone are located on or near the SOFAR channel (SOFAR = *so*und *f*ixing *a*nd *r*anging), where the velocity has a minimum. Seismological methods are also very useful; in underwater explosions, the coupling between the explosive charge and the solid earth is as a rule better than for explosions in other media.

3. The atmosphere, i.e. atmospheric explosions. These produce an atmospheric pressure wave and hence are best observed by means of microbarographs. But seismological methods are also of great significance, even though only a very small fraction of the total explosive energy is transmitted into seismic wave energy. Atmospheric explosions have been detonated both on or near the surface as well as at altitudes of a few kilometres.

4. The ionosphere, i.e. explosions at an altitude of a few hundred kilometres above the earth's surface. These are best recorded by suitable magnetic instruments. Figure 116 shows such a case, obtained by an instrument for recording short-period variations of the earth's magnetic field. The magnetic method is very sensitive, and the reaction is practically instantaneous. On the other hand, this method–at least as it operated at Uppsala–proved to be insensitive to explosions in the lower atmosphere–even to the very strongest of them. Magnetic methods are apparently among the only ones useful for ionospheric explosions. No seismic effects are observed for such explosions.

The detection methods outlined here are augmented by a number of other observations, for instance of radioactive fallout, especially from atmospheric tests but also from leaking underground explosions. Direct observations by means of satellites also play a great role.

In the partial test-ban treaty of 1963, all nuclear explosions were forbidden except the underground ones. It was then considered that suf-

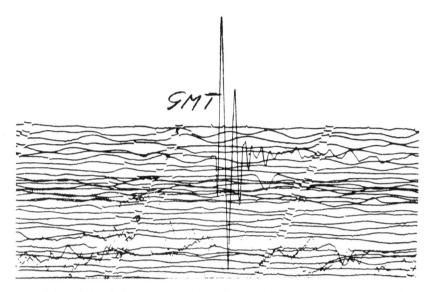

Fig. 116. An explosion of 1.4 megaton at an altitude of about 400 km above Johnston Island in the Pacific on July 9, 1962, as recorded by a magnetic instrument at Uppsala ($\Delta = 103°1/2$).

ficiently reliable control methods were available for all tests except for those made underground. For this reason efforts in detection work have since then been concentrated on underground explosions and primarily on seismological methods. While the treaty mentioned was signed by the USSR, the USA and Great Britain, to name a few, it has not been signed by among others France and China. The latter two have also conducted atmospheric nuclear tests after 1963.

11.4 *Detection: Position, Depth, Origin Time*

The fact that instrumental seismological observations still cover too short a time interval to guarantee against surprises (see Chapter 5), has an important application to the identification of unknown events (whether explosions or earthquakes). If an event is recorded, which can be localized to a suspected area, e.g. a known test area, and in which no seismic activity is known, this is certainly a reason to investigate more carefully the nature of the source, if possible. But the very circumstance that earthquakes occur from time to time in areas, where they are hitherto unknown, is a warning against too hastily drawn conclusions. On the other hand, it is naturally much easier to 'hide' an underground test by performing it within a seismic area.

An increased precision in the calculation of epicentral locations could contribute to an improvement of the possibilities of deciding the nature of a recorded event. To wit, it is highly improbable that an underground explosion be set off under the ocean bottom or very near the frontier of another country.

Seismicity studies have clearly been considerably stimulated because of these new aspects. For instance, the seismic conditions in the Soviet Union have attracted great interest. The Soviet Academy of Sciences in Moscow has also contributed efficiently to these studies by publishing very comprehensive tables and maps regarding the seismic conditions in the USSR for the period 1911–57, a classical work in its field.

The problem of calculating the depth of a seismic event is of great importance for the identification of the nature of the source. Even though the precision of depth determinations is sufficiently great to establish structures as shown in Figures 54 and 55, seismic depth calculations often suffer from errors of ±25 km. To date, the deepest underground explosions have been made at depths of 1–2 km. If depth calculations could be made with higher precision, events with calculated depths of say 5–10 km or more could be referred to the earthquake category, and closer investigations could be limited to more superficial phenomena. From this point of view it is obvious that increased precision of depth calculations appears as one of the most important contributions to the supervision of a test-ban treaty for underground explosions. For this to be accomplished, increased precision of seismic measurements will be necessary, comprising for one thing reliable timing to 0.1 sec or better, and for another thing applications of time corrections because of local structural deviations both for the source area and for every single seismograph station. A comprehensive investigation would be needed for the calculation of these corrections. The most reliable depth calculations are based on the arrival-time difference pP–P. In Chapter 4, we have seen how this method can be applied with higher accuracy by simultaneous inspection of records from a seismograph network. Spectral methods offer even more powerful techniques, applying the destructive interference between P and pP (see Fig. 95 and Section 8.4). This method is particularly useful for the shallow depths of explosions, where clear separation of P and pP may be difficult directly in the records.

Also the time of an event may provide some guide in detection work. It is well known that in most test series explosions are touched off on a full hour or on a full minute, and they are frequently limited to certain hours of the day. For instance, underground explosions in Kazakh (Semipalatinsk) are carried out in our early morning hours, always on a full minute. Quite obviously, such occurrences cannot by themselves serve as reliable identification arguments: it may happen both that an earthquake occurs on an exact minute and that explosions are made at other times. However, together with other indications, observations of the origin time may constitute a contribution of some value and interest.

11.5 Detection: Source Mechanism and Seismic Waves

Detection of nuclear tests by means of information on location, depth and time, as outlined in the preceding section, is possible only in favourable cases. As an example, I might mention the very large record obtained on January 15, 1965, from an explosion at the Semipalatinsk test site. The records we obtained at Uppsala were of considerably larger amplitude than for any earlier underground explosion, and our immediate reaction was that the record was due to an ordinary earthquake. We calculated the location from the Uppsala records and found excellent agreement with the known test area at Semipalatinsk. But earthquakes could not be excluded for this area. However, the records exhibited typical features suggesting an explosive origin, such as large amplitudes of P and PP on short-period records, as opposed to comparatively weak surface waves on the long-period records. But we know that this is a characteristic feature also of deep earthquakes. A study of available seismicity maps showed definitely that no deep earthquakes had ever been recorded from this area. But still the earthquake hypothesis could not be completely excluded: we could be faced with a seismological surprise, i.e. an event for which existing observational series are too short to show a parallel case. The possibility of this appeared small, and it became vanishingly so when we also considered that the event had occurred exactly on a full hour, a usual explosion time at Semipalatinsk.

In the case mentioned the considerations described were sufficient to decide the nature of the phenomenon. Similar arguments are usually effective in most cases when we have to deal with longer series of tests performed within a smaller area. In the case of single tests, especially those of smaller strength, this line of argument is frequently not enough. Then recourse has to be taken to more detailed studies of the seismic records, including comparison between records of known origin, both explosions and earthquakes. Even after such a careful study, it is not always possible to reach one hundred percent reliability.

The problem of using different properties of the records in order to decide the nature of the source can in general be illustrated as in Figure 117. If some property proves to be completely different for earthquakes and explosions (Fig. 117a), the discrimination is one hundred percent reliable

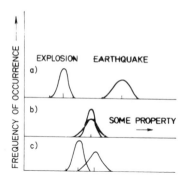

Fig. 117. Schematic comparison of the frequency distributions for different properties for explosions and earthquakes: a) Complete separation, b) no separation at all, c) partial separation.

on the basis of this property. However, if the frequency curves coincide (Fig. 117b), the corresponding property cannot be used for discrimination. As a rule, frequency curves for explosions and earthquakes overlap partially (Fig. 117c), and such a property may not permit one hundred percent reliability in the discrimination. Then, it will be necessary to study as many different properties on the records as possible. If they all give the same indication, this naturally strengthens the suspicions based on just one such property. In Figure 117 we have on purpose made the variational width broader for earthquakes than for explosions. This is a generally observed fact, depending on the greater variability of the source mechanisms of earthquakes.

Earlier (Chapter 6) we studied the mechanism of earthquakes. As distinct from earthquakes the mechanism of an explosion is spherically symmetric, at least theoretically, radiating a longitudinal wave of equal amplitude in all directions. This corresponds to the ideal case. However, this does not happen in practice, because of the inhomogeneous structure of the earth's crust, particularly its different structure in different directions. Therefore, explosions also exhibit a certain lack of symmetry, even though this is not as pronounced as in earthquakes. The observed fact that explosions also generate transverse waves, even though such waves would not be expected in the ideal case, is another expression of the asymmetry.

The differences in mechanism with ensuing differences in the radiation of seismic waves can be used for identification of the source in uncertain cases. But in order to apply this procedure successfully, observations are needed in as many different directions from the source as possible for simultaneous inspection. This requirement would be met by a system providing seismic observations in at least 36 different directions from the source and with an even distribution, i.e. one observation for every $10°$ of azimuth and all observations within a comparable distance range, say 3000–4000 km or less. Such observational material would offer good opportunities to decide on the source nature, however, still not always with one hundred percent reliability. The number of observation points could be diminished with the availability of array-station records.

Another related factor of great significance to the appearance of the records concerns the extent in time and space of the source action. Earthquakes–especially those of larger magnitude–always exhibit a certain extension both in time and space. Often the rupture starts at one end of the fault and propagates with a velocity around 3 km/sec along the fault to its opposite end. This entails that a comprehensive spectrum of seismic waves is generated and that records are obtained both on short-period and long-period seismographs. For an underground explosion the source effect is strictly limited both in space and time. As a consequence, mainly short-period (high-frequency) waves are created, and records are obtained primarily with short-period instruments (of *P*, *PP*, *Pn*, etc.). On long-period instruments, short-period surface waves may be recorded (especially of higher-order modes, if the path is purely continental), while surface waves of longer period (fundamental mode) certainly may exist, but are much underdeveloped in comparison with short-period P. Comparison of spectra both of P-waves and of Rayleigh waves for underground explosions and earthquakes has proved to provide a reliable discrimination method. Atmospheric explosions give rise to seismic waves through the impact of the atmospheric pressure wave on the ground surface (or ocean surface) below the explosion point. In this case we have again a source of seismic waves that is extended both in space and time. The seismic source consists of an approximately circular area on the surface just under the explosion, and this source area may have an extent of several kilometres. Similarly, it takes a certain time for the atmospheric pressure wave to propagate over this area. As a consequence, atmospheric explosions produce seismic records which resemble those of earthquakes much more than those of underground explosions. This means that the atmospheric test records are dominated by longer-period waves, especially surface waves, and are thus best recorded by long-period seismographs. Our discussion of the respective seismic records can be summarized in the following two points:

1. Atmospheric explosions resemble earthquakes at normal depth.

2. Underground explosions resemble earthquakes at greater depth.

This scheme may serve as a starting point and there are several modifications to it in detail. Whether the similarity between records of underground explosions and deep earthquakes–especially the dominant occurrence of high frequencies and of relatively simple wave forms–may be taken as an indication of a more concentrated source for the latter (e.g. by means of phase changes; see Chapter 6) is a problem of great significance. What has been said above for underground explosions, is also applicable to underwater explosions. The results suggest that once the identification of an explosion by means of seismic records has been made, it is generally no greater problem to decide if the explosion took place above or below the ground surface. The atmospheric explosions discussed here refer to those made at altitudes of a few kilometres above the surface. Atmospheric explosions detonated on or near the surface are, just as the underground explosions, strongly limited in time and space, with regard to their seismic effect. Therefore, their records are more similar to those of underground tests.

There are also a number of discrepancies between records of atmospheric explosions and of earthquakes at shallow depth. Surface-wave amplitudes are more important in relation to body-wave amplitudes for atmospheric explosions than for earthquakes. A striking comparison which illustrates this fact can be seen in Figure 118. In addition, records of atmospheric explosions are 'smoother', i.e. they are lacking much of the higher-frequency motion of an earthquake record. Detailed frequency analysis of different waves may provide powerful means to discriminate between different phenomena. Moreover, atmospheric explosions exhibit only insignificant motion perpendicular to the plane of propagation as expected. Almost all the motion on the seismogram is contained in the plane of propagation, i.e. it consists of P, the SV-component of S and of R. However, some poorly developed transverse motions are found, but these are considerably weaker than the motion along the plane of propagation. The atmospheric explosions–for instance, those performed at Novaya Zemlya–did not show any traces of SH or of Love waves at Swedish stations; on the other hand, we have found traces of higher-mode Love waves. As a consequence of inhomogeneities in the earth's crust, transverse motion is more common in records of underground explosions than of atmospheric tests. Clear Love waves of fundamental mode were thus recorded from the underground test at Novaya Zemlya on October 27, 1966 (Fig. 114) as well as from the underground test in Nevada on December 20, 1966, just to mention a couple of examples.

In addition to frequency analysis (spectral analysis) of recorded waves, a detailed study of wave forms (in the time domain) may be of great assistance. It has been found that short-period records of P-waves from underground explosions consistently show a first displacement (compression) followed by a second swing (dilatation) about two to three times as large as the first one (Fig. 114). If a record is found with this typical appearance, this feature may indicate an explosive origin. But this indication alone is rather uncertain, as many earthquakes are able to give rise to exactly the same wave form. In order to utilize the wave forms better, simultaneous access to as many records as possible at widely different directions from the source is necessary, which has already been emphasized above. On the other hand, it can be maintained that a wave form, which deviates significantly from the one described, suggests that the origin has been an earthquake and not an explosion.

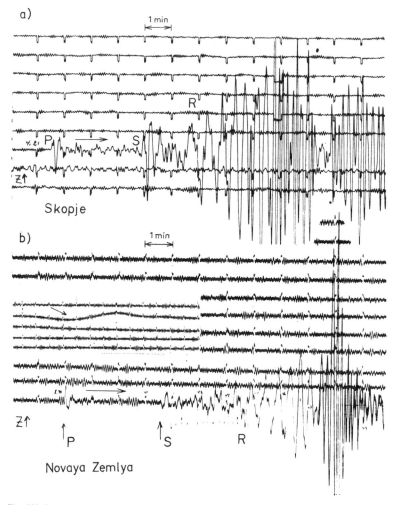

Fig. 118. Long-period vertical-component records at Uppsala:
a) The Skopje earthquake, July 26, 1963, $\Delta = 18°$, $M = 6.0$.
b) The Novaya Zemlya atmospheric explosion, October 30, 1961, $\Delta = 19.5°$, $M = 5.4$ (the largest explosion ever). The larger M for a) than for b) is due in part to different amplitude response of the two records.
Inserted: the atmospheric pressure wave from this explosion, arriving about 2 hours after the explosion.

In the original efforts to discriminate between explosions and earthquakes, proposed in Geneva in 1958, great emphasis was laid on the direction of the first motion of the *P*-wave. This constitutes only a very small fraction of the whole record, and in addition may be difficult to assess with reliability, especially with low signal-to-noise ratio. Therefore, it has been emphasized quite correctly that it is necessary to extract information from the whole record, not only from the direction of the first swing. Besides *P*-waves, also surface waves have been used with considerable success. From surface waves it has been possible to deduce source functions. These in turn provide valuable information about the nature of the source.

Estimates of the height above ground for atmospheric tests are of importance in connection with warnings for radioactive fallout. Unfortunately, it is hardly possible to make any reliable absolute height estimate from seismic records alone. On the other hand, there are some possibilities to make relative height estimates for a series of tests in the same area, as for Novaya Zemlya tests as recorded at Swedish stations. In this work we apply a principle which is clear from our discussion above, namely that the higher an explosion is made, the more will long-period components dominate over short-period ones in a seismic record. Amplitude ratios between different parts of the surface-wave train as well as between surface waves and *P*-waves can be used for such relative height estimates. More accurate and absolute height calculations could be made by a combination of seismic and electromagnetic observations. Seismic records provide an information about the moment when the pressure wave hits the surface, whereas electromagnetic observations tell us the shot moment itself. The difference between these two times could provide a relatively accurate piece of information about the height.

Tables 26 and 27 summarize some properties of records of underground explosions (Nevada) and of earthquakes, as obtained at Pasadena in southern California. These results deviate in part from those obtained at Uppsala from records of Semipalatinsk explosions. An essential reason is to be found in the different source distances involved. While the distance Semipalatinsk–Uppsala is about 3800 km, the epicentral distances in Tables 26 and 27 are only about one-tenth of this. This demonstrates

Table 26. Generation of long-period surface waves ($T > 5$ sec) from earthquakes and explosions[1].

Magnitude M_L	M	Surface waves
< 3.9	< 3.0	No surface waves observed.
3.9—4.4	3.0—3.8	25% of cases exhibit surface waves.
> 4.4	> 3.8	Surface waves generally observed, at least up to 3000 km distance.

[1] After F. PRESS, G. DEWART and R. GILMAN (1963).

Table 27. Some properties which can be used for source identification[1].

Property	Earthquake (q)		Explosion (x)
Dominating Love-wave period	≥ 13 sec		< 8 sec
Area of surface-wave envelope		$q > x$	
Amplitude ratio $Lg(\text{H}): Rg(\text{Z})^2$	2.13		0.77
Energy ratio $(Sg + 4 \text{ sec})/(Sg - 4 \text{ sec})$	41.2		8.4 (underground)
			4.1 (atmospheric)

[1] After F. PRESS, G. DEWART and R. GILMAN (1963).
[2] H = horizontal component, Z = vertical component.

that in giving various diagnostic properties it is very important to state also for what distance range they are applicable.

Some comments are necessary concerning the magnitudes in Table 26. By later installations of highly sensitive long-period seismographs, it has been possible to decrease the magnitude limits still further. Thus, surface waves have been recorded from events with magnitudes M less than 3.0. Recent installations of long-period seismographs with a maximum magnification of 60 000 and more will still further improve the possibilities of recording surface waves from weak events.

11.6 *Magnitude and Energy*

It is equally important to be able to calculate magnitude and energy of the source as it is to be able to identify its nature. Calculations of energy and magnitude have earlier been developed for earthquakes (Chapter 4). But when we have to deal with explosions, caution is required in the application of such methods.

It is still common practice to calculate magnitudes for explosions in exactly the same way as for earthquakes, also to use the same equation as for earthquakes in calculations of elastic wave energy. However, if we consider the conditions at an explosion (limiting the discussion to underground explosions), we will find that this procedure could be improved. As we have seen, records of distant underground explosions exhibit only short-period P, sometimes also short-period PP, but no S and not any or only weak surface waves. In addition, explosions are made practically at the surface level, and the duration of the wave train is generally considerably shorter than for earthquakes. The result of these effects is that for a given distance and for a given ratio amplitude/period (a/T) the wave energy from an explosion amounts only to about one-tenth of its value from an earthquake.

This circumstance can easily be included in our calculations. If by a subscript x we denote quantities which refer to an explosion (underground), we are able to calculate the total seismic wave energy from the same equation as (5) in Chapter 4, i.e.

$$\log E_x = 4.78 + 2.57 m_x \qquad (1)$$

where, however, m_x is expressed in a magnitude scale for underground explosions. The relation between m_x and the usual m (defined for earthquakes) is simply the following (Båth, 1966):

$$m_x = m - 0.4 \qquad (2)$$

This may at least serve as a first approximation to a magnitude scale for underground explosions. For atmospheric explosions the same procedure as for earthquakes may serve as a first approximation.

However, calculations of magnitude or wave energy are often not sufficient. In addition, we need calculations of yields. As concerns calculation of the yield from a knowledge of E_x or m_x, we are faced with the problem of estimating the fraction of the total explosive energy that has been transferred into seismic wave energy. This fraction depends upon a number of different factors and varies within wide limits.

From the political and military side as well as from news agencies the question is often raised as to how far from a source of a given yield it is possible to get a seismic record. A unique answer can hardly be given to this question, as the same factors are also of importance here. The most important of these are the following:

1. Yield. Figure 98 shows a relation between maximum recording distance (for the most sensitive seismograph stations, excluding array stations) and yield. It demonstrates that, for instance, an underground explosion of 1 kiloton could be recorded to distances of 3000–4000 km under favourable circumstances. For surface explosions, about 1000 times as large a yield is required for the same recording range.

2. Depth of the explosion below the surface. The deeper the explosion is emplaced, the better is the conversion into seismic energy. That is at least true for very shallow depths, until a depth of complete coverage is reached.

3. Nature of the medium in which the explosion takes place. From the USA, experience has been gained on explosions in various media, such as salt, volcanic tuff, granite. Of these, granite and salt are the most favourable for the transfer of explosive energy into seismic energy.

4. Degree of coupling between the charge and adjacent medium. Especially in the USA, comprehensive investigations, both theoretical and experimental, have been made concerning decoupling of an explosive charge. These indicate that it would be possible in this way to lower the seismic output several hundred times.

A few examples will be given which demonstrate in a very striking way the enormous variation there is in the possibility of recording an underground explosion. The so-called Gnome test on December 10, 1961, in New Mexico, USA, which corresponded to 3.5 kilotons and was made at a depth of 800–900 m in a salt layer, was very well recorded in Sweden. On the other hand, the Sedan explosion on July 6, 1962, in Nevada, which amounted to 100 kilotons and was performed at a depth of about 200 m, was not recorded at all in Sweden. The latter test was a crater experiment: the top of the mountain was blown off by the explosion and too little energy propagated downwards to give any record at our distances.

The following equation gives an approximate relation between the magnitude m_x, the yield Y in kilotons and the seismic coupling factor α, i.e. the ratio of seismic energy to total explosive energy:

$$m_x = 5.4 + 0.4 \log \alpha Y \qquad (3)$$

This formula is applicable to all kinds of explosions, not only the underground ones, provided that the correct value of the coupling factor is used. The yield Y is generally expressed either in kilotons ($1 \text{ kt} = 10^3$ tons) or in megatons ($1 \text{ Mt} = 10^6$ tons), by which we mean that the total explosive energy is the same as if that amount (kt or Mt) of TNT were used. For calculation of seismic energy we also need to know the conversion factor between kt and ergs: 1 kt corresponds to an energy of 4×10^{19} ergs.

A number of different calculations of seismic coupling factors have been published, of which we present a short summary in Table 28, referring

Table 28. Seismic coupling, i.e. the fraction of the total explosion energy which is converted into seismic wave energy, for different locations of the explosion in relation to the earth's surface (according to American data).

Location of explosion		Coupling factor (α)	
	10 km altitude		1×10^{-5}
	1 km altitude		3×10^{-5}
	surface		1×10^{-4}
	300 m underground		1×10^{-3}
	30 m under water		5×10^{-3}
	100 m under water		2×10^{-2}
	500 m under water		4×10^{-2}

to an explosion of $Y = 20$ kt. The coupling may vary considerably from case to case. The values in Table 28 have been determined by American scientists. Our own research in this field fully confirms the low coupling factors for atmospheric explosions, whereas for explosions below the surface, especially in water, we have found reason to suspect considerably larger coupling factors than those given in Table 28.

It is internationally agreed (in 1968) that existing station networks are able to detect underground explosions in the northern hemisphere down to 20 kt, if the explosion has taken place in hard rock, like granite, and that the source can be located to within 10–40 km. From the American side it is maintained in 1971 that underground explosions as small as 1 kt can be detected to distances exceeding 6000 km. It is of interest to compare this statement with my graph of 1958, presented in Figure 98. Using this graph and equation (3) above, we have in this case: distance = 6000 km = 54°

(maximum range in Fig. 98), corresponding to $m = 5.4$, $m_x = 5.0$ from equation (2), $Y = 1$ kt (given), which gives $\alpha = 10^{-1}$ by equation (3). This is higher than the corresponding values of Table 28, but in better agreement with our own experience of favourable coupling.

For atmospheric explosions the seismic output is very poor indeed. If the total energy of the 58-megaton bomb on October 30, 1961, at Novaya Zemlya had been converted into seismic energy, it would have corresponded to an earthquake of magnitude $M = 8.4$, i.e. in the same class as the largest earthquakes. However, the conversion into seismic energy was only about 5×10^{-5} of the total energy of the bomb. The rest of the energy appeared in other forms, especially as an atmospheric pressure wave. Therefore, from a seismological point of view, this bomb corresponded to an earthquake with a magnitude of only $M = 5.4$, as far as energy is concerned. Figure 119 also demonstrates that the bomb was relatively insigni-

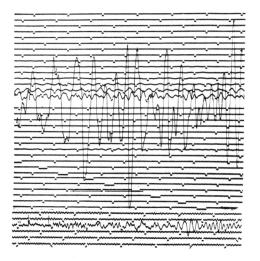

Fig. 119. Records at Uppsala showing part of the surface waves from the Iranian earthquake of September 1, 1962 (upper figure) and from the Novaya Zemlya explosion of October 30, 1961 (lower figure). The records are written by the same instrument, and therefore the trace amplitudes are directly comparable.

ficant as a seismic phenomenon. This picture shows two records from the same instrument at Uppsala, one from the earthquake in Iran on September 1, 1962 (magnitude 7.1), the other from the bomb on October 30, 1961.

The relation between body-wave magnitude m and surface-wave magnitude *M*, which we have discussed earlier (Chapter 4), is valid for earthquakes. On account of the different focal mechanism for earthquakes and explosions, the relation *m–M* will be different for underground explosions. Figure 120 illustrates how a given body-wave magnitude *m* corresponds to a remarkably lower surface-wave magnitude *M* for explosions than it does for earthquakes. This circumstance has sometimes been considered as the most reliable criterion for discriminating underground explosions from earthquakes. However, a similar difference exists between deep and shallow earthquakes, and therefore the criterion has to be combined with some information on focal depth.

There is now a large collection of observational material, including source information, available for underground explosions made by the USA, but we are still lacking corresponding information regarding the Asiatic

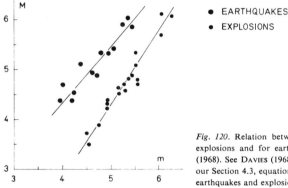

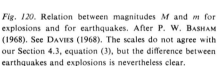

Fig. 120. Relation between magnitudes *M* and *m* for explosions and for earthquakes. After P. W. BASHAM (1968). See DAVIES (1968). The scales do not agree with our Section 4.3, equation (3), but the difference between earthquakes and explosions is nevertheless clear.

continent. The continents vary in their structure, even in different parts of the same continent. This may have a considerable influence on the distance range to which seismic recordings can be obtained as well as on other properties of obtained records. Obviously, experience from the North American continent cannot simply be applied to the Asiatic continent. In order to provide the necessary basis for the disarmament discussions in Geneva and for a possible test-ban treaty, including underground explosions, it would be of great significance if information on a series of tests on the Asiatic continent could be made available.

11.7   Some Statistics on Nuclear Explosions

Table 29. Number of nuclear tests up to 1976, inclusive, which have been reliably recorded by the Swedish network (Table 2).

Explanation of notation:

A Novaya Zemlya	1 Underground
B Semipalatinsk	2 Underwater
C Caspian Sea-Ural region	3 Surface
D USA, particularly Nevada	4 Atmosphere
E Aleutian Islands	
F Pacific Ocean	
G Sahara	
H Sinkiang	
I India	

Year	A1	A2	A4	B1	C1	D1	E1	F2	F3	F4	G1	G3	H1	H4	I1	Sum
	USSR					USA					France		China		India	
1954									1							1
55			1					1								2
56									2							2
57			4													4
58			11			1				2[1]						14
59																0
60												1				1
61		2	18			1										21
62			24	1		4				10[1]	1					40
63						3					1					4
64	2			4		2										8
65				9		4	1				2					16
66	1			11	1	9					1					23
67	1			14		11								1		27
68	1			11	1	20				2[2]				1		36
69	1			10	4	20	1						1	1		38
70	1			8	4	21				2[2]				1		37
71	1			11	5	7	1			1[2]						26
72	1			11	5	4										21
73	3			7	4	9								1		24
74	2			13	2	5								1	1	24
75	3			8		13										24
76	2			10	1	9										(22[3])
Sum	19	2	58	128	27	143	3	1	3	17	5	1	1	6	1	415

[1] USA.
[2] France.
[3] Probably incomplete.

Table 29 gives a summary of all nuclear tests which have been recorded by the Swedish seismograph station network (Fig. 18) up to 1976, inclusive. Nuclear explosions were certainly started by 1945, but it took several years until the explosions were large enough to be recorded seismically at greater distance and also until our own station network had been created and equipped with instruments of high sensitivity (see Table 2). Table 29 gives an idea of the extent of the material we have to offer on this point and at the same time clearly demonstrates the number of tests which have been large enough to be recorded at greater distance. Military and political aspects (for instance, the partial test-ban treaty in 1963) are also reflected in the material. In addition, there is a large number of smaller tests which has not been recorded by us. As a rough estimate, the total number of our recordings (415) is hardly more than 40% of the total number of tests performed. Our statistics (Table 29) cuts the top and is therefore still representative of the "seismopolitical" development.

The right-hand column in Table 29 shows clearly how the interval can be divided into four periods, of which the first three show an increasing frequency of explosions:

1. 1954–8, before the first test ban.

2. 1960–2, dominated by the large number of atmospheric explosions.

3. 1963–70, after the partial test-ban treaty in 1963, exhibiting a steady increase of the number of underground explosions. Each of the three years, 1968–70, has nearly the same total number, which might suggest that some degree of saturation has been reached, technically, economically, politically or military.

4. 1971–6, with a constant and somewhat lower frequency of explosions. Notable features from this period are the continued atmospheric tests by France and China, besides underground tests, also that India conducts its first nuclear test (1974) and that the USSR expands activity on Novaya Zemlya to include two test areas from 1973.

Table 30 summarizes the largest underground explosions which have been made. The list includes those explosions which have given a body-wave magnitude (m) at the Swedish network of 6.5 and over.

Even though we have based the magnitudes on Swedish records only and even though there are still certain problems connected with magnitude determinations, the list can be considered as relatively homogeneous, es-

Table 30. The largest underground nuclear explosions ($m \geqq 6.5$) up to 1976, incl. *American tests*[1]:

Date	Location	Magnitude (*m*)	Remark
1968 January 19	Nevada	7.0	About 1 megaton at 980 m depth
April 26	Nevada	6.5	1.2 megaton at 1160 m depth
December 19	Nevada	6.5	1.1 megaton at 1400 m depth
1969 September 16	Nevada	6.5	800—900 kiloton at 1160 m depth
October 2	Aleutians	7.0	1.2 megaton at 1220 m depth
1970 March 26	Nevada	6.7	Over 1 megaton at 1210 m depth
1971 November 6	Aleutians	7.2[2]	5 megaton at 1791 m depth
1975 May 14	Nevada	6.5	765 m depth
October 28	Nevada	6.6	1265 m depth
1976 March 14	Nevada	6.6	0.5—1 megaton

Russian tests:

Date	Location	Magnitude	Remark
1965 January 15	Semipalatinsk	7.0	
1966 February 13	Semipalatinsk	6.6	
March 20	Semipalatinsk	6.8	
October 19	Semipalatinsk	6.5	
October 27	Novaya Zemlya	6.6	
December 18	Semipalatinsk	6.7	
1967 February 26	Semipalatinsk	6.7	
1968 July 1	North of Caspian Sea	6.6	
November 7	Novaya Zemlya	6.5	
1969 October 14	Novaya Zemlya	6.5	
November 30	Semipalatinsk	7.0	
December 28	Semipalatinsk	6.6	
1970 October 14	Novaya Zemlya	6.9[3]	
December 12	East of Caspian Sea	6.6	
December 23	East of Caspian Sea	6.5	
1971 April 25	Semipalatinsk	6.7	
September 27	Novaya Zemlya	6.7	
December 22	North of Caspian Sea	6.8	
1972 August 28	Novaya Zemlya	6.7	
November 2	Semipalatinsk	6.9	
December 10	Semipalatinsk	7.0	
1973 July 23	Semipalatinsk	7.1	
September 12	Novaya Zemlya	6.7	

Russian tests (continued):

1973	October 27	Novaya Zemlya	7.1
	December 14	Semipalatinsk	6.7
1974	May 31	Semipalatinsk	6.9
	August 29	Novaya Zemlya	6.5
	November 2	Novaya Zemlya	7.0
1975	April 27	Semipalatinsk	6.7
	August 23	Novaya Zemlya	6.6
	October 29	Semipalatinsk	6.7
	December 25	Semipalatinsk	6.7
1976	July 4	Semipalatinsk	6.7
	July 29	North of Caspian Sea	6.5
	August 28	Semipalatinsk	6.6
	November 23	Semipalatinsk	6.8
	December 7	Semipalatinsk	6.7

[1] For further detailed information on recent American tests, I refer to SPRINGER and KINNAMAN (1971, 1975).
[2] The largest magnitude for any explosion so far, partly due to favourable coupling.
[3] Estimated to correspond to 6 megatons by the American Atomic Energy Commission, and rated as the strongest underground test so far performed.

pecially as it is based on the same instruments, which have worked with unchanged characteristics for the whole observation period. The magnitudes have been revised to conform to the Zurich recommendations (Chapter 4) and they are averages from Uppsala and Kiruna short-period P-wave records. On account of the uncertainties of any transformation from magnitude to yield, I have refrained from making any such calculation, and the information on yield and depth given for the American tests is taken from American sources.

From Table 30 we see that large underground tests have not been made until during the last few years–since 1965. For magnitudes of 6.5 and above there are up to 1976, inclusive, 37 Russian and 10 American underground tests. This result is not exclusively explained by an overweight of strong Russian explosions. It is also to some extent depending upon the fact that our magnitudes for Russian tests, especially those in the Semipalatinsk area, tend to be somewhat high, probably due to favourable wave propagation properties. For the American explosions we see that most yields are about 1 megaton and the depths around 1 km. There is good reason to assume that most Russian tests listed have had similar values for yield and depth.

For further detailed discussion of "seismopolitical" aspects of nuclear explosions, the reader is referred to an excellent book by BOLT (1976).

11.8. Peaceful Application of Nuclear Explosions

Although originally developed for military purposes, there is no doubt that there are many peaceful applications of nuclear devices. In the USA the Atomic Energy Commission has developed a special programme, termed Plowshare, to incorporate such projects. Several such explosions have also been performed by the Americans. The so-called Sedan explosion in July 1962 is a good example of a cratering experiment. The explosion corresponded to 100 kt and was carried out at a depth of about 200 m. It produced a crater which was about 365 m across and about 100 m deep. In another American experiment, five 1-kiloton bombs were arranged along a line and simultaneously detonated. As a result, a linear crater was produced. Such experiments open up the possibility of using this method for the construction of large canals. Other fields of application include dam construction. Nuclear explosions carried out at greater depth can be used efficiently for excavation purposes, for instance, for construction of underground chambers for storage of oil and natural gas. Other applications involve stimulation of natural gas formations and oil reservoirs, etc. For instance, for gas stimulation an experiment was made in Colorado, USA, on May 17, 1973. Three charges, each of 30 kiloton, were placed vertically above each other in the depth range 1780 m to 2040 m. Scientific applications, especially seismological, have already been emphasized above.

It has been maintained that nuclear explosions provide a means to carry out large engineering works more efficiently and above all more economically than by conventional methods. On the other hand, even peaceful application of nuclear explosions entails a number of problems. Radioactivity, seismic shock and acoustic waves tend to limit the application to sparsely populated areas. Consideration of such effects, also of possible violation of a test-ban treaty, has caused much opposition to the use of nuclear explosions at all.

There is an abundance of relevant information from the USA. From the USSR there is some related information, but much more sparse. There seems to be a Soviet programme in many respects similar to the American Plowshare programme.

In 1970, the USSR described plans for peaceful application of large (nuclear) explosions in a report to the International Atomic Energy Agency

Table 31. Soviet explosions within the area of the Caspian Sea and Ural Mountains up to 1976, incl. Magnitudes are calculated as in Table 30.

Date	Location		Magnitude (*m*)	Remark
1954 September 14	62°N, 58°E	Urals	4.4	Chemical explosion.
1956 April		Urals		Chemical explosion, 174 tons (USSR). Not recorded in Sweden.
1957 December 7	60°N, 60°E	Urals	4.3	Chemical explosion, 432 tons (USSR).
1958 March 25	60°N, 60°E	Urals	5.3	Chemical explosion, 3.1 kilotons (USSR).
1966 April 22	48°N, 48°E	North of Caspian Sea	6.1	
September 30	40°N, 64°E	Uzbekistan	5.9	Nuclear (USSR).
1968 July 1	48°N, 48°E	North of Caspian Sea	6.6	Nuclear according to USA. Cavity in salt dome.
1969 September 2	57°N, 55°E	Urals	5.6 ⎫	Oil field
September 8	57°N, 55°E	Urals	5.6 ⎭	experiments.
September 26	46°N, 43°E	West of Caspian Sea	6.1	Gas stimulation.
December 6	44°N, 55°E	East of Caspian Sea	6.2	
1970 February 21	60°N, 59°E	Urals		Depth 1 km.
June 25	52°N, 56°E	Urals	5.6	
December 12	44°N, 55°E	East of Caspian Sea	6.6	
December 23	44°N, 55°E	East of Caspian Sea	6.5	
1971 March 23	61°N, 56°E	Urals	5.9	
July 2		Urals		
July 10	64°N, 55°E	Urals	5.3	
October 22	52°N, 54°E	Urals	5.8	
December 22	48°N, 48°E	North of Caspian Sea	6.8	
1972 August 20	49°N, 48°E	North of Caspian Sea	6.2	
September 21	52°N, 52°E	North of Caspian Sea	5.3	

Date		Location	Magnitude (*m*)	Remark
	October 3	47°N, 45°E	Northwest of Caspian Sea	6.4	
	November 24	53°N, 51°E	North of Caspian Sea		
	November 24	52°N, 64°E	Southern Ural region	5.6	
1973	August 15	43°N, 67°E	Uzbekistan	5.9	
	August 28	50°N, 68°E	Southeast of Urals		
	September 30	52°N, 55°E	Southern Urals	5.7	
	October 26	54°N, 55°E	Southern Urals	4.9	
1974	August 14	69°N, 76°E	Northeast of Urals	5.8	
	August 29	67°N, 62°E	Urals	5.4	
1976	July 29	48°N, 48°E	North of Caspian Sea	6.5	

in Vienna, among other things for regulation of the water system of the Caspian Sea. According to records at our stations, explosions partly with this purpose have been carried out already for several years. In Table 31, I have summarized all those explosions, known to us, which could be of interest in this connection. Coordinates, given to the nearest full degree, have in most cases been determined by the NEIS (USCGS), whereas magnitudes (*m*) are based on Swedish records. The explosions in Table 31, which were carried out during the 1950's, were all chemical, according to Russian reports. As we see from Table 31, the explosions during the 1960's are considerably larger, and most or all of them may be nuclear. The relation between many of the listed explosions and the Caspian Sea and its water system is quite obvious. In addition to these cases, there are also others, both east and west of the Ural Mountains. The Russians have an active programme for the application of nuclear explosions to extraction of ores, oil and natural gas, as indicated for some of the cases listed in Table 31.

The performance of large nuclear explosions, especially for peaceful purposes, is no doubt a major technical achievement. But it appears to require a far greater achievement also to have such explosions under control politically.

11.9 Chemical Explosions

Sometimes the question has been raised as to what extent it is possible to distinguish nuclear and chemical explosions by seismic means. Quite obviously, this problem is much more difficult than to discriminate explosions from earthquakes. The magnitude of an event may give some clue, especially if an explosion proves to be very large, in which case a nuclear origin would appear to be likely. The volume of the charge is considerably larger for a chemical explosion than for a nuclear explosion of corresponding yield. This could possibly influence minor details in a record. However, no noteworthy progress hitherto seems to have been made as far as deciding the exact nature of an explosion is concerned. We have probably to be content with the possibility of deciding that an event is an explosion and not an earthquake. It is in this connection, among others,

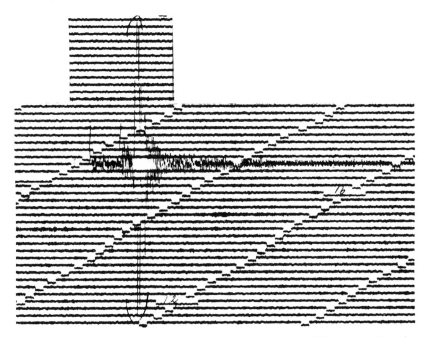

Fig. 121. Submarine mine explosion at Sandhamn (Baltic Sea) on August 6, 1957, as recorded by a short-period vertical-component Benioff seismograph at Uppsala. In order of arrival, the waves are *Pg1*, *Sg1* and *Rg*, the latter with the largest amplitudes.

that the request for direct inspection at the explosion site has been discussed. For instance, a chemical explosion would not give any radioactive fallout.

Smaller explosions of conventional nature are frequently recorded and they are also of interest from the seismological point of view. A few examples may be given. Figure 121 shows the record at Uppsala of the detonation of a submarine mine near Sandhamn in the Baltic Sea in August 1957. The distance is 100 km. The amplitudes (especially of the *Rg*-wave) are considerable, and a weak record was obtained as far as our station at Kiruna. The explosion occurred under water, which is the reason for the very good coupling to the solid earth.

The explosions in the Kiruna iron-ore mines are regularly recorded at the Kiruna station. It is true that to a certain extent they disturb our records, but in general we have no difficulties in separating the explosions from earthquakes in these cases. Figure 122 shows a typical record at Kiruna of one such explosion; the distance is about 10 km. Besides *P*- and *S*-waves (denoted *Pg2* and *Sg2*), there is a wave about 10 sec after *P*

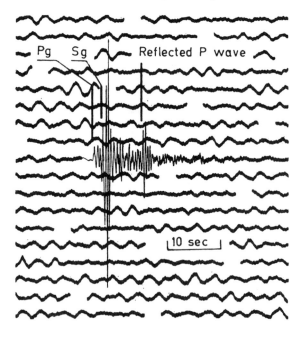

Fig. 122. Typical record of an explosion in the Kiruna iron-ore mines written by the Grenet seismograph at the Kiruna seismograph station on October 23, 1951, at 17.53 GMT. The waves *Pg2*, *Sg2* and the *P*-wave reflected at the Mohorovičić discontinuity are specially marked.

which has been reflected at the base of the earth's crust at a depth of 30–35 km.

Especially since the 1960's, the Russians have performed a long and intensive series of large chemical explosions on the Kola Peninsula. The charges are of the order of 50–100 tons of dynamite and records are as a rule obtained at our stations at Kiruna, Umeå and Skalstugan, in exceptional cases all over our network. The explosions are probably connected with mining activities in the area, but also construction of power plants, river regulation, etc., are possible reasons. Other explosions are frequently carried out in the Soviet border regions, especially in the Lake Ladoga area and around the coasts of Esthonia.

A comparative study of our records of chemical explosions in and near Fennoscandia and of earthquakes in the same region would no doubt yield valuable information for discrimination purposes. Because of the short distances involved, such research would hardly be of any value for detections by means of records at greater distance. On the other hand, such a study would be of essential importance in seismicity studies in Fennoscandia, considering the large number of explosions carried out in the area. On the average, the Swedish network can locate around 50 events per month within Scandinavia and its immediate surroundings. Of these, no less than 80% are explosions. Under these circumstances reliable discrimination criteria appear as very significant for seismicity studies in such areas. The following preliminary results may be mentioned:

1. Clear Rg-waves are recorded on short-period vertical-component seismographs up to distances around 500 km from explosions, but not from earthquakes, unless these are very shallow, i.e. with focal depths less than 2–3 km (as rockbursts).

2. The amplitude ratio Pgl/Sgl (or between any other pair of P- and S-waves) is generally larger for explosions than for earthquakes.

3. Combination of travel times of different waves through the crustal layers may provide information about the focal depth, which in turn may serve as a discriminating factor.

There are exceptions to these rules, and in several cases discrimination may be very difficult or impossible.

Besides discrimination problems, we note also the extensive use of controlled explosions for investigation both of path properties (Section 7.2) and of source properties. Thus, numerous investigations have been made on the generation of seismic waves from different controlled sources, e.g. from underwater explosions at varying water depths and with varying charges.

Chapter 12

Planetary Seismology

12.1 *Purpose of Research and Earlier Results*

Since the first satellite was launched by the Russians in October 1957, development has been very rapid in the exploration of outer space. This development is of great interest also in seismological research and in geophysical research in general. In this chapter, we shall consider geophysics as a whole and not just restrict ourselves to the seismological aspects. Furthermore, we shall concentrate our attention onto the moon, as this is the first heavenly body that has given any direct results in this connection.

Geophysics and moon research–this sounds like a contrast between, on the one hand, a research branch which is concerned with the earth, especially the physical properties of its interior, and, on the other hand, research which is directed outwards from the earth to outer space. However, there is really no such contrast. Instead, geophysics can expect very useful impulses and results from such direct planetary observations, as have now been realized in the case of the moon. The closer investigation of the moon and the planets means nothing else than a new epoch in the development of geophysics–for the first time, man is provided with bodies comparable to the earth in size for direct inspection. It may perhaps seem as if this would not diminish but rather increase our problems. To a certain extent this is true, but on the other hand, the new observations will provide a better check on our theories about the earth itself. It may be easy to formulate hypotheses, when observations are lacking. But we are not satisfied with hypotheses alone–we want to know the truth, and then observations from other planets will be of great significance also for geophysics.

There are several examples demonstrating how increased knowledge about the moon could influence and enhance our knowledge about our own earth. Here, I shall limit myself to a few of these. The base of the earth's crust, the Mohorovičić discontinuity, is generally considered to be a chemical discontinuity between overlying basic and underlying ultrabasic rocks. However, according to another hypothesis, this discontinuity would correspond to a phase change and not to a chemical discontinuity. There appeared to be good possibilities of getting some contribution to the solution of this problem from observations on the moon. If it should correspond to some chemical separation, possibly arisen during some cooling process, then the depth to this surface is expected to be about the same in the moon as in the earth (Chapter 7). However, if instead it were a phase change, the surface should be found in the moon at the same pressure and temperature as are prevailing at the Moho in the earth. And this may occur at quite a different depth in the moon. Seismic methods, directly applied to the moon, would thus be able to assist in the solution of this problem. This appeared to be a well-defined problem, until observations were really obtained. As we shall see in Section 12.4, the moon's crust is about twice as thick as the earth's crust. Nevertheless, due to differences in composition and in the stage of planetary development, no clear-cut answer to the question could still be obtained.

Astronomers have already for a long time supplied us with accurate information about the orbit of the moon–this is testified to by the accuracy with which eclipses of the sun and moon can be predicted as well as the accuracy of the propagation paths of the moon satellites. In addition, the mass, volume, density, etc., of the moon are known with good accuracy (see Table 32). Information of this kind, which is also available for the

Table 32. Some data for the earth, the moon, Venus and Mars.

Planet	Mass	Mean radius	Volume	Mean density	Gravity at surface
	10^{12} megaton[1]	km	10^9 km^3	g/cm^3	cm/sec^2
Earth	5975	6371	1083	5.52	981
Moon	73.5	1738	22	3.34	163
Venus	4860	6100	950	5.12	874
Mars	644	3385	163	3.95	389

[1] 1 megaton $= 10^6$ tons.

planets with high accuracy, has already been utilized by seismologists in comprehensive calculations for comparison with the conditions in the earth. The Australian seismologist BULLEN, notable among others, has published an extensive series of such investigations. These have already been accomplished facts for many years–BULLEN's papers in this field extend over more than 25 years back in time. On the other hand, access to direct observations on or near other heavenly bodies has become a reality only in the last decade.

Our present knowledge about the moon's internal constitution can be compared with our knowledge about the earth's interior around the turn of the century, before seismic recordings were available to a greater extent. It is probably correct to say, however, that the basis is more favourable in the case of the moon's exploration, especially as seismological experience from the earth can be utilized in the planning of experiments on the moon. On the whole, we could expect a repetition of the development of seismology during the present century–but now instead concerning the moon. With all probability this development will be considerably faster than in the case of the earth.

Guided by data as collected in Table 32 and theoretical considerations, certain ideas have already been formulated regarding the interior of the moon. A usual model for the moon's density assumes a surface value of 3.28 g/cm^3 and a value of 3.41 g/cm^3 at the centre. This is the simplest model derived from the compression of the material under increasing pressure. Then it is also assumed that the composition of the moon is chemically homogeneous and that no phase changes occur. Also spherical symmetry is assumed. However, other facts indicate that there are deviations from hydrostatic equilibrium and that density variations exist laterally and not only in a vertical direction. The pressure at the centre of the moon has been calculated to be slightly less than 5×10^{10} dynes/cm^2, approximately the same as the pressure at a depth of only about 150 km in the earth. These and other calculations await verification by direct observations on the surface of the moon. Among these, seismic observations play a significant role.

12.2 *The Moon's Magnetic Field, Topography and Origin*

Whether or not the moon has a central core, corresponding to the earth's core, is still an unsolved problem. Sometimes, it has been suspected that a central core is lacking. This has been based partly upon the relatively low mean density of the moon (3.34 g/cm^3), partly upon the large rigidity of the lunar body. The latter is proved by the shape of the moon, which is elongated in the direction to the earth. This figure has arisen in an earlier epoch as a consequence of gravity effects and still remains. Whereas remanent magnetization has been found in rock samples from the moon, as well as in recent observations in the vicinity of the moon, it appears as if the moon lacks a present magnetic field of any significance. This is a very interesting observation which seems to indicate the absence of a moon core and at the same time lends support to the theory that the earth's magnetism has its origin in the fluid core. However, the question about the existence or non-existence of a moon core can probably not be fully settled until we have obtained the corresponding seismic records from the moon.

An essential contribution to the study of the topography of the moon surface was made by the Russians in 1959. Using a very much developed technique they succeeded for the first time, by means of Lunik III, in taking pictures of the rear side of the moon, which up to that time had been completely unknown to us. Among the most striking discoveries was a moon crater about 300 km in diameter, i.e. considerably larger than the largest crater known up to that time (about 200 km in diameter). After the publication of the Russian moon pictures, it was advocated that the moon had been the scene of large catastrophes–violent collisions with other heavenly bodies which had caused lava flows over the moon's surface, this in turn creating the so-called seas. At first, it was believed that the rear side of the moon was much smoother and had fewer contours than the front side. But further photographic work, both by Russians and Americans, revealed that the rear side resembles the front side in all its properties. Numerous photographs have been taken on later moon flights, especially by the American expedition in July 1971 (Apollo 15).

Before landings on the moon were undertaken, careful investigations were made of the surface structure. Among others, a 'moon-scraper' was used to test the properties of the surface. It had been feared that large parts of the moon might be covered by soft material of volcanic origin. People and instruments could then be in danger of sinking through this material to unknown depth. However, both the results from the moon-scraper and the placing of heavy photographic equipment on the moon as well as the American Apollo landings on the moon (beginning in July 1969), all testified that the surface is solid and able to support the loads applied, at least at the places investigated.

An interesting observation concerns rifts on the moon, which could suggest a seismic activity. In May 1967, it was communicated from the Jet Propulsion Laboratory in Pasadena in California that the American moon satellite Lunar Orbiter 4 had photographed a moon rift, which was 318 km long and 16 km wide. It was considered that this rift may have been created by a moonquake and that lava had poured out from within the moon's crust. Similar rifts had also been noticed earlier on moon photographs. As an alternative explanation, the possibility has been suggested that a meteor has grazed the lunar surface and thus caused a rift. Some scientists are of the opinion that the fissures on the moon's surface form a regular pattern, which could be explained by convection currents in the moon's interior–possibly in an earlier epoch. In some cases, features have been found on the moon which bear strong resemblance to certain features on the earth's surface, e.g. the fissure occupied by the Red Sea. On the other hand, it has also been emphasized that mountain ranges do not exist on the moon in the same sense as on the earth. The nature of possible tectonic phenomena on the moon is still an open question.

Regarding the origin of the moon, we can distinguish between three different hypotheses:

1. Separation from the earth.
2. Origin simultaneous with the earth as a double planet.
3. Capture from outer space by the earth.

According to the old idea, the moon had been separated from the earth and left the Pacific Ocean as its mark. The circumstance mentioned above, that a lunar core is probably lacking, has been thought to support this idea. It has been envisaged that a mighty tidal wave, amplified by resonance, was able to break loose from the earth and to form the moon. This theory has been almost completely abandoned because of several difficulties, especially that friction would have limited the height of the tidal wave too greatly; furthermore that the formation would have taken place when the earth was liquid, and then no trace, like the Pacific Ocean, could have been left. As alternative theories one has considered the possibilities that the earth and the moon were formed simultaneously as a kind of double planet or that the moon was captured from outer space by the earth. However, Professor BULLEN in Sydney, with collaborators, has proved that phase changes in the earth's interior with reasonable assumptions would be able to generate sufficient energy to eject the moon from the earth. In a way, this throws us back to the old hypothesis, even though many problems remain to be solved.

This idea about the moon's possible origin from the earth, which was proposed by BULLEN in 1951, seems to be relatively unknown but deserves to be briefly described in this connection. According to certain newer ideas, the earth's core inside 2900 km depth, has the same chemical composition as the surrounding mantle (silicates), but by a phase transition under high pressure and high temperature, the core material is in a so-called metallic phase (Chapter 7). A planet with such a core of very small size is in an unstable state, and the stable state corresponds to the normal (molecular) phase. The phase change to the stable state entails liberation of energy, which under certain circumstances could be enough to eject the moon from the earth. The Indian scientist DATTA showed in 1954 that this is possible quantitatively, if we start from an original earth-moon with three phases, a normal phase, X, and two high-pressure phases, Y and Z. The disappearance of the phase Z would be connected with enough energy liberation to eject the moon from the earth, provided that the radius of the phase Z amounted to at least 1500 km and that its density were at least 18.5 g/cm³. These assumptions appear to be reasonable, but cooperation is required with some resonance effect which could bring the body over a certain potential barrier.

Age determinations of lunar rock samples, collected during the Apollo landing in July 1969, gave surprisingly high values, about 3.5×10^9 years. Age determinations of samples collected during later Apollo expeditions have yielded even higher values, up to about 4.6×10^9 years. Such results will be considerably amplified by investigation of rock samples from other localities on the lunar surface as well as from beneath it, through planned drilling to about 3 m depth. The presently most-favoured idea about the moon's origin is that it was captured by the earth about 4×10^9 years ago.

12.3   *Selenophysical Phenomena: Volcanic Eruptions, Moonquakes, etc.*

In this section we shall essentially restrict our attention to observations of selenophysical phenomena from the time prior to the first moon landing (July 1969). A question of the greatest interest is whether the moon is physically dead or alive, i.e. if volcanic eruptions, moonquakes, etc., exist or not. The surface of the moon is mainly characterized by extensive lava fields (maria), surrounded by mountain ranges, and an enormous number of craters, possibly formed by meteor impacts in an early epoch. For comparison, it should be emphasized that similar craters have been found on the earth's surface, as in Arizona and Canada, probably with the same origin. To discover such craters on the earth is considerably more difficult, due in part to the cover of soil, partly to the weathering and eroding effect of wind and water. The exact origin of moon craters has been very much debated, and the ideas vary from an exclusively volcanic origin to an exclusively meteoritic origin. Many scientists hold an intermediate position and consider some craters to be volcanic and others to have been formed by meteor impacts. Probably only a direct investigation of the moon's surface will lead to an answer to this question. In the event that the hypothesis of meteor impacts be verified, then the further exploration of the moon's surface will contribute to our knowledge of meteors also.

Many times it has been maintained that changes have occurred on the moon's surface, such as the disappearance of some crater or the appearance of a new one, which could suggest volcanic eruptions or moonquakes. But many of these observations, especially those which are visual and not photographic, have to be taken with caution, because of the great influence of different lights and shadows. The best-known case is the disappearance of the crater Linné which could suggest a lunar catastrophe some time between 1843 and 1866. Such observations do not in general provide very reliable evidence either for,' or against any selenophysical activity. We have to bear in mind that even large moonquakes could occur without leaving any surface trace visible from the earth. Just as on the earth, we have to expect possible moonquakes to occur at many kilometres depth below the surface.

The temperature variation on the surface of the moon amounts to about 280° between its day- and night-side. There is no doubt that this will cause heavy stresses in the moon's crust, which could provoke breaks and displacements. Observation of radiofrequency radiation demonstrates a temperature variation of only about 110°. This corresponds to conditions at a depth of about 40 cm below the moon's surface. Likewise, the Apollo 15 heat flow measurements reveal that the diurnal temperature variation is almost negligible at 50 cm depth. These observations show not only that the superficial layer consists of material with very low heat conductivity, but also that the stresses produced must be restricted to the very top layer. Nevertheless, it has been found that a large number of small moonquakes must be ascribed to the diurnal temperature variation on the moon's surface. These quakes, so-called *thermal moonquakes,* are explained as slumping of lunar soil triggered by diurnal thermally induced stresses.

It has been reported repeatedly that haze or fog has been observed on the moon, especially in some crater. Most likely this consists of gas emanating from the moon's interior and is not a result of condensation in any thin atmosphere. Of particular interest are the much discussed observations of presumed volcanic activity on the moon that were made from the Crimea Observatory, for the first time in November 1958, and since then on several other occasions. The question of the possible existence of a very thin moon atmosphere has not yet been answered definitely. It does not appear impossible that a thin upper atmosphere and an ionosphere could exist, to some extent resembling conditions on the earth. A few observations of light flashes, interpreted as due to meteorites passing through the outer moon atmosphere, could be taken as an indication of this. Ionosphere scientists and meteorologists will probably also have an interest in closer investigation of the moon, even though some of the planets, such as Venus or Mars, will have more to offer for these branches.

12.4   *Instrumental Observations and Models of the Moon*

In 1957, I mentioned the importance of seismograph installations on the moon to the leaders of the American and Russian committees for the International Geophysical Year project. At both places, the proposal was received with the greatest interest and sympathy and was considered to be unique and of considerable value, perhaps only raised too early. Since then, the development of suitable seismographic equipment for the moon has advanced considerably, especially in the USA. In the beginning of the 1960's, the Americans launched three rockets with seismograph equipment for automatic installation and operation on the moon, but these efforts failed and at least one of them missed its target. It was with the American astronaut landing on the moon in July 1969 (Apollo 11), that seismometers first were mounted on the moon's surface. The recording was done on the earth by means of radiowave transmission. This equipment consisted of four seismometers: one 3-component long-period system and a short-period vertical-component apparatus. In principle, they were of a capacitor type (Chapter 2) with magnifications reaching 1.6 million. During the lapse of 21 days, these recorded several events suggesting moon slides (probably a temperature effect on the moon surface) and a few events whose origin could not be fully ascertained: moonquakes or meteor impacts. A clear velocity dispersion could be observed, suggesting a layering in the moon's crust.

The absence of water and, practically, of atmosphere on the moon implies absence of noise (microseisms) on seismic records. This means that considerably higher magnification could be used than is possible for any station on the earth. On the other hand, we cannot exclude the possibility that temperature variations in the superficial layer with ensuing motions could disturb the records. The continuous impacts of meteorites against the moon are another and perhaps even more important source of noise. Experience has shown, however, that none of these effects are able to produce any significant background noise, at least not at the sites investigated. In fact, the background seismic noise has been found to be extremely low, far below the noise level at any known earth site. As a consequence, it has been found possible to detect motions on the moon's surface as small as $1 \text{ Å} = 10^{-8} \text{ cm} = 10^{-4}$ microns. In the study of records of possible moonquakes we will be faced with the problem of distinguishing between meteor impacts and real quakes, a problem which very much resembles that of distinguishing explosions from earthquakes on the earth (Chapter 11).

Recordings of meteor impacts on the moon may give very valuable information about their frequency of occurrence. Previously, only moon photographs could provide any material for such studies. From careful studies of these it has been found that the number of craters increases exponentially with their decreasing size, and that the whole surface of the moon has been covered by craters over and over again. Such results have been of significance in connection with studies of craters on the earth as well as for a better apprehension of conditions in space, of importance for future space flights among other things.

By means of smaller explosions, a method is offered to generate seismic waves both regionally and through the whole of the moon, whose records would clarify the layering in its interior, the possible existence of a lunar core, etc. By means of records of moonquakes, the seismic conditions on the moon could ultimately be mapped.

An interesting seismological experiment was carried out in connection with the moon landing of Apollo 12 in November 1969. After takeoff from the moon, the ascent stage of the lunar module was shot against the moon's surface. This generated an artificial moonquake which gave strong records on the seismometers placed 75.9 km from the impact point. Most remarkable in this connection is the long duration of the record–no less than 55 minutes–which gave rise to many discussions (Fig. 123). With

Fig. 123. Seismic signal received on the lunar long-period vertical-component seismometer from the Apollo 12 lunar-module impact. After G. LATHAM et al. (1970). 1 nm = 1 nanometre = 10^{-9} metre = 1 millimicron.

a mass of 2383 kg and an impact velocity of 1.68 km/sec, the kinetic energy of the lunar module corresponds to a body-wave magnitude $m=5.0$ at 100% conversion into seismic wave energy [using equations (1) and (2) in Chapter 11], to $m=4.6$ at 10% energy conversion, and to $m=4.2$ at 1% energy conversion. However, it is estimated that the conversion factor may be much smaller, around 0.001%, i.e. 10^{-5} of the kinetic energy. For such a conversion we get $m=3.0$. The ascent stage, shot down from Apollo 13 in the spring of 1970, had a larger mass (13 925 kg) and a higher impact velocity (2.58 km/sec). The seismic magnitude m is calculated to be 5.4, 5.0, 4.6 and 3.5, corresponding to the four cases, respectively, just dealt with. A 4-hour-long seismic record was then obtained at a distance of 115 km. Later experiments during the following Apollo flights gave similar results. For comparison, on February 4, 1967, an earthquake occurred in Värmland,

Sweden, with a magnitude $m=4.6$, which gave records on short-period instruments of about 6 minutes duration both at Uddeholm (distance 53 km) and at Uppsala (distance 245 km). On the moon, we have to expect longer duration, partly because of the considerably larger seismometer magnifications used and partly because of considerably smaller or absent background noise. But it appears that in addition we need to have recourse to some structural difference in order to explain the long duration of the moon record. A number of structural properties has been suggested in speculation about the long duration of the moon record which we can summarize in the following points:

1. Existence of a much greater number of homogeneous (competent) layers in the moon, which would make numerous reverberations possible.

2. Existence of a sharp discontinuity at shallow depth in the moon, which to a large extent would preserve vibrations within the top layer ('blanket effect').

3. Existence of blocks (perhaps of meteoric origin) with loose coupling to the rest of the moon, and that impact and seismometers were located on the same block.

4. A high quality factor Q (see Chapter 7) of the moon material, of the order of 3000–5000, or low absorption coefficient, because of degassing, dehydration and possibly compression.

5. Scattering of seismic waves in a highly heterogeneous medium.

6. A spray of secondary ejecta around the seismometer, i.e. lunar dust particles which rain down on the lunar surface close to the seismometer. If true, this explanation for the long duration of the record naturally detracts much of the seismological interest in the experiment.

Some of these explanations are obviously contradictory, and only repeated experiments will permit more reliable conclusions. The fact that similarly long durations were obtained at other places on the moon in 1970 and 1971 suggests that it is a moon-wide phenomenon and not of local nature.

The seismometers, placed on the moon in connection with the Apollo 12 landing in November 1969, continued for a long period to record about one event per day. These were found to be due to two causes, on the one hand, lunar tides (attraction mainly from the earth) and, on the other hand, meteor impacts. This means that both reasons for the moonquakes were exterior to the moon and any interior activity, corresponding to that known for the earth had so far not been discovered. The observation of moonquakes due to lunar tides is interesting because it sheds light on an old problem concerning the earth, i.e. whether earthquakes are released by tidal triggering action or not. The experience from the moon seems to justify the assumption of such an effect, even though it is generally hidden by other more important phenomena on the earth. For the moon it did not appear clear whether the tides act as generators of quakes, i.e. causing both strain and its release, or if they act simply as triggers, i.e. releasing strain existing in the moon for other reasons.

It is of considerable interest to note that moon recordings generally do not resemble those regularly obtained on the earth, except for those due to such phenomena as volcanic tremors, microseisms, landslides, etc. Distinct phases as observed in earthquake records do not show up on the moon in the time-compressed record of Figure 123. With better resolution, however, separate phases have been found, corresponding to P, S and surface waves, even though these are much less conspicuous than on the earth. If any quake activity should exist, similar to that on the earth, then the time of observation would have been long enough to record many of these. At present, we can speculate either that tectonic activity is absent on the moon, at least it is not at all comparable to the one on earth, or that the moon's interior is of such a nature as to prevent wave propagation to greater distances. The latter idea obviously conflicts with suggestions of high quality.

With the installation of the seismograph equipment of Apollo 15 in July, 1971, the first triangular array was established, further developed by the installations of Apollo 16 and 17. This naturally enhanced the possibilities of locating seismic events. The most remarkable finding was moonquakes at depths of 700–1200 km with magnitudes of 2 to 3. While generally no larger moonquakes have been observed, the depth is quite remarkable. Among other events we note swarms (Section 6.4) of moonquakes with one shock every second hour and lasting for several days, but still with unknown reason. It has also been possible to locate specially active zones – one of which is responsible for about 80% of the moon's seismic activity.

With the much more abundant data which gradually became available, it was possible also to conduct some statistical studies of moonquakes. For instance, in the frequency–magnitude relation $\log N = a - bM$, cf. Section 5.2, it was found that $b \cong 2$ for moonquakes as distinct from its value around 1 for tectonic earthquakes. Such a high value of the b-coefficient is characteristic for volcanic earthquakes, and could suggest microfracturing, possibly due to thermal stresses.

Guided by our seismological knowledge about the earth, quite a number of theoretical calculations have been made regarding the seismic conditions of the moon. Among such calculations, we may mention determination of preliminary travel times for P- and S-waves, assuming constant velocities of 8.0 km/sec and 4.6 km/sec, respectively, for the entire moon. On the basis of these assumptions, it will take a little more than 7 min and $12\frac{1}{2}$ min for a P-wave and an S-wave, respectively, to pass straight through the moon. Later, these calculations have been extended to various layered models of the moon, also to a number of body waves in addition to P and S (as pP, PP, sS, SS). See NAKAMURA and LATHAM (1970) and DERR (1970). Other theoretical calculations concern surface waves. Among other things, dispersion curves for Rayleigh waves with periods between 10 and 100 sec have been drawn for two moon models with different layering in the moon's crust. Similarly, the free vibrations of the moon have been calculated under the assumption of certain models. Observations of spheroidal moon vibrations are expected to be able to discriminate between different moon models. Such calculations, of which just a few examples have been mentioned here, are mostly theoretical and speculative. But they will certainly become of great importance by comparison with direct seismological observations of the moon.

More recently, it has become possible to place considerations of moon models on a more reliable basis thanks to direct observations. See Figure 124 and Table 33. The velocity structure is based on travel times and amplitudes of recorded seismic waves, especially P, together with information from theoretical seismograms and laboratory measurements of seismic wave velocities in rock samples from the moon. A two-layered crust with total thickness of no less than about 65 km is suggested, overlying a lithosphere which extends to depths around 1200 km. The existence of remarkably deep moonquakes, at depths of 700–1200 km, suggests that the interior is solid enough to such depths to hold stresses. The innermost part, astheno-

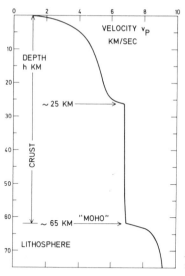

Fig. 124. P-wave velocity profile in the crust and uppermost mantle of the moon. After TOKSÖZ et al. (1972), modified.

sphere, extending from about 1200 km depth to the moon's centre, is probably partially molten, as indicated by stronger attenuation of S-waves and absence of moonquakes. There is no correspondence to the earth's core, unless one would like to look upon the inner asthenospheric globe as a certain counterpart, even though it better corresponds to the earth's lower mantle. Comparing the gross features of the earth's and the moon's interior, we could say that the moon exaggerates the outer parts of the earth, but

Table 33. Model of the moon's internal structure.

Layer	Depth range km	P-wave velocity km/sec	Remark
Crust {	0—25	0.1—6.8	Increasing velocity due to compaction and chemical change.
	25—65	6.8	Practically constant velocity.
Lithosphere	65—1200	9.0 at 75 km depth	Quakes located to depths of 700—1200 km.
Asthenosphere	1200—1738		Partially molten. S-waves attenuated. Possibly decreasing P-wave velocity.

suppresses the inner parts. For instance, taken in relation to the respective radii of the moon and the earth, both the crustal thickness and the depth to the deepest quakes are 6 to 7 times as large for the moon as for the earth.

Other geophysical–or more correctly selenophysical–measurements should also be conducted on the moon, such as heat flow and radioactive measurements, observations of moon tides, gravity measurements and selenodetic measurements. Several of these are already included in the manned lunar landing projects, besides the collection of lunar rock samples. Selenothermal measurements ought to give information about the thermal history of the moon, and thus contribute to the solution of the problem of the origin of the moon and our planetary system. The shape of the moon should be accurately determined, and also an exact coordinate network (longitudes and latitudes) will be necessary.

A recent observation on the moon concerns so-called *mascons*, i.e. mass concentrations, which exhibit themselves by increased gravity. Their origin has been much discussed, and the suggestions vary from meteoritic origin to an origin from the moon's interior. Besides the existence of mascons, there are several other indications of relatively low internal temperatures, probably below the melting point, within the moon, e.g. results from the Apollo surface magnetometer experiment, furthermore the shape of the moon and the existence of very deep quakes, which both testify to considerable internal strength. Also direct heat flow measurements on the moon's surface give similar results. The heat flow has been measured as 30 ergs/cm^2 sec, i.e. $0.7 \cdot 10^{-6}$ cal/cm^2 sec or only about half of the earth's average value, estimated as $1.5 \cdot 10^{-6}$ cal/cm^2 sec. Only preliminary estimates of the moon's internal temperature have been made, all of them lying below 3000 °C and some even below 1000 °C.

Observations of cosmic radiation, not to mention astronomical observations, both optical and in the radiofrequency range, could be made with advantage from the moon as well as all observations for which the earth's atmosphere is a disturbing factor. Among these we may mention all kinds of radiation measurements, such as radiation from the sun and from the earth. These and other observations have been made with rockets, which, however, are unable to provide continuous observations over a longer time span. Satellites, however, provide such a possibility. For instance, for observations of corpuscular radiation from the sun, distances on the earth are too short to provide sufficiently long base lines. But simultaneous observations from the earth and the moon would be able to furnish important information about the propagation and distribution of this radiation. The moon contains such a record of the sun's development as is impossible to find in meteorites or in any substance on the earth. For 4.6×10^9 years, the moon has been exposed to an uninterrupted solar wind. This means that particles from the sun can be found on the moon and corresponding samples have now been collected for laboratory investigation on the earth. Also a number of observations of our own earth could be made with advantage from the moon, e.g. of the tail of leaking atmosphere that sometimes has been suggested to exist. At present, it is mainly a question of technical facilities, whether many of these observations are better made on the moon's surface or from a space laboratory. The low gravity and the practically complete absence of air resistance make the moon very suitable as a platform for continued satellite and rocket experiments on our way towards outer space.

A laser reflector placed on the moon in July 1969 (Apollo 11) has made it possible to measure the earth-moon distance to an accuracy of about 15 cm. This increased accuracy has great significance for many studies, such as of the moon's motion and of continental drift and polar wandering on the earth. Therefore, a network of observing stations around the earth would provide accurate means for testing hypotheses as expressed in the new global tectonics (Chapter 6).

Even though interest at present is naturally focussed on the moon, preliminary calculations, similar to the ones described above for the moon, are being made for the planets next in line for manned landings, i.e. Venus and Mars.

As the mass and the mean density of Venus are nearly the same as for the earth (Table 32), these two planets have presumably almost the same composition. Observations on Venus may therefore have a special relevance to observations on our own earth. It is considered on good grounds that Venus has a core, which is metallic and liquid, like the outer core of the earth. Accurate determinations of the radius of the Venusian core would make a discrimination possible between the two most important hypotheses for the boundary of the core: whether this is a chemical discontinuity or represents a phase change (see Chapter 7). We see that this problem is analogous to the problem concerning the Moho discontinuity and the moon.

Mars is like Venus a terrestrial planet, and seismological information can be expected in the future which will resemble that from Venus. The existence of a Martian core is more uncertain than a Venusian core. BULLEN has estimated that the radius of the Martian core amounts to 900 km at most, but direct measurements are needed to test the hypothesis. The American chemist UREY has suggested that Mars is chemically homogeneous. This too can be tested by direct measurements of wave velocities on Mars. In July and September, 1976, the Americans succeeded in placing instruments on the surface of Mars with the unmanned expeditions Viking I and Viking II, respectively. They both brought seismographs, of which Viking I was damaged on landing, but of which Viking II in January, 1977, gave the first positive indication of a marsquake. Otherwise, this activity has concerned taking lots of photographs and some meteorological observations, and, above all, searching for indications of possible life – now or in the past – on Mars.

Finally, let us look at the development from a historical perspective- back in time but essentially into the future. Throughout all time, man has been attracted towards exploration of the unknown. Quite clearly, it is easiest to move in horizontal directions along the earth's surface, both over continents and oceans. This was done during the time of the great geographic discoveries in past centuries, when new continents and new oceans were found. In our time, this expansion corresponds to man's exploration of outer space–the enormously increased problems connected with such journeys are now being solved. In this chapter, we have been dealing mainly with the moon, but exploration of other planets will also have great repercussions on geophysical research. In the present rapid development, geophysics occupies a rather strange position. The object for its study–the interior of the earth–is at the same time so close to us and so inaccessible. During the Apollo 11 expedition in July 1969, samples of moon rocks were for the first time carried to the earth for direct investigation in the laboratory. Thus, it has proved to be easier to get a rock sample from the moon than to get a corresponding sample from only 20 km depth in the earth. Another example: our knowledge about the nature of the core-mantle boundary only 2900 km beneath our feet is less certain than our knowledge about the outer parts of the sun.

The recent direct observations on the moon have stimulated an enormous literature and also raised a number of new problems not envisaged before. But there is no doubt that with these observations new pathways are opened up to science, which will lead to a more reliable apprehension of the universe and its development.

Chapter 13

Seismological Education and Practice

13.1 *Seismology Curriculum at Uppsala University*

University institutes in seismology and in geophysics now exist in many countries, where both research and education are carried on. Just as an example, I will quote below the present curriculum in seismology followed at the Seismological Institute at Uppsala, Sweden. Seismology here became a university subject with an established curriculum in 1961, but only on the postgraduate level. This means that students should first have a degree corresponding to B. Sc. or M. Sc. and then start with seismology which would lead to a degree corresponding to Ph. D.

Recently, the Swedish degree system has been reorganized, such that the basic university degree (*filosofie kandidat*, roughly equivalent to B. Sc.) can be attained in three years after high school. This is then followed by a four-year programme of education leading to the new doctor's degree (*filosofie doktor*, roughly equivalent to Ph. D.). The programme of education in seismology still exists only on the latter, postgraduate level. I reproduce here in full a translated version of the present (1977) curriculum in seismology with brief explanations inserted where needed.

Program of Studies for Graduate Education in Seismology
at Uppsala University

This program of studies is formalized by the Office of the Chancellor of the University March 27, 1973. Graduate studies are provided to the extent that available resources allow.

1. *Subject Description and Educational Objectives*

1.1 *Subject Description*

Seismology, literally translated, means the study of earthquakes. The central problem in seismological research is the generation, transmission, and registration of elastic waves. Registration of seismic waves by a world-wide network of seismograph stations makes it possible to determine the location of earthquakes as well as their energy, mechanism, etc. and thereby gives valuable insight into the global tectonics. The properties of seismic waves (velocity, amplitude, etc.) depend upon the medium through which they pass. Thereby, studies of the physical properties of the Earth's interior are made possible, and seismic methods present a more certain and meaningful determination of these properties than any other methods heretofore presented. Seismology has many practical connections as well, among which prospecting, seismic protection of various types of construction projects, and nuclear weapons detection should be especially pointed out.

Seismology is an applied mathematical and physical science and, like geophysics in general, it is broadly based upon physical observations, laboratory experiments (seismological modelling, etc.), and theoretical studies.

1.2 *Educational Objectives*

The educational goal is to provide the student with deepened knowledge and capability in the methodology of research, leading to the ability to carry out independent seismological research, including planning and supervision.

2. *Requirements for Admission*

To be considered for admission to graduate studies in seismology, the applicant must have a bachelor's degree, as well as at least 40 points credit in the area of mathematics and at least 60 points credit in physics or mathematical statistics. (In the Swedish university system, a passing grade in a one-semester course normally corresponds to 20 points which is equivalent to 1 point per week).

It is recommended that this preparation includes a course in computer programming (for example, 10 points in computational physics as part of a physics course, or in information processing, especially numerical analysis).

Applicants who, in other programs inside or outside Sweden, have acquired capabilities corresponding basically to the above will also be considered for admission. In addition, a person who has not been formally admitted as a graduate student has the right to undergo examinations and defend a thesis for the doctor's degree, if he fulfills the general admission requirements.

3. *Educational Program*

3.1 *General*

The doctor's degree program is planned so that it normally requires four years' full-time study. The course-related work corresponds to approximately 40 points and is mainly concentrated in the first two years. This includes both elements which are common to all students and elements which are directed towards the scientific studies particular to the individual dissertation.

At the beginning of the program, an individual study plan which the student intends to follow is defined. This plan is worked out between the student and his advisor.

3.2   *Courses*

3.2.1 *Methodology and introduction*, 8 points comprising:

a) Physical geology (features and formations of the surface of the Earth, constitution of the crust, external and internal processes, corresponding, for example, to HOLMES: *Principles of Physical Geology*).

b) Statistical processing of observations (distributions, means, correlations, significance calculations, spectral theory, least-square method).

c) Applied mathematics (contour integration, conformal mapping, stationary-phase method, series integration, Bessel functions, Legendre functions, integral transforms, matrix algebra, calculus of variations, integral equations; BÅTH: *Mathematical Aspects of Seismology*).

d) Computer programming, in case this is not included in the student's preparation (the ability to write the necessary programs may be best acquired by auditing one of the programming courses given in other areas).

e) Spherical trigonometry and map projections (familiarity with spherical trigonometric formulas, especially sine and cosine relations, as well as the most common map projections).

f) Introduction to seismology (BÅTH: *Introduction to Seismology*) and to areas related to seismology, such as gravimetry, Earth's magnetism, tectonics, geothermometry, etc. (HOWELL: *Introduction to Geophysics*).

3.2.2 *Theoretical seismology*, 20 points (wave equation, transmission through layered media, surface and body waves, effects of gravity, curvature of the Earth, and viscosity upon wave transmission, vibrations of plates and cylinders, transmission through media with variable velocity, free oscillations of the Earth; EWING, JARDETZKY, PRESS: *Elastic Waves in Layered Media*, BULLEN: *Theory of Seismology*, BÅTH: *Mathematical Aspects of Seismology*).

3.2.3 *Seismicity and general seismology*, 3 points (seismic condition of the Earth, earthquake processes, direct observations, magnitude and intensity, more important earthquakes and their effects, calculations from records, etc; RICHTER: *Elementary Seismology*, GUTENBERG, RICHTER: *Seismicity of the Earth*).

3.2.4 *Tectonics, internal structure of the Earth*, 3 points (interior of the Earth, the crust, the mantle, the cores, physical properties in the Earth's interior, inelastic processes, tectonophysical processes; GUTENBERG: *Internal Constitution of the Earth*, GUTENBERG: *Physics of the Earth's Interior*, JEFFREYS: *The Earth*).

3.2.5 *Instrument and seismograph theory*, 3 points (theory of seismographs, methods for determining constants, theory as well as practice; BYERLY: *Seismology*, BENIOFF: *Earthquake Seismographs and Associated Instruments*).

3.2.6 *Seismic prospecting*, 3 points (theory of seismic prospecting, practical aspects, participation in field courses and field work; HEILAND: *Geophysical Exploration*, DIX: *Seismic Prospecting for Oil*, SLOTNICK: *Lessons in Seismic Computing*, DOBRIN: *Introduction to Geophysical Prospecting*).

The above-mentioned literature is meant only to specify approximately the level of each topic and may be replaced by course literature of similar value after agreement with the supervisor.

The courses are terminated by written and/or oral examinations.

One or more of the courses may, with agreement of the research supervisor, be replaced or modified, whereupon the supervisor shall determine how course credits shall be arranged. The substituted course can also be selected from subject areas other than, but related to, seismology.

3.3 *Dissertation*

For the doctor's degree it is also required that the candidate write and publicly defend a scientific presentation. Specifications of the form of the dissertation and of the public defence are found in §§ 59–69 Declaration (1969:50) on education in the philosophical faculties.

The dissertation shall be of such quality that it is judged to fulfil recognized, reasonable requirements to be accepted, as a whole or in summary, for publication in an international scientific journal of high quality.

The choice of subject for the thesis is normally made in agreement with the advisor.

4. *Instruction*

The Seismological Institute provides instruction for the basic courses for the doctor's degree to the extent that instructional staff is available. Beyond this, lectures and seminars in specialized areas as well as guest lectures are arranged.

5. *Advisors*

Candidates accepted to graduate studies are entitled to supervision by an advisor, full-time students for four years, and part-time students with university positions as assistants or technicians for five years. Other part-time students are entitled to research supervision of corresponding extent. Upon entrance into the program, each student is assigned an advisor who is continuously available to supply guidance concerning the course of studies and the selection of course literature, etc. Later, usually when the student has completed some of the basic courses and begun to consider the nature of the dissertation, the student will work with a research advisor who need not be the advisor first assigned and in the normal case directs the special studies and research leading to the dissertation. Normally, this research advisor is also an examiner for the doctoral dissertation.

It is the responsibility of the student to keep his supervisors informed of the progress of his studies and research.

A supervisor who is not himself appointed to be an examiner for the doctoral examination is requested to reach a final decision in agreement with someone who is a qualified examiner in those cases mentioned in points 3 and 6 in this Program of Studies.

6. *Examinations*

Examinations for the doctor's degree in seismology cover the basic courses by means of written or oral examinations, both at the terminations of courses and at other times when required.

Those who wish to carry out studies and dissertation work without being accepted as a doctoral candidate are entitled to undergo examinations for the doctor's degree.

Students are entitled to transfer credit for examination passed in graduate studies in a philosophical faculty even at another university. This also applies to examinations in graduate studies at colleges within Sweden.

7. *Miscellaneous Information*

Because of the international nature of the research community, all instruction (lectures, seminars, etc.) is given as a rule in English. The students also use English in their seminar talks, which seems to give the students practice for participation in international scientific conferences.

13.2 *Exercises*

In this section we give a number of exercises, by which the reader will get an opportunity to test his knowledge acquired from the reading of the present book. All the problems are relatively simple and of such a nature as a seismologist is confronted with in his daily work with records.

1. Determine the epicentre and the origin time for the following event:

		h m s			h m s
Uppsala	i	22 50 57.0	Skalstugan	i*Pn*	22 48 11.2
	i*Sg*	22 51 03.7		i	22 48 34.5
				i*Sg*	22 48 57.2
Kiruna	i*Pg*	22 48 16.9			
	i	22 48 30.1	Umeå	e*Pn*	22 48 24
	i*Sg*	22 49 00.7		i*Pg*	22 48 37.0
				i*Sn*	22 49 07.5
				i*S**	22 49 22.2
				i*Sg*	22 49 33.0

Hint: Use the Jeffreys-Bullen travel-time tables and a map of the Nordic countries, preferably of scale $1:5 \times 10^6$.

Answer: West coast of Norway at 66.6 °N, 13.6 °E. Origin time $= 22$ 47 22.

2. Same problem as 1 for the following event:

		h m s			h m s
Kiruna	i*Pn*	00 09 48	Umeå	i*Sn*	00 11 28.1
	i*Sn*	00 10 42.1		i*S**	00 11 43.7
	i*Sg*	00 11 05.9		i*Sg*	00 12 03.0
Skalstugan	e*Sg*	00 13 35			

Answer: Northwest Russia, 68.2 °N, 31.6 °E. Origin time $= 00$ 08 37.

3. Same problem as 1 for the following event:

		h m s			h m s
Uppsala	i*Pg*	04 16 56	Göteborg	e*Pn*	04 16 17
	i*Sn*	04 17 34		e*Sg*	04 16 57
	i*Sg*	04 17 53			
			Karlskrona	i*Pg*	04 15 40
Skalstugan	e*Sn*	04 19 00		i*Sg*	04 15 48
	i*Sg*	04 19 49			

Answer: Southern Baltic Sea, 55.6 °N, 15.0 °E. Origin time $= 04$ 15 28.

Comment: Somewhat more accurate results are obtained by using regional travel-time tables, e.g. those of BÅTH (1971), Chapter 3.

4. Are the following phases possible and, if so, how do they propagate: *PcPPKP, PPPKP, SKSP, SKKKP, SKPSKP, PKIKKIKP, sPcP, pSKP, ScSScS*?

5. Suppose we have a given earthquake and a given station, such that a) both *S* and *SKS* are recorded; b) both *P* and *PKP* are recorded. Explain how this is possible and within what distance ranges a and b can happen.

6. Suppose we have records in three components (E, N, Z). Which of the following waves can be used for determination of the direction to the epicentre: *P, PP, S, PcS, ScS, SKS, L, R*?

7. Given the following earthquake: November 9, 1963, 21 15 30.4 GMT, 9.0 °S, 71.5 °W, western Brazil, depth about 600 km, magnitude $m=7.1$. Calculate the following quantities for the station at Uppsala:
a) Epicentral distance to 0.1° by means of spherical trigonometry.
b) Azimuth of the epicentre as seen from the station and vice versa to 0.1° by means of spherical trigonometry.
c) Theoretical arrival times of *P, pP, PP, SKS, S, PKKP, P'P'* using both the Jeffreys-Bullen tables and the Gutenberg-Richter tables.
d) Compare the results from c with observed arrival times at Uppsala.
e) Angle of incidence (or angle of emergence) at the focus and at the station of *P, pP, PP, S*.
f) Depth of deepest penetration of *P, S* and *PP* rays.
Hint: For the solution of a and b, knowledge of spherical trigonometry is required (as a rule, the cosine and the sine theorems are sufficient for all seismological purposes). For part d, the Uppsala readings are communicated:

i*P*	21 28 01.5
i*pP*	21 30 06.5
i*PP*	21 32 18.6
i*SKS*	21 37 42
i*S*	21 38 36
i*PKKP*	21 44 28.2
i*P'P'*	21 52 51.9

8. Repeat the calculations under c, d, e and f in problem 7, but instead under the assumption of zero focal depth, i.e. with a focus located on the earth's surface.

9. Calculate the magnitude for the earthquake of September 1, 1962, 19 20 38.7 GMT, 35.6 °N, 49.9 °E, Iran, depth 20 km, by means of the following readings at Uppsala and Kiruna:

		microns	sec				microns	sec
Uppsala	P E	8.9	10	Kiruna	P	E	12	7
	P N	5.4	6		P	N	8.9	7
	P Z	6.5	6		P	Z	26	9
	P Z'	0.4	0.6		P	Z'	2.4	1.0
	S N	2.7	4		PP	Z	14	6
	R E	230	18		S	E	17	11
	R N	250	17		S	N	27	11
	R Z	330	18		R	E	200	13
					R	N	210	17
					R	Z	430	17

Hint: Use graphs and other means from RICHTER's book *Elementary Seismology.*
Answer: $m = 6.9$, $M = 7.1$.

10. Same problem as 9 for the following earthquake: December 7, 1962, 14 03 37.0 GMT, 29.2 °N, 139.2 °E, Bonin Islands, depth 440 km.

		microns	sec				microns	sec
Uppsala	P E	1.4	1	Kiruna	P	E	3.3	5
	P N	1.8	3		P	Z	5.4	5
	P Z'	0.6	0.5		P	Z'	0.4	0.5
	PP E	1.5	3		PP	Z	3.2	4
	PP N	2.1	3		S	E	17	7
	PP Z	3.8	3		S	N	19	7
	S E	11	3		R	E	7.0	15
	S N	12	3		R	N	4.1	16
	R E	4.8	17		R	Z	6.9	17
	R N	6.7	19					
	R Z	6.1	16					

Answer: $m=6.8$, $M=6.2$ (M not corrected for focal depth).

11. Given all Uppsala records for the following earthquake: July 26, 1963, 04 17 16.7 GMT, 42.1 °N, 21.5 °E, Skopje, surface focus assumed.

a) Determine epicentral distance from the records and compare with trigonometrically calculated distance.

b) Repeat the problem a for the azimuth of the epicentre.

c) Determine origin time from Uppsala records and compare with given origin time.

d) Determine the magnitude.

The Uppsala readings follow here:

iP 04 21 22
iS 04 24 44

		microns	sec
P	N	2.3	3
P	Z	1.3	3
P	Z'	0.2	0.6
S	E	3.9	6
S	N	3.0	5
L, R	E	45	15
L, R	N	34	10
R	Z	25	9

Hint: Problem b supposes an access to all Uppsala records of this earthquake. However, as an exercise, the problem can be applied to any other station and to any well-recorded earthquake.

12. Arrival times of the P-wave are given for a number of stations according to the following table from U.S. Coast and Geodetic Survey:

ADK	iP	16 27 18.3	NEW	eP	16 34 09.0
BIG	eP	16 30 29.0	BMO	eP	16 34 22.8
TNN	eP	16 31 12.0	HHM	$e(P)$	16 34 24.3
SCM	eP	16 31 18.0	ALE	eP	16 34 32.0
COL	eP	16 31 26.0	BOZ	eP	16 34 46.7
PJO	eP	16 31 28.0	EUR	iP	16 34 54.5
MBC	eP	16 33 19.2	WDY	iP	16 34 58.0

LF4	e*P*	16 35 00.4	FBC	e*P*	16 35 55.0
DUG	e*P*	16 35 06.1	WMO	e*P*	16 36 35.0
UBO	e*P*	16 35 20.5	CPO	e*P*	16 37 20.2
TFO	e*P*	16 35 41.8	SHL	i*P*	16 37 36.6
GOL	e*P*	16 35 42.0			

Verify the solution of the USCGS: origin time = 16 26 34.1 GMT, epi-central coordinates 51.2 °N, 179.0 °E, Aleutian Islands, focal depth = 36 km. Make also an independent solution.

Hint: A table of stations and their coordinates has to be used as well as travel-time tables.

13.3 *Practical Application of Seismology*

It is undoubtedly of very great value to students to be acquainted with the possibilities for practical application and employment in their field after a degree has been obtained. Apart from employment as high-school teachers in physics, we have to state that the possibilities are more restricted than in 'big' subjects, like physics and chemistry. But in spite of this, it is very satisfying to see that seismology has gained more and more practical significance in the community in recent time. In this section, I shall give a review of various practical applications of seismology. On the whole, when a study of seismic waves enters a problem, the applications aim at increased knowledge of the source, of the transmitting medium or of the receiving end.

1. Seismic prospecting. In this branch, seismic methods are applied to exploration for occurrences of oil, salt, ore bodies, minerals, etc. As only economically profitable resources are of interest, the seismic investigations are limited to relatively shallow depths. In principle, the methods are the same as used in seismic soundings of the earth's crust (Chapter 7). Just as in that case, there are principally two methods which are applied: the reflection method and the refraction method. In prospecting work, however, the measurements have to be done on another scale: the investigated depths are smaller, and smaller explosions and shorter profiles are enough. But on the other hand, the timing accuracy has to be increased to a corresponding degree to yield sufficiently accurate results.

It ought to be emphasized that it is not possible to detect oil, etc., by seismic or any other geophysical means. It is only possible to discover geological *formations* where there may be reason to suspect the occurrence of oil, etc. When such a suspected location has been found, the drilling of bore holes can be started. Oil prospecting has recently received renewed stimulus. Intensive prospecting for oil is thus made in several different parts of the earth, as for instance, in the North Sea, the Baltic Sea, in the Sahara, and other places.

Geophysical firms who conduct prospecting work are now in existence in many countries. Depth-to-bedrock investigations (soil depth, etc.) is another branch of applied seismology, also made by such firms and of importance in connection with various constructional problems.

2. Vibration measurements. As a consequence of an ever increasing extent of built-up areas (cities, etc.), the effect of vibrations on various structures as well as on human beings is a problem of increasing importance. This includes vibrations from explosions in connection with various kinds of works and from mining blasts as well as from modern traffic. Therefore, vibration measurements have become of great practical importance, and they are usually made at a large number of points in exposed areas. As a rule, such measurements are made with accelerographs. One example is the frequent construction of underground railways in cities with old buildings, which one wants to keep intact (Stockholm provided such an example). Vibration problems in towns in the immediate vicinity of large mines have attracted special interest. Kiruna, which is located near the iron-ore mines, provides such an example where vibration measurements have been made on several occasions.

3. Stress measurements. Measurement of absolute stress has application within several different fields: in investigation of strength of building materials, in investigation of stability in mines, etc. In Sweden, such measurements are of great importance and Professor N. HAST, Stockholm, has developed a very sensitive method for these (Chapter 2). He and his collaborators have conducted measurements both in the Nordic countries and elsewhere, as in Canada, Egypt, Zambia, Iceland, etc. Professor HAST has coordinated and studied his observations from a geophysical point of view. There is no doubt that such measurements in the solid earth will be of great significance to the problem of earthquake prediction (Chapter 10). In the planning of various constructions, as dams and power plants, it is

also of the greatest significance to have an accurate knowledge about the stress conditions and possible fracture zones in the solid rock.

4. Engineering seismology or earthquake engineering. This is such an important branch of applied seismology, that it is represented by special institutes and professional research workers in a number of countries, especially seismic countries. This field is concerned with the effects of earthquakes on all kinds of building structures. Thorough studies are now usually made after every destructive earthquake, frequently by special missions sent out by UNESCO in Paris. In addition, laboratory investigations on models play a very great role in this field. It is on the basis of such studies, combined with seismicity research, that so-called building codes can be formulated, which regulate the way of building in seismic areas.

Obviously, these problems are of the greatest importance in earthquake countries. But they are of concern not only to the inhabitants of such countries, but to everyone who is going to build something in such a country. Therefore, many firms in non-seismic countries, like Sweden, who are involved in large construction works in seismic areas, are very much concerned with seismic conditions and possible damage in such areas. This is an important field where the seismologist can be of great help, thanks to the information he can provide on seismicity around the world.

Sometimes very severe requirements have to be made on a site, which should be free from vibrations of seismic or other origin. This concerns, for instance, accelerator installations and more recently, nuclear power plants. Then even very small shocks have to be taken into account as well as the rock structure and in great detail.

5. Earthquake insurance. The seismological part of this problem is closely related to the one described under point 4. From the seismological side, information is provided on the seismic risk, i.e. the intensities (magnitudes) and their frequency of occurrence to be expected for a location under interest. Information on seismic risk has then to be converted into insurance premiums. A recent investigation of this problem is presented in a paper by KULHÁNEK and BÅTH (1976), cf. Chapter 4, where methods are developed and applied to selected areas in Central and South America.

6. Earthquake prediction. That this represents a field of the greatest practical importance is clear beyond any doubt. We touched upon this

problem under 3 above and for a more detailed account we refer to Chapter 10.

7. Recording of seismic waves to establish the nature of the source. This field comprises several parts, of which we give the most significant examples.

a) Nuclear test detection. Detection of nuclear explosions has now to be considered as an important branch of applied seismology. The details of the problem have been described in Chapter 11. Research on this problem as well as practical detection are being conducted both by government agencies and by seismological institutes in several countries.

b) Seismic detection in connection with military operations. In such operations, it has been customary for a long time already to listen in various media, such as in the atmosphere and in water (hydrophones). More recently, these methods have been supplemented by listening methods applied to the ground, using seismic detectors.

c) Seismic detection of rockbursts. Rockbursts (Section 10.8) are small ruptures which can be located seismically by recording with a number of receivers. Such recordings are of significance as they can locate possible crack formation in tunnels, underground chambers, mines, etc. They may therefore be of importance for safety measures against major ruptures.

8. Seismic-glaciological research. Every line of work, where seismic methods are being employed, can be considered as a branch of applied seismology. Seismic-glaciological research offers one such example. Investigations have been made of the internal structure of glaciers in many areas, by the application of seismic refraction techniques.

On the whole, it can be said that there is an increasing demand for seismologists all over the world with a good theoretical and experimental background. Technically highly skilled seismologists are required for the operation of seismograph station networks and of array stations as well as for the reading of records and other handling of data. Universities need well-qualified people to carry on teaching in seismology on an advanced level. This together with the practical side, of which a number of examples have been given above, tend to make the prospects for seismologists brighter than ever before.

Literature Review

The intention of the following literature review, arranged according to chapter, is both to give the main sources upon which the present book is based as well as to give suggestions for further reading. On the other hand, the intention is not to give any complete coverage of the seismological literature, which would be practically impossible with the great abundance of published papers. The list is rather to be looked upon as a selection of papers suitable for further reading. In the papers listed there are often extensive references to further literature. As a considerable part of the material in this book is based upon observations and studies at the Seismological Institute, Uppsala, there is a corresponding overweight of papers from this institute. In case publications refer to more than one chapter, they are generally listed only on the first of these chapters. Books given already in Chapter 13 are in general not repeated here.

Chapter 1

BÅTH, M., *Seismology in the Upper Mantle Project*, Tectonophysics *1*, 261—271 (1964).

BÅTH, M., *Seismology*, Acta Univ. Ups., Uppsala University 500 Years, Fac. Sci. *8*, 181—191 (Math. & Physics), and *10*, 207—217 (Earth & Life Sci.) (1976).

BOLT, B. A. (Ed.), *Methods in Computational Physics*, Vol. 13: *Geophysics* (Acad. Press, New York 1973), 473 pp.

CIVETTA, L., GASPARINI, P., LUONGO, G., and RAPOLLA, A. (Editors), *Physical Volcanology* (Elsevier, Amsterdam 1974), 333 pp.

GUTENBERG, B. (Ed.), *Handbuch der Geophysik*, Vol. 4 (Borntraeger, Berlin 1932), 1202 pp.

GUTENBERG, B., *Seismology*, Geol. Soc. Amer., 50th Ann. Vol., 439—470 (1941).

JEFFREYS, H., and JEFFREYS, B. S. (Editors), *Collected Papers of Sir Harold Jeffreys on Geophysics and Other Sciences*, Vol. 1—6 (Gordon and Breach, London 1971—7).

LOVE, A. E. H., *A Treatise on the Mathematical Theory of Elasticity* (Dover, New York 1944), 643 pp.

PARASNIS, D. S., *Principles of Applied Geophysics* (Chapman and Hall, London 1972), 214 pp.

PRESS, F., and SIEVER, R., *Earth* (Freeman, San Francisco 1974), 945 pp.

SCHEIDEGGER, A. E., *Foundations of Geophysics* (Elsevier, Amsterdam 1976), 238 pp.

SHARMA, P. V., *Geophysical Methods in Geology* (Elsevier, Amsterdam 1976), 428 pp.

Chapter 2

ANDERSON, J. A., and WOOD, H. O., *Description and Theory of the Torsion Seismometer*, Bull. Seism. Soc. Amer. *15*, 1—72 (1925).

BÅTH, M., *Development of Instrumental Seismology in Sweden in 1949—1958*, Geofis. pura e appl. *43*, 108—130 (1959).

BENIOFF, H., *A New Vertical Seismograph*. Bull. Seism. Soc. Amer. *22*, 155—169 (1932).

BENIOFF, H., *A Linear Strain Seismometer*. Bull. Seism. Soc. Amer. *25*, 283—309 (1935).

BENIOFF, H., *Earthquake Seismographs and Associated Instruments*, Advances in Geophys. *2*, 219—275 (1955).

BENIOFF, H., *Long-Period Seismographs*, Bull. Seism. Soc. Amer. *50*, 1—13 (1960).

BORG, H., and BÅTH, M., *The Uppsala Seismograph Array Station*, Pure and Appl. Geophys. *89*, 19—31 (1971).

COULOMB, J., and GRENET, G., *Nouveaux principes de construction des séismographes électromagnétiques*, Ann. Phys., Sér. 11, *3*, 321—369 (1935).

DE BREMAECKER, J. CL., DONOHO, P., and MICHEL, J. G., *A Direct Digitizing Seismograph*, Bull. Seism. Soc. Amer. *52*, 661—672 (1962).

GUTENBERG, B. (Ed.), *Handbuch der Geophysik*, Vol. 4 (Borntraeger, Berlin 1932), 1202 pp.

HAST, N., *The Measurement of Rock Pressure in Mines*, Swedish Geological Survey, Yearbook 52, 183 pp. (1958).

HAST, N., *The State of Stresses in the Upper Part of the Earth's Crust*, Engineering Geology *2*, 5—17 (1967).

WILLMORE, P., *The Detection of Earth Movements*. Methods and Techniques in Geophysics (S. K. RUNCORN, Ed.), Interscience, *1*, 230—276 (1960).

Chapter 3

ANSELL, J. H., *Observation of the Frequency-Dependent Amplitude Variation with Distance of P waves from 87° to 119°*, Pure and Appl. Geophys. *112*, 683—700 (1974).

BARBER, N. F., *Fourier Methods in Geophysics*, Methods and Techniques in Geophysics (S. K. RUNCORN, Ed.), Interscience, *2*, 123—204 (1966).

BÅTH, M., *An Investigation of the Uppsala Microseisms* (Almqvist & Wiksell, Uppsala 1949), 168 pp.

BÅTH, M., *Comparison of Microseisms in Greenland, Iceland, and Scandinavia*, Tellus *5*, 109—134 (1953).

BÅTH, M., *A Study of T Phases Recorded at the Kiruna Seismograph Station*, Tellus *6*, 63—72 (1954).

BÅTH, M., *Ultra-long-period Motions from the Alaska Earthquake of July 10, 1958*, Geofis. pura e appl. *41*, 91—100 (1958).

BÅTH, M., *Channel Waves in the Earth's Continental Crust*, Scientia *56*, 8 pp. (1962).

BÅTH, M., *Propagation of Sn and Pn to Teleseismic Distances*, Pure and Appl. Geophys. *64*, 19—30 (1966).

BÅTH, M., *An Earthquake with Exceptionally Strong Higher-mode Surface Waves*, Pure and Appl. Geophys. *66*, 16—24 (1967).

BÅTH, M., *Observations of Teleseismic Pn Phases*, Pure and Appl. Geophys. *66*, 30—36 (1967).

BÅTH, M., *Average Crustal Structure of Sweden*, Pure and Appl. Geophys. *88*, 75—91 (1971).

BÅTH, M., *Short-Period Rayleigh Waves from Near-Surface Events*, Phys. Earth Planet. Interiors *10*, 369—376 (1975).

BÅTH, M., and LOPEZ ARROYO, A., *Attenuation and Dispersion of G Waves*, J. Geophys. Res. *67*, 1933—1942 (1962).

BÅTH, M., and LOPEZ ARROYO, A., *Pa and Sa Waves and the Upper Mantle*, Geofis. pura e appl. *56*, 67—92 (1963).

BÅTH, M., and CRAMPIN, S., *Higher Modes of Seismic Surface Waves — Relations to Channel Waves*, Geophys. J. *9*, 309—321 (1965).

BÅTH, M., and SHAHIDI, M., *T-phases from Atlantic Earthquakes*, Pure and Appl. Geophys., *92*, 74—114 (1971).

BÅTH, M., et al., *A Seismic Refraction Investigation of Superficial Granitic Layering,* Seism. Inst., Uppsala, Rep. No. 7—76, 21 pp., 3 tables, 7 figs (1976).

BENIOFF, H., *Long Waves Observed in the Kamchatka Earthquake of November 4, 1952,* J. Geophys. Res. *63,* 589—593 (1958).

BOLT, B. A. (Ed)., *Methods in Computational Physics,* Vol. 11: *Seismology, Surface Waves and Earth Oscillations* (Acad. Press, New York 1972), 309 pp.

BOLT, B. A. (Ed.), *Methods in Computational Physics,* Vol. 12: *Seismology, Body Waves and Sources* (Acad. Press, New York 1972), 391 pp.

BROWN, R. J., *Azimuthally Varying P-Wave Travel-Time Residuals in Fennoscandia and Lateral Inhomogeneity,* Pure and Appl. Geophys. *105,* 741—758 (1973).

BROWN, R. J., and ENAYATOLLAH, M. A., *Comparison of Short- and Long-Period P-Wave Travel Times and Arrival Angles,* Pure and Appl. Geophys. *109,* 1638—1652 (1973).

CRAMPIN, S., *Higher Modes of Seismic Surface Waves — Preliminary Observations,* Geophys. J. *9,* 37—57 (1964).

CRAMPIN, S., *Higher Modes of Seismic Surface Waves — Phase Velocities Across Scandinavia,* J. Geophys. Res. *69,* 4801—4811 (1964).

CRAMPIN, S., *Higher Modes of Seismic Surface Waves — Propagation in Eurasia,* Bull. Seism. Soc. Amer. *56,* 1227—1239 (1966).

DERR, J. S., *Free Oscillation Observations Through 1968,* Bull. Seism. Soc. Amer. *59,* 2079—2099 (1969).

DONN, W. L., *Microseisms,* Earth-Sci. Rev. *1,* 213—230 (1966).

ENAYATOLLAH, M. A., *Travel Times of P-Waves for the Swedish-Finnish Seismograph Network,* Pure and Appl. Geophys. *94,* 101—135 (1972).

GUTENBERG, B., *Effects of Low-Velocity Layers,* Geofis. pura e appl. *28,* 1—10 (1954).

GUTENBERG, B., *Channel Waves in the Earth's Crust,* Geophysics *20,* 283—294 (1955).

GUTENBERG, B., and RICHTER, C. F., *Materials for the Study of Deep-Focus Earthquakes,* Bull. Seism. Soc. Amer. *26,* 341—390 (1936).

HADDON, R. A. W., HUSEBYE, E. S., and KING, D. W., *Origins of Precursors to P'P',* Phys. Earth Planet. Interiors *14,* 41—70 (1977).

HERRIN, E., et al., *1968 Seismological Tables for P Phases,* Bull. Seism. Soc. Amer. *58,* 1193—1352 (1968).

HUSEBYE, E. S., *Correction Analysis of Jeffreys-Bullen Travel Time Tables,* Bull. Seism. Soc. Amer. *55,* 1023—1038 (1965).

JEFFREYS, H., and BULLEN, K. E., *Seismological Tables,* Brit. Assoc. Adv. Sci., 50 pp. (1967).

KULHÁNEK, O., and BÅTH, M., *Power Spectra and Geographical Distribution of Short-Period Microseisms in Sweden,* Pure and Appl. Geophys. *94,* 148—171 (1972).

LEONG, L. S., *Sp Converted Phases, Synthetic Long-Period S Waveforms and Crustal Structure at Umeå, Sweden,* Seism. Inst., Uppsala, Rep. No. 4—76, 13 pp., 4 figs (1976).

PAYO SUBIZA, G., and BÅTH, M., *Core Phases and the Inner Core Boundary,* Geophys. J. *8,* 496—513 (1964).

SAVARENSKY, E., *Seismic Waves* (Mir Publishers, Moscow 1975), 349 pp.

SHAHIDI, M., *Variation of Amplitude of PKP Across the Caustic,* Phys. Earth Planet. Interiors *1,* 97—102 (1968).

STAMOU, P., and BÅTH, M., *The Caustic and Other Properties of SKP,* Phys. Earth Planet. Interiors *8,* 317—331 (1974).

TRYGGVASON, E., *Crustal Thickness in Fennoscandia from Phase Velocities of Rayleigh Waves*, Ann. Geofis. *14*, 267—293 (1961).

WHITE, J. E., *Seismic Waves — Radiation, Transmission, and Attenuation* (McGraw-Hill, New York 1965), 302 pp.

Chapter 4

ANONYMOUS, *Norma sismorresistente*, Inst. Geograf. y Catastral, 173 pp. (1968).

BÅTH, M., *Earthquake Energy and Magnitude*, Phys. and Chem. Earth *7*, 115—165 (1966).

BÅTH, M., *Handbook on Earthquake Magnitude Determinations*. 2nd rev. ed., Seism. Inst., Uppsala, 158 pp., 1969 (VESIAC Spec. Rep. 7885—36—X; this book contains a bibliography on magnitude and energy calculations, comprising about 410 items).

BÅTH, M., *Seismicity of the Tanzania Region*, Tectonophysics *27*, 353—379 (1975).

BÅTH, M., KULHÁNEK, O., van ECK T., and WAHLSTRÖM, R., *Engineering Analysis of Ground Motion in Sweden*, Seism. Inst., Uppsala, Rep. No. *5—76*, 37 pp., 8 tables, 11 figs (1976).

BEN-MENAHEM, A., and BÅTH, M., *A Method for Determination of Epicenters of Near Earthquakes*, Geofis. pura e appl. *46*, 37—46 (1960).

FREEMAN, J. R., *Earthquake Damage and Earthquake Insurance* (McGraw-Hill, New York 1932), 904 pp.

GUTENBERG, B., *SV and SH*, Trans. Amer. Geophys. Union *33*, 573—584 (1952).

GUTENBERG, B., and RICHTER, C. F., *Magnitude and Energy of Earthquakes*, Ann. Geofis. *9*, 1—15 (1956).

GUTENBERG, B., and RICHTER, C. F., *Earthquake Magnitude, Intensity, Energy, and Acceleration*, Bull. Seism. Soc. Amer. *46*, 105—145 (1956).

HOUSNER, G. W., *The Design of Structures to Resist Earthquakes*, Dept. Nat. Res. California, Div. of Mines, Bull. *171*, 271—277 (1955).

HUSEBYE, E. S., *A Rapid, Graphical Method for Epicenter Location*, Gerl. Beitr. Geophys. *75*, 383—392 (1966).

KULHÁNEK, O., and BÅTH, M., *Earthquake Insurance Coefficients with Application to Some South-Central American Capitals*, Seism. Inst., Uppsala, Rep. No. *8—76*, 28 pp., 12 tables, 2 figs (1976).

LOMNITZ, C., *Global Tectonics and Earthquake Risk* (Elsevier, Amsterdam 1974), 320 pp.

LOMNITZ, C., and ROSENBLUETH, E. (Editors), *Seismic Risk and Engineering Decisions* (Elsevier, Amsterdam 1976), 425 pp.

MEDVEDEV, S. V., SPONHEUER, W., and KÁRNÍK, V., *Seismic Intensity Scale, MSK 1964*, Acad. Sci. USSR, Sov. Geophys. Comm., 13 pp. (1965).

NEWMARK, N. M., and ROSENBLUETH, E., *Fundamentals of Earthquake Engineering* (Prentice-Hall, Englewood Cliffs, New Jersey 1971), 640 pp.

RICHTER, C. F., *An Instrumental Earthquake Magnitude Scale*, Bull. Seism. Soc. Amer. *25*, 1—32 (1935).

SHAPIRA, A., and BÅTH, M., *Location of Teleseisms from P-Wave Arrivals at the Swedish Stations*, Seism. Inst., Uppsala, Rep. No. *9—76*, 14 pp., 4 tables, 4 figs (1976).

WYSS, M., and BRUNE, J. N., *Seismic Moment, Stress, and Source Dimensions for Earthquakes in the California-Nevada Region*, J. Geophys. Res. *73*, 4681—4694 (1968).

Chapter 5

BÅTH, M., *Seismicity of Fennoscandia and Related Problems*, Gerl. Beitr. Geophys. *63*, 173—208 (1954).

BÅTH, M., *Earthquakes, Large, Destructive*, Int. Dict. Geophys. *1*, 417—424 (1967).

BÅTH, M., and DUDA, S. J., *Strain Release in Relation to Focal Depth*, Geofis. pura e appl. *56*, 93—100 (1963).

BENIOFF, H., *Orogenesis and Deep Crustal Structure — Additional Evidence from Seismology*, Bull. Geol. Soc. Amer. *65*, 385—400 (1954).

BENIOFF, H., *Seismic Evidence for Crustal Structure and Tectonic Activity*, Geol. Soc. Amer., Spec. Paper, *62*, 61—74 (1955).

DUDA, S. J., *Secular Seismic Energy Release in the Circum-Pacific Belt*, Tectonophysics *2*, 409—452 (1965).

DUDA, S. J., *Regional Seismicity and Seismic Wave Propagation from Records at the Tonto Forest Seismological Observatory, Payson, Arizona*, Ann. Geofis. *18*, 365—397 (1965).

HEEZEN, B. C., *The Rift in the Ocean Floor*, Sci. Amer., 1—15 (Oct. 1960).

HEEZEN, B. C., and EWING, M., *The Mid-Oceanic Ridge and its Extension Through the Arctic Basin*, Geology of the Arctic, 622—642 (1961).

OLIVER, J., RYALL, A., BRUNE, J. N., and SLEMMONS, D. B., *Microearthquake Activity Recorded by Portable Seismographs of High Sensitivity*, Bull. Seism. Soc. Amer. *56*, 899—924 (1966).

ROTHÉ, J. P., *The Seismicity of the Earth 1953—1965*, UNESCO, Earth Sciences *1*, 336 pp. (1969).

SCHEIDEGGER, A. E., *Physical Aspects of Natural Catastrophes* (Elsevier, Amsterdam 1975), 289 pp.

SCHNEIDER, G., *Erdbeben, Entstehung, Ausbreitung, Wirkung* (Enke, Stuttgart 1975), 406 pp.

Chapter 6

BÅTH, M., *Lateral Inhomogeneities of the Upper Mantle*, Tectonophysics *2*, 483—514 (1965).

BÅTH, M., and BENIOFF, H., *The Aftershock Sequence of the Kamchatka Earthquake of November 4, 1952*, Bull. Seism. Soc. Amer. *48*, 1—15 (1958).

BÅTH, M., and DUDA, S. J., *Earthquake Volume, Fault Plane Area, Seismic Energy, Strain, Deformation and Related Quantities*, Ann. Geofis. *17*, 353—368 (1964).

BELOUSSOV, V. V., *Against the Hypothesis of Ocean-Floor Spreading*, Tectonophysics *9* 489—511 (1970).

BENIOFF, H., *Earthquakes and Rock Creep*, Bull. Seism. Soc. Amer. *41*, 31—62 (1951).

BENIOFF, H., *Circum-Pacific Tectonics*, Publ. Dom. Obs. Ottawa, *20(2)*, 395—402 (1957).

BENIOFF, H., *Movements on Major Transcurrent Faults*, Int. Geophys. Ser., Acad. Press *3*, 103—134 (1962).

BENIOFF, H., *Source Wave Forms of Three Earthquakes*, Bull. Seism. Soc. Amer. *53*, 893—903 (1963).

BENIOFF, H., *Earthquake Source Mechanisms*, Science *143*, 1399—1406 (1964).

BEN-MENAHEM, A., *Radiation of Seismic Surface Waves from Finite Moving Sources*, Bull. Seism. Soc. Amer. *51*, 401—435 (1961).

BONNIN, J., and DIETZ, R. S. (Editors), *Present State of Plate Tectonics*, Tectonophysics *38*, 168 pp. (1977) (contains papers presented at a symposium in Grenoble in 1975).

CAREY, S. W., *The Expanding Earth — An Essay Review*, Earth Sci. Rev. *11*, 105—143 (1975).

COX, A., *Plate Tectonics and Geomagnetic Reversals* (Freeman, San Francisco 1973), 702 pp.

EVISON, F. F., *On the Occurrence of Volume Change at the Earthquake Source*, Bull. Seism. Soc. Amer. *57*, 9—25 (1967).

ISACKS, B., OLIVER, J., and SYKES, L. R., *Seismology and the New Global Tectonics*, J. Geophys. Res. *73*, 5855—5899 (1968).

KHATTRI, K., *Earthquake Focal Mechanism Studies — A Review*, Earth Sci. Rev. *9*, 19—63 (1973).

LENSEN, G. J., *Principal Horizontal Stress Directions as an Aid to the Study of Crustal Deformation*, Publ. Dom. Obs. Ottawa *24 (10)*, 389—397 (1960).

LE PICHON, X., *Sea-Floor Spreading and Continental Drift*, J. Geophys. Res. *73*, 3661—3697 (1968).

OLIVER, J., SYKES, L., and ISACKS, B., *Seismology and the New Global Tectonics*, Tectonophysics *7*, 527—541 (1969).

OLSSON, R., *Strain Energy Release, Deformation Energy Release and Related Problems*, Riv. Ital. Geofis. *22*, 341—352 (1973).

OLSSON, R., *Earthquake Activity in the Kamchatka-Kurile Islands-Japan Region November 6, 1958—November 30, 1970*, Riv. Ital. Geofis. *23*, 43—56 (1974).

OLSSON, R., *Some Aftershock Sequences in the Japan—Kamchatka Region*, Seism. Inst., Uppsala, Rep. No. *3—76*, 12 pp., 2 tables, 2 figs (1976).

ROUSE, G. E., and BISQUE, R. E., *Global Tectonics and the Earth's Core*, The Mines Mag., Mining Eng., 8 pp., March 1968.

SCHEIDEGGER, A. E., *The Geometrical Representation of Fault-Plane Solutions of Earthquakes*, Bull. Seism. Soc. Amer. *47*, 89—110 (1957).

SCHEIDEGGER, A. E., *Recent Advances in Geodynamics*, Earth-Sci. Rev. *1*, 133—153 (1966).

STEFÁNSSON, R., *The Use of Transverse Waves in Focal Mechanism Studies*, Tectonophysics *3*, 35—60 (1966).

STEFÁNSSON, R., *Methods of Focal Mechanism Studies with Application to Two Atlantic Earthquakes*, Tectonophysics *3*, 209—243 (1966).

UYEDA, S. (Ed.), *Subduction Zones, Mid-Ocean Ridges, Oceanic Trenches and Geodynamics*, Tectonophysics *37*, 246 pp. (1977) (contains papers presented at a symposium in Grenoble in 1975).

WILSON, J. T. (Ed.), *Continents Adrift*, Readings from Sci. Amer. 1952—72 (Freeman, San Francisco 1972), 172 pp.

Chapter 7

ANDERSON, D. L., *Recent Evidence Concerning the Structure and Composition of the Earth's Mantle*, Phys. and Chem. Earth *6*, 1—131 (1966).

ANONYMOUS, *Geodynamics Project: Development of a U.S. Program*, Trans. Amer. Geophys. Union *52*, 396—405 (1971).

BÅTH, M., *Seismic Exploration of the Earth's Crust — Recent Developments*, Geol. Fören. Förhandl. *80*, 291—308 (1958).

BÅTH, M., *Die Conrad-Diskontinuität*, Freiberger Forsch.-Hefte *C 101*, 5—34 (1961).

BÅTH, M., *Crustal Structure in Iceland and Surrounding Ocean*, ICSU Review *4*, 127—133 (1962).

BÅTH, M., *An Analysis of the Time Term Method in Refraction Seismology*, Seism. Inst., Uppsala, Rep. No. *10—76*, 22 pp., 2 figs (1976).

BÅTH, M., et al., *A Seismic Refraction Investigation of Superficial Granitic Layering*, Seism. Inst., Uppsala, Rep. No. *7—76*, 21 pp., 3 tables, 7 figs (1976).

BROWN, R. J., *Lateral Inhomogeneity in the Crust and Upper Mantle from P-Wave Amplitudes*, Pure and Appl. Geophys. *101*, 102—154 (1972).

BULLEN, K. E., *The Earth's Density* (Chapman and Hall, London 1975), 420 pp.

BULLEN, K. E., and HADDON, R. A. W., *Evidence from Seismology and Related Sources on the Earth's Present Internal Structure*, Phys. Earth Planet. Interiors *2*, 342—349 (1970).

COOK, K. L., *The Problem of the Mantle-Crust Mix: Lateral Inhomogeneity in the Uppermost Part of the Earth's Mantle*, Advances in Geophys. *9*, 295—360 (1962).

FINDLAY, D. C., and SMITH, C. H., *Drilling for Scientific Purposes*, Geol. Surv. Canada, Paper *66—13*, 264 pp. (1966).

GURWITSCH, I. I., *Seismische Erkundung* (Geest & Portig, Leipzig 1970), 699 pp.

GUTENBERG, B., *PKKP, P'P', and the Earth's Core*, Trans. Amer. Geophys. Union *32*, 373—390 (1951).

HADDON, R. A. W., and BULLEN, K. E., *An Earth Model Incorporating Free Earth Oscillation Data*, Phys. Earth Planet. Interiors *2*, 35—49 (1969).

IBRAHIM, A. K., *Leaking and Normal Modes as a Means to Determine Crust-Upper Mantle Structure for Different Paths to Sweden*, Bull. Seism. Soc. Amer. *59*, 1695—1712 (1969).

IBRAHIM, A. K., *Effects of a Rigid Core on the Reflection and Transmission Coefficients from a Multi-Layered Core-Mantle Boundary*, Pure and Appl. Geophys. *91*, 95—113 (1971).

IBRAHIM, A. K., *The Amplitude Ratio PcP/P and the Core-Mantle Boundary*, Pure and Appl. Geophys., *91*, 114—133 (1971).

KANAI, K., *On the Group Velocity of Dispersive Surface Waves*, Bull. Earthq. Res. Inst. Tokyo *29*, 49—60 (1951).

KULHÁNEK, O., and BROWN, R. J., *P-Wave Velocity Anomalies in the Earth's Mantle from the Uppsala Array Observations*, Pure and Appl. Geophys. *112*, 597—617 (1974).

LEE, W. H. K., and TAYLOR, P. T., *Global Analysis of Seismic Refraction Measurements*, Geophys. J. *11*, 389—413 (1966).

LUBIMOVA, E. A., *Thermal History of the Earth with Consideration of the Variable Thermal Conductivity of its Mantle*, Geophys. J. *1*, 115—134 (1958).

MASON, B., *Principles of Geochemistry* (Wiley & Sons, New York 1952), 276 pp.

MÜLLER, G., MULA, A. H., and GREGERSEN, S., *Amplitudes of Long-Period PcP and the Core-Mantle Boundary*, Phys. Earth Planet. Interiors *14*, 30—40 (1977).

PRESS, F., *Earth Models Consistent with Geophysical Data*, Phys. Earth Planet. Interiors *3*, 3—22 (1970).

SANTÔ, T., *Lateral Variation of Rayleigh Wave Dispersion Character; I. Observational Data*, Pure and Appl. Geophys. *62*, 49—66 (1965).

SANTÔ, T., *Lateral Variation of Rayleigh Wave Dispersion Character; II. Eurasia*, Pure and Appl. Geophys. *62*, 67—80 (1965).

SANTÔ, T., *Lateral Variation of Rayleigh Wave Dispersion Character; III. Atlantic Ocean, Africa and Indian Ocean*, Pure and Appl. Geophys. *63*, 40—59 (1966).

TOKSÖZ, M. N., CHINNERY, M. A., and ANDERSON, D. L., *Inhomogeneities in the Earth's Mantle*, Geophys. J. *13*, 31—59 (1967).

Chapter 8

AL-SADI, H. N., *Dependence of the P-Wave Amplitude Spectrum on Focal Depth*, Pure and Appl. Geophys. *104*, 439—452 (1973).

BÅTH, M., *Futur développement des réseaux de stations séismologiques*, Scientia *99*, 184—191 (1964).

BÅTH, M., *Seismic Recording Possibilities in Sweden*, FOA4, Stockholm, *A 4466—4721*, 34 pp. (1965).

BÅTH, M., *Underground Measurements of Short-Period Seismic Noise*, Ann. Geofis. *19*, 107—117 (1966).

BÅTH, M., *Methods to Improve Seismograph Data*, Scientia *102*, 304—317 (1967).

BÅTH, M., *Spectral Analysis in Geophysics* (Elsevier, Amsterdam 1974), 563 pp.

BEAUCHAMP, K. G. (Ed.), *Exploitation of Seismograph Networks*, NATO Advanced Study Inst., *E 11* (Noordhoff, Leiden 1975), 647 pp.

BROWN, R. J., *Slowness and Azimuth at the Uppsala Array, Part 1: Array Calibration and Event Location*, Pure and Appl. Geophys. *105*, 759—769 (1973).

BROWN, R. J., *Slowness and Azimuth at the Uppsala Array, Part 2: Structural Studies*, Pure and Appl. Geophys. *109*, 1623—1637 (1973).

CRAMPIN, S., and BÅTH, M., *Higher Modes of Seismic Surface Waves: Mode Separation*, Geophys. J. *10*, 81—92 (1965).

ENAYATOLLAH, M. A., *Continental-Array Measurements of P-Wave Velocities in the Mantle*, Pure and Appl. Geophys. *94*, 136—147 (1972).

FARNBACH, J. S., *The Complex Envelope in Seismic Signal Analysis*, Bull. Seism. Soc. Amer. *65*, 951—962 (1975).

FROSCH, R. A., and GREEN, P. E., JR., *The Concept of a Large Aperture Seismic Array*, Proc. Roy. Soc. *A 290*, 368—384 (1966).

HUSEBYE, E. S., and JANSSON, B., *Application of Array Data Processing Techniques to the Swedish Seismograph Stations*, Pure and Appl. Geophys. *63*, 82—104 (1966).

JANSSON, B., and HUSEBYE, E. S., *Application of Array Data Processing Techniques to a Network of Ordinary Seismograph Stations*, Pure and Appl. Geophys. *69*, 80—99 (1968).

KORKMAN, K., *Aftershock P-wave Spectra and Dynamic Features of the Aleutian Islands Earthquake Sequence of February 4, 1965*, Tectonophysics *5*, 245—266 (1968).

KULHÁNEK, O., *P-wave Amplitude Spectra of Nevada Underground Nuclear Explosions*, Pure and Appl. Geophys. *88*, 121—136 (1971).

KULHÁNEK, O., *Signal and Noise Coherence Determination for the Uppsala Seismograph Array Station*, Pure and Appl. Geophys. *109*, 1653—1671 (1973).

KULHÁNEK, O., *Introduction to Digital Filtering in Geophysics* (Elsevier, Amsterdam 1976), 168 pp.

LEONG, L. S., *Crustal Structure of the Baltic Shield beneath Umeå, Sweden, from the Spectral Behavior of Long-Period P Waves*, Bull. Seism. Soc. Amer. *65*, 113—126 (1975).

SUTTON, G. H., MCDONALD, W. G., PRENTISS, D. D., and THANOS, S. N., *Ocean-Bottom Seismic Observations*, Proc. IEEE *53*, 1909—1921 (1965).

WHITCOMB, J. H., *Array Data Processing Techniques Applied to Long-Period Shear Waves at Fennoscandian Seismograph Stations*, Bull. Seism. Soc. Amer. *59*, 1863—1887 (1969).

WHITEWAY, F. E., *The Use of Arrays for Earthquake Seismology*, Proc. Roy. Soc. *A 290*, 328—342 (1966).

Chapter 9

ANDERSON, O. L., and LIEBERMANN, R. C., *Sound Velocities in Rocks and Minerals*, VESIAC Rep. *7885—4—X*, 182 pp. (1966).

ANONYMOUS, *Symposium on Seismic Models*, Liblice near Prague, November 9—12, 1965 (a collection of papers presented at the symposium), Stud. Geophys. Geodaet. *10*, 239—400 (1966).

BRACE, W. F., *Laboratory Studies Pertaining to Earthquakes*, Trans. New York Acad. Sci., Ser. II *31*, 892—906 (1969).

DUDA, S. J., *The Stress Around a Fault According to a Photoelastic Model Experiment*, Geophys. J. *9*, 399—410 (1965).

LAVIN, P. M., and HOWELL, B. F., JR., *Model Studies of Effects of Near-Source Velocity Discontinuities on First-Motion Patterns*, Bull. Seism. Soc. Amer. *53*, 933—954 (1963).

PRESS, F., *A Seismic Model Study of the Phase Velocity Method of Exploration*, Geophysics *22*, 275—285 (1957).

PRESS, F., *Elastic Wave Radiation from Faults in Ultrasonic Models*, Publ. Dom. Obs. Ottawa *20 (2)*, 271—277 (1957).

PRESS, F., OLIVER, J., and EWING, M., *Seismic Model Study of Refractions from a Layer of Finite Thickness*, Geophysics *19*, 388—401 (1954).

RAMBERG, H., *Gravity, Deformation and the Earth's Crust* (Academic Press, New York 1967), 214 pp.

VANĔK, J., *Revised Amplitude Curves of Seismic Body Waves for the Region of South-Eastern Europe*, Stud. Geophys. Geodaet. *13*, 173—179 (1969).

Chapter 10

ADAMS, R. D., *The Haicheng, China, Earthquake of 4 February, 1975: The First Successfully Predicted Major Earthquake*, Bull. New Zealand Nat. Soc. Earthq. Eng. *9*, 32—42 (1976).

ANONYMOUS, *ESSA Symposium on Earthquake Prediction*, U.S. Dept. Commerce, 167 pp. (1966).

ANONYMOUS, *Proposal for a Ten-Year National Earthquake Hazards Program*. Ad hoc Interagency Working Group for Earthquake Research, Fed. Council Sci. & Tech., Washington, D.C., 81 pp. (1968).

BÅTH, M., *Earthquake Prediction*, Scientia *101*, 234—243 (1966).

BÅTH, M., *Artificial Release of Earthquakes*, Scientia *105*, 19 pp (1970).

BÅTH, M., and WAHLSTRÖM, R., *A Rockburst Sequence at the Grängesberg Iron Ore Mines in Central Sweden*, Seism. Inst., Uppsala, Rep. No. *6—76*, 30 pp., 3 tables, 1 appendix, 11 figs (1976).

BLOT, C., *Origine profonde des séismes superficiels et des éruptions volcaniques*. IUGG XII Gen. Ass. Seismology, 1963.

BOUCHER, G., RYALL, A., and JONES, A. E., *Earthquakes Associated with Underground Nuclear Explosions*, J. Geophys. Res. *74*, 3808—3820 (1969).

CALOI, P., and SPADEA, M. C., *Principali risultati conseguiti durante l'osservazione geodinamica, opportunamente estesa nel tempo, di grandi dighe di sbarramento, e loro giustificazione teoriche*, Ann. Geofis. *19*, 261—286 (1966).

CARDER, D. S., *Seismic Investigations in the Boulder Dam Area, 1940—1944, and the Influence of Reservoir Loading on Local Earthquake Activity*, Bull. Seism. Soc. Amer. *35*, 175—192 (1945).

COMNINAKIS, P., DRAKOPOULOS, J., MOUMOULIDIS, G., and PAPAZACHOS, B., *Foreshock and Aftershock Sequences of the Cremasta Earthquake and their Relation to the Waterloading of the Cremasta Artificial Lake*, Ann. Geofis. *21*, 39—71 (1968).

DUDA, S. J., *Phänomenologische Untersuchung einer Nachbebenserie aus dem Gebiet der Aleuten-Inseln*, Freiberger Forsch.-Hefte *C 132*, 9—90 (1962).

DUDA, S. J., *Strain Release in the Circum-Pacific Belt: Chile 1960*, J. Geophys. Res. *68*, 5531—5544 (1963).

DUDA, S. J., and BÅTH, M., *Strain Release in the Circum-Pacific Belt: Kern County 1952, Desert Hot Springs 1948, San Francisco 1957*, Geophys. J. *7*, 554—570 (1963).

FEDOTOV, S. A., et al., *Long- and Short-Term Earthquake Prediction in Kamchatka*, Tectonophysics *37*, 305—321 (1977).

FOWLER, R. A., *Earthquake Prediction from Laser Surveying*, NASA *SP-5042*, 32 pp. (1968).

GUPTA, H. K., and RASTOGI, B. K., *Dams and Earthquakes* (Elsevier, Amsterdam 1976), 229 pp.

GUTENBERG, B., *Seismology*, Geol. Soc. Amer., 50th Ann. Vol., 439—470 (1941).

HAGIWARA, T., and RIKITAKE, T., *Japanese Program on Earthquake Prediction*, Science *157*, 761—768 (1967).

KNOPOFF, L., *Earth Tides as a Triggering Mechanism for Earthquakes*, Bull. Seism. Soc. Amer. *54*, 1865—1870 (1964).

MEAD, T. S., and CARDER, D. S., *Seismic Investigations in the Boulder Dam Area in 1940*, Bull. Seism. Soc. Amer. *31*, 321—340 (1941).

MJACHKIN, V. I., BRACE, W. F., SOBOLEV, G. A., and DIETERICH, J. H., *Two Models for Earthquake Forerunners*, Pure and Appl. Geophys. *113*, 169—181 (1975).

PAGE, R. (Ed.), *Proceedings of the Second United States-Japan Conference on Research Related to Earthquake Prediction*, Nat. Sci. Foundation and Japan Soc. Prom. Sci., 106 pp. (1966).

PRESS, F., *Earthquake Prediction*, Sci. Amer. *232*, 14—23 (1975).

PRESS, F., and BRACE, W. F., *Earthquake Prediction*, Science *152*, 1575—1584 (1966).

RIKITAKE, T., *A Five-Year Plan for Earthquake Prediction Research in Japan*, Tectonophysics *3*, 1—15 (1966).

RIKITAKE, T. (Ed.), *Earthquake Prediction*, Tectonophysics *6*, 1—87 (1968).

RIKITAKE, T. (Ed.), *Earthquake Mechanics*, Tectonophysics *9*, 97—300 (1970) (contains papers presented at a symposium in Madrid in September 1969).

RIKITAKE, T. (Ed.), *Focal Processes and the Prediction of Earthquakes*, Tectonophysics *23*, 219—318 (1974) (contains papers presented at a symposium in Lima in 1973).

RIKITAKE, T., *Earthquake Prediction* (Elsevier, Amsterdam 1976), 357 pp.

RYALL, A., and SAVAGE, W. U., *A Comparison of Seismological Effects for the Nevada Underground Test Boxcar with Natural Earthquakes in the Nevada Region*, J. Geophys. Res. *74*, 4281—4289 (1969).

SAVARENSKY, E. F. (Ed.), *Prediction of Earthquakes* (in Russian) (Publ. House Mir, Moscow 1968), 213 pp. (a collection of papers translated from English).

SAVARENSKY, E. F., and RIKITAKE, T. (Editors), *Forerunners of Strong Earthquakes, Tectonophysics 14,* 175—348 (1972) (contains papers presented at a symposium in Moscow in 1971).

SIMON, R. B., *The Denver Earthquakes, 1962—1967,* Earthquake Notes *39,* 37—40 (1968).

SUYEHIRO, S., *Difference Between Aftershocks and Foreshocks in the Relationship of Magnitude to Frequency of Occurrence for the Great Chilean Earthquake of 1960,* Bull. Seism. Soc. Amer. *56,* 185—200 (1966).

TAMRAZYAN, G. P., *On the Periodicity of Seismic Activity in the Last Fifteen Hundred to Two Thousand Years (as revealed in Armenia),* Izv. Akad. Nauk SSR, Ser. Geofiz. *1,* 76—85 (1962).

TARAKANOV, R. Z., *Aftershocks of the Earthquake from November 4, 1952,* Trud. Sakhalin Compl. Sci. Res. Inst. *10,* 112—116 (1961).

TSUBOI, C., WADATI, K., and HAGIWARA, T., *Prediction of Earthquakes,* Earthq. Res. Inst. Tokyo, 21 pp. (1962).

WHITCOMB, J. H., GARMANY, J. G., and ANDERSON, D. L., *Earthquake Prediction: Variation of Seismic Velocities before the San Fernando Earthquake,* Science *180,* 632—635 (1973).

Chapter 11

ANONYMOUS, *The Detection and Recognition of Underground Explosions,* U.K. Atomic Energy Authority, Spec. Rep., 118 pp. (1965).

BATES, C. C., *VELA UNIFORM, the Nation's Quest for Better Detection of Underground Nuclear Explosions,* Geophysics *26,* 499—507 (1961).

BÅTH, M., *Seismic Records of Explosions — Especially Nuclear Explosions.* III, FOA4, Stockholm, *A 4270—4721,* 116 pp. (1962).

BÅTH, M., *Earthquake Energy and Magnitude,* Phys. and Chem. Earth *7,* 115—165 (1966).

BÅTH, M., *Short-Period Rayleigh Waves from Near-Surface Events,* Phys. Earth Planet. Interiors *10,* 369—376 (1975).

BÅTH, M., and TRYGGVASON, E., *Deep Seismic Reflection Experiments at Kiruna,* Geofis. pura e appl. *51,* 79—90 (1962).

BOLT, B. A., *Nuclear Explosions and Earthquakes — The Parted Veil* (Freeman, San Francisco 1976), 309 pp.

BULLARD, E., *The Detection of Underground Explosions,* Sci. Amer. *215,* 19—29 (1966).

BULLEN, K. E., *Seismology in Our Atomic Age,* UGGI IASPEI, CR *12* (Toronto 1957), 19—35 (1958).

BULLEN, K. E., and BURKE-GAFFNEY, T. N., *Diffracted Seismic Waves Near the PKP Caustic,* Geophys. J. *1,* 9—17 (1958).

CRAMPIN, S., *Higher Mode Seismic Surface Waves from Atmospheric Nuclear Explosions over Novaya Zemlya,* J. Geophys. Res. *71,* 2951—2958 (1966).

DAVIES, D. (Ed.), *Seismic Methods for Monitoring Underground Explosions* (SIPRI, Stockholm 1968), 130 pp.

EVERNDEN, J. F., *Identification of Earthquakes and Explosions by Use of Teleseismic Data,* J. Geophys. Res. *74,* 3828—3856 (1969).

FARNBACH, J. S., *The Composition of P Waves from Underground Explosions,* Seism. Inst., Uppsala, Rep. No. *6—75,* 25 pp., 6 tables, 7 figs (1975).

GORDON, D. W., *Travel Time Curve*, ESSA Symp. Earthq. Pred., 141—145 (1966).

GRIGGS, D. T., and PRESS F., *Probing the Earth with Nuclear Explosions* (Univ. of Calif., Lawrence Radiation Lab., 1959), 31 pp.

KOGAN, S. D., *Travel Times of Body Waves from Surface Focus*, Izv. Akad. Nauk SSR, Ser. Geofiz. *3*. 371—380 (1960). Also in IASPEI, Trav. Sci. *A 21*, 15—22 (1961).

KORKMAN (KOGEUS), K., *A Synthesis of Short-Period P-wave Records from Distant Explosion Sources*, Bull. Seism. Soc. Amer. *58*, 663—680 (1968).

KULHÁNEK, O., *Source Parameters of Some Presumed Semipalatinsk Underground Nuclear Explosions*, Pure and Appl. Geophys. *102*, 51—66 (1973).

MEYER, K., *Secondary Pressure Pulses from Underwater Explosions*, Seism. Inst., Uppsala, Rep. No. *9—75*, 17 pp., 5 tables, 7 figs (1975).

MEYER, K., and BÅTH, M., *First-Arrival Seismic Waves Generated by Underwater Explosions*, Seism. Inst., Uppsala, Rep. No. *2—75*, 13 pp., 6 tables, 5 figs (1975).

MOROKHOV, I. D. (Ed.), *Nuclear Explosions for Peaceful Purposes*, Transl. from the Russian Atomnye vzryvy v mirnykh tselyakh, Atomizdat, Moscow, 124 pp. (1970), U.S. Atomic Energy Commission, UCRL-Trans-*10517*, 190 pp. (1971).

NORDYKE, M. D. (Ed.), *French and Soviet Papers Presented at the Second Panel on the Peaceful Uses of Nuclear Explosions, Vienna, Austria, 18—22 January 1971*, U.S. Atomic Energy Commission, UCRL-Trans-*10543*, 116 pp. (1971). (There are similar publications also on later conferences in Vienna on this topic.)

PERSEN, L. N., *Rock Dynamics and Geophysical Exploration* (Elsevier, Amsterdam 1975), 276 pp.

PRESS, F., DEWART, G., and GILMAN, R., *A Study of Several Diagnostic Techniques for Identifying Earthquakes*, J. Geophys. Res. *68*, 2909—2928 (1963).

ROBINSON, E. A., *Statistical Communication and Detection with Special Reference to Digital Data Processing of Radar and Seismic Signals* (Griffin, London 1967), 362 pp.

RODEAN, H. C., *Understanding and Constructively Using the Effects of Underground Nuclear Explosions*, Rev. Geophys. *6*, 401—445 (1968).

RODEAN, H. C., *Nuclear-Explosion Seismology* (U.S. Atomic Energy Commission, 1971), 156 pp.

SPRINGER, D. L., and KINNAMAN, R. L., *Seismic Source Summary for U.S. Underground Nuclear Explosions, 1961—1970*, Bull. Seism. Soc. Amer. *61*, 1073—1098 (1971).

SPRINGER, D. L., and KINNAMAN, R. L., *Seismic Source Summary for U.S. Underground Nuclear Explosions, 1971—1973*, Bull. Seism. Soc. Amer. *65*, 343—349 (1975).

Chapter 12

ALFVÉN, H., *Origin of the Moon*, Science *148*, 476—477 (1965).

BASTIN, J. A., *The Diffusion of Lunar Seismic Energy*, Phys. Earth Planet. Interiors *4*, 218—221 (1971).

BOWIN, C., SIMON, B., and WOLLENHAUPT, W. R., *Mascons: A Two-Body Solution*, J. Geophys. Res. *80*, 4947—4955 (1975).

BULLEN, K. E., *Origin of the Moon*, Nature *167*, 29 (1951).

BULLEN, K. E., *On the Internal Constitutions of the Planets*, Proc. Astr. Soc. Australia *1*, 1—6 (1967).

CHANG, G. K., GUNTHER, P., and JAMES, D. B., *A Secondary Ejecta Explanation of a Lunar Seismogram*, J. Geophys. Res. *75*, 7426—7438 (1970).

COOPER, M. R., KOVACH, R. L., and WATKINS, J. S., *Lunar Near-Surface Structure,* Rev. Geophys. and Space Phys. *12,* 291—308 (1974).

DATTA, A. N.. *On the Energy Required to Form the Moon,* Monthl. Not. Roy. Astr. Soc., Geophys. Suppl. *6(9),* 535—539 (1954).

DERR, J. S., *Travel Times, Variational Parameters, and Love Numbers for Moon Models,* Bull. Seism. Soc. Amer. *60,* 697—716 (1970).

DUENNEBIER, F., and SUTTON, G. H., *Thermal Moonquakes,* J. Geophys. Res. *79,* 4351—4363 (1974).

DUENNEBIER, F., and SUTTON, G. H., *Meteoroid Impacts Recorded by the Short-Period Component of Apollo 14 Lunar Passive Seismic Station,* J. Geophys. Res. *79,* 4365—4374 (1974).

EWING, M., et al., *Seismology of the Moon and Implications on Internal Structure, Origin and Evolution,* Highlights of Astronomy (De Jager, Ed.), 155—172 (1971).

KOVACH, R. L., and ANDERSON, D. L., *The Interiors of the Terrestrial Planets,* J. Geophys. Res. *70,* 2873—2882 (1965).

KOZYREV, N. A., *Volcanism on the Planets,* Tectonophysics *1,* 451—454 (1964).

LATHAM, G., EWING, M., PRESS, F., and SUTTON, G., *The Apollo Passive Seismic Experiment,* Science *165,* 241—250 (1969).

LATHAM, G. V., et al., *Apollo 11 Passive Seismic Experiment,* Proc. Apollo 11 Lunar Science Conf. *3,* 2309—2320 (1970). In abbreviated form also in Science *167,* 455—457 (1970).

LATHAM, G. V., MCDONALD, W. G., and MOORE, H. J., *Missile Impacts as Sources of Seismic Energy on the Moon,* Science *168,* 242—245 (1970).

LATHAM, G., et al., *Moonquakes,* Science *174,* 687—692 (1971).

MIDDLEHURST, B. M., *An Analysis of Lunar Events,* Rev. Geophys. *5,* 173—189 (1967).

NAKAMURA, Y., and LATHAM, G. V., *Travel Times of Body Waves in the Moon,* Bull. Seism. Soc. Amer. *60,* 63—78 (1970).

NASA (National Aeronautics and Space Administration), *Apollo 15, Preliminary Science Report,* Spec. Publ. *SP-289,* 526 pp. (1972).

O'KEEFE, J. A., *Origin of the Moon,* J. Geophys. Res. *74,* 2758—2767 (1969), and *75,* 6565—6574 (1970).

SUTTON, G. H., and STEINBACHER, R., *Surveyor Seismograph Experiment,* J. Geophys. Res. *72,* 841—844 (1967).

TOKSÖZ, M. N., et al., *Velocity Structure and Properties of the Lunar Crust,* The Moon, Reidel Publ. Co., *4,* 490—504 (1972).

TOKSÖZ, M. N., DAINTY, A. M., SOLOMON, S. C., and ANDERSON, K. R., *Velocity Structure and Evolution of the Moon,* Geochim. et Cosmochim. Acta, Suppl. 4, *3,* 2529—2547 (1973).

Subject Index

412

418

Author index

Geographical index